Analysis of Medical Modalities for Improved Diagnosis in Modern Healthcare

T0321382

Analysis of Medical Modalities for Improved Diagnosis in Modern Healthcare

Edited by
Varun Bajaj
G. R. Sinha

CRC Press
Taylor & Francis Group
Boca Raton London New York

CRC Press is an imprint of the
Taylor & Francis Group, an **informa** business

First edition published 2021
by CRC Press
6000 Broken Sound Parkway NW, Suite 300, Boca Raton, FL 33487-2742

and by CRC Press
2 Park Square, Milton Park, Abingdon, Oxon, OX14 4RN

CRC Press is an imprint of Taylor & Francis Group, LLC

ISBN: 978-0-367-70536-7 (hbk)
ISBN: 978-0-367-70537-4 (pbk)
ISBN: 978-1-003-14681-0 (ebk)

Typeset in Times
by codeMantra

Dedicated to my late father Mahendra Bajaj and family members.

Varun Bajaj

Dedicated to my late grand parents, my teachers
and Revered Swami Vivekananda

G. R. Sinha

Contents

Preface

In modern healthcare, various medical modalities play an important role in improving the diagnostic performance in healthcare systems for various applications, such as prosthesis design, surgical implant design, diagnosis and prognosis, and detection of abnormalities in the treatment of various diseases. This book also discusses the uses of analysis, modeling, and manipulation of modalities, such as EEG, ECG, EMG, PCG, COP, EOG, MRI, and FMRI for an automatic identification and classification for diagnosis of any disorder and physiological states. For addressing a wide range of modalities of medical imaging, their analysis and applications for post-processing and diagnosis are much-needed topics for a number of researchers and faculty members all across the world attempting to conduct research in this area.

Therefore, this book emphasizes the real-time challenges in medical modalities for a variety of applications for analysis, classification, and identification of different states for the improvement of healthcare systems. Each chapter starts with the introduction, need and motivation of the medical modality, and a number of applications for the identification and improvement of healthcare systems. Moreover, the chapters can be read independently by research scholars, graduate students, faculty members, and R&D engineers who wish to explore research in the field of computer sciences, electronics, medical sciences, and biomedical engineering.

The chapter-wise description of this book as follows: Chapter 1 presents the dual-tree complex wavelet transform (DTCWT)-based approach for the classification of alertness and drowsiness states utilizing EEG signals. The DTCWT decomposes the EEG signal into different band-limited sub-bands. The characteristics of DTCWT provide sub-bands that are extracted in terms of several time-domain measures. Further, the measures are tested over several machine learning algorithms to effectively classify the two important states as alertness state and drowsiness state of EEG signals. In Chapter 2, stochastic event synchrony (SES) is studied as an approach of quantifying synchrony between two time series. The time-frequency transform of each signal is approximated as a sum of half-ellipsoid basis functions, referred to as "bumps." Each bump is considered as an event on the time-frequency plane. Method is applied to combat related post-traumatic stress disorder (PTSD) EEG signals, in addition to healthy controls, and trauma-exposed none-PTSD veterans. EEG signals are collected in two resting states (eyes-open and closed). In Chapter 3, the authors have carried out an in-depth survey of all the prevalent and nonprevalent healthcare chatbot systems, and discussed in detail regarding a suitable dataset and an underlying deep learning model for the said chatbot application. Chapter 4 presents a framework composed of the three main modules. In the first module, EMG signals are denoised by using MSPCA denoising technique. In the second module, the coefficients of wavelet-based time-frequency methods are calculated for each category of EMG signal, and then statistical values of each sub-band are computed. In the last module, the extracted features are employed as an input to a classifier to diagnose different neuromuscular disorders. In Chapter 5, the upper-limb prostheses and their control are discussed. The results of some experiments are further provided showing that such undersampled signals could be used for various applications required in advanced prosthesis control, e.g., force prediction, elbow angle prediction, movement detection, and time and frequency parameter extraction using undersampled sEMG signals. Finally, a low-cost controller for BRUNEL HAND 2.0 from Open Bionics is designed to link low-cost recording and prosthesis. In Chapter 6, current diagnosis based on clinical symptoms, laboratory tests, real-time reverse transcription polymerase chain reaction (RT-PCR) test, and chest CT imaging are discussed. Although the real-time RT-PCR test is considered as the gold standard method for the diagnosis of SARS-CoV-2 infection, this test presents some limitations, which may lead to the delay in detecting the disease. Accurate, noninvasive, and easy-to-use tools are on high demand for a rapid and timely diagnosis of the disease. Since fever is a common symptom in COVID-19 patients, noncontact infrared thermometers have been extensively used to measure the body temperature rapidly and noninvasively as an early diagnostic tool. However, this approach could be misleading and may affect the effectiveness of SARS-CoV-2 detection. In Chapter 7, statistical features are extracted from the segmented breast region for breast cancer prognosis. Machine learning algorithms like support vector machine (SVM), k-nearest neighborhood (kNN), naïve Bayes, and logistic regression without principal component analysis (PCA) and with PCA as a precursor are applied to the extracted data. We fit various classifiers on the PCA-transformed data to classify thermograms as malignant or benign. This work indicates that thermal imaging is capable of predicting breast pathologies coupled with machine learning algorithms.

Chapter 8 explores different techniques for image acquisition, followed by image pre-processing, image segmentation to find the region of interest (ROI), feature extraction of the segmented image, and classification techniques. Chapter 9 focuses on enhancing requirements of the current voice command and speaking technologies with consideration of emotional states, so that the interaction with computers and smart devices may become more human. In Chapter 10, a thinking process is suggested to result in a different pattern of the brain signals. Hence, BCI can be viewed as a pattern recognition system. Brain signals will have different information in each activity. Different channels capture these signals. The foremost aspects of feature extraction are dimensions' reduction, then classification, and finally statistical analysis. Chapter 11 further explores the BCI systems based on different brain signal patterns, such as slow cortical potential (SCP), sensorimotor rhythms (SMR), P300 event-related potentials (ERP), and steady-state evoked potentials (SSEPs) with the help of existing literature. Techniques to improve signal quality and feature extraction are also reviewed. Chapter 12 gives an idea of type of deep models called as deep neural networks (DNNs). According to the research, DNNs can directly work on raw inputs to automate the functions. This chapter explains the identification process for human behavior (users) using activities, actions, and behavior patterns of intra- and interactive contents. Chapter 13 suggests ECG as a noninvasive technique, which helps to analyze the functioning of the heart. It isn't just a tedious task; it's often prone to error. The rare occurrences of abnormal heartbeats represent the most significant barrier for heart disease detection, and even more are some unique typologies that are not adequately represented in signal datasets. To overcome such obstacles, we implement a Stockwell transform (ST)-based random forest classifier.

MATLAB® is a registered trademark of The MathWorks, Inc. For product information, please contact:

The MathWorks, Inc.
3 Apple Hill Drive
Natick, MA 01760-2098 USA
Tel: 508-647-7000
Fax: 508-647-7001
E-mail: info@mathworks.com
Web: www.mathworks.com

Acknowledgments

Dr Bajaj expresses his heartfelt appreciation to his mother Prabha, wife Anuja, and daughter Avadhi, for their wonderful support and encouragement throughout the completion of this important book on **Analysis of Medical Modalities for Improved Diagnosis in Modern Healthcare**. His deepest gratitude goes out to his mother-in-law and father-in-law for their constant motivation. This book is an outcome of sincere efforts that could be given to this book only due to great support of the family. He also thanks Prof Sanjeev Jain Director of PDPM IIITDM Jabalpur for his support and encouragement.

Dr Sinha too expresses his gratitude and sincere thanks to his family members, his wife Shubhra, daughter Samprati, parents, and teachers.

We would like to thank all our friends, well-wishers, and all those who keep us motivated in doing more and more, better and better. We sincerely thank all the contributors for writing the relevant theoretical background and real-time applications of analysis of medical modalities for improved diagnosis in modern healthcare.

We express our humble thanks to Dr Gagandeep Singh, Publisher (Engineering), and all editorial staff of CRC Press for their great support, necessary help, appreciation, and quick responses. We also wish to thank Lakshay Gaba and the publication team of CRC Press for the support, and the publishing house for giving us this opportunity to contribute on some relevant topics with reputed publisher. Finally, we want to thank everyone, in one way or another, who helped us to edit this book.

Last but not least, we would also like to thank God for showering us in his blessings and strength to do this type of novel and quality work.

Varun Bajaj
G. R. Sinha

Editors

Varun Bajaj (Ph.D., MIEEE 16 SMIEEE20) has been working as a Faculty in the discipline of Electronics and Communication Engineering, at Indian Institute of Information Technology, Design and Manufacturing (IIITDM) Jabalpur, India, since 2014. He worked as a Visiting Faculty in IIITDM Jabalpur from September 2013 to March 2014. He worked as an Assistant Professor at Department of Electronics and Instrumentation, Shri Vaishnav Institute of Technology and Science, Indore, India, during 2009–2010. He received B.E. degree in Electronics and Communication Engineering from Rajiv Gandhi Technological University, Bhopal, India, in 2006, and M.Tech. degree with Honors in Microelectronics and VLSI design from Shri Govindram Seksaria Institute of Technology and Science, Indore, India, in 2009. He received his Ph.D. degree in the Discipline of Electrical Engineering, at Indian Institute of Technology, Indore, India, in 2014.

He is an Associate Editor of *IEEE Sensor Journal* and Subject Editor-in-Chief of *IET Electronics Letters*. He served as a Subject Editor of *IET Electronics Letters* from November 2018 to June 2020. He is a Senior Member IEEE June 2020 and also contributing as active technical reviewer of leading international journals of IEEE, IET, and Elsevier, etc. He has authored more than 100 research papers in various reputed international journals/conferences like IEEE Transactions, Elsevier, Springer, IOP etc. He has edited *Modelling and Analysis of Active Biopotential Signals in Healthcare* - Volume 1, 2 published in IOP books. Currently, he is editing a book in CRC Press (Taylor & Francis Group). The citation impact of his publications is around 2048 citations, *h* index of 20, and i10 index of 47 (Google Scholar Sep 2020). He has guided 6 (03 completed and 3 in-process) Ph.D. scholars and 6 M.Tech. scholars. He is a recipient of various reputed national and international awards. His research interests include biomedical signal processing, image processing, time-frequency analysis, and computer-aided medical diagnosis.

G. R. Sinha is an Adjunct Professor at International Institute of Information Technology Bangalore (IIITB) and currently deputed as a Professor at Myanmar Institute of Information Technology (MIIT), Mandalay, Myanmar. He obtained his B.E. (Electronics Engineering) and M.Tech. (Computer Technology) with Gold Medal from National Institute of Technology Raipur, India. He received his Ph.D. in Electronics & Telecommunication Engineering from Chhattisgarh Swami Vivekanand Technical University (CSVTU), Bhilai, India. He has been a Visiting Professor (Honorary) in Sri Lanka Technological Campus Colombo for one year (2019–2020).

He has published 259 research papers, book chapters, and books at an international level that includes Biometrics published by Wiley India, a subsidiary of John Wiley; Medical Image Processing published by Prentice Hall of India; and 07 Edited books on Cognitive Science-Two Volumes (Elsevier), Optimization Theory (IOP), Biometrics (Springer) and Modelling of Bio-potential Signals (IOP), and Assessment of Learning Outcomes (IGI). He is currently editing 06 more books on biomedical signals; brain and behavior computing; modern sensors and data deduplication with Elsevier, IOP, and CRC Press. He is an Associate Editor of three SCI journals: *IET-Electronics Letters*, *IET-Image Processing Journal,* and *IEEE Access-Multidisciplinary Open Access Journal.* He has teaching and research experience of 22 years. He has been the Dean of Faculty and Executive Council Member of CSVTU and currently a member of Senate of MIIT. Dr Sinha has been delivering ACM lectures as an ACM Distinguished Speaker in the field of DSP since 2017 across the world. His few more important assignments include Expert Member for Vocational Training Program by Tata Institute of Social Sciences (TISS) for two Years (2017–2019); Chhattisgarh Representative of IEEE MP Sub-Section Executive Council (2016–2019); and Distinguished Speaker in the field of Digital Image Processing by Computer Society of India (2015). He served as a Distinguished IEEE Lecturer in IEEE India Council for Bombay section.

He is a recipient of many awards and recognitions like TCS Award 2014 for outstanding contributions in Campus Commune of TCS, Rajaram Bapu Patil ISTE National Award 2013 for Promising Teacher in Technical Education by ISTE New Delhi, Emerging Chhattisgarh Award 2013, Engineer of the Year Award 2011, Young Engineer Award 2008, Young Scientist Award 2005, IEI Expert Engineer Award 2007, ISCA Young Scientist Award 2006 Nomination, and Deshbandhu Merit Scholarship for 05 years. He is a Senior Member of IEEE, Fellow of Institute of Engineers India, and Fellow of IETE India.

He has delivered more than 50 Keynote/Invited Talks and Chaired many Technical Sessions in International Conferences across the world such as Singapore, Myanmar, Sri Lanka and India. His Special Session on "Deep Learning in Biometrics" was included in IEEE International Conference on Image Processing 2017. He is also a member of many national professional bodies like ISTE, CSI, ISCA, and IEI. He is a member of various committees of the university and has been a Vice President of Computer Society of India for Bhilai Chapter for two consecutive years. He is a Consultant of various skill development initiatives of NSDC, Government of India. He is a regular Referee of Project Grants under DST-EMR scheme and several other schemes of Government of India. He received few important consultancy supports as grants and travel support.

Dr Sinha has supervised 08 Ph.D. scholars and 15 M.Tech. scholars, and has been supervising 01 more Ph.D. scholar. His research interest includes biometrics, cognitive science, medical image processing, computer vision, outcome-based education (OBE), and ICT tools for developing employability skills.

Contributors

Golnaz Amiri
Biomedical Engineering Department,
   Faculty of Engineering
University of Isfahan
Isfahan, Iran

R. N. Awale
Department of Electrical Engineering
Veermata Jijabai Technological Institute
Mumbai, India

Mohan Awasthy
Department of Electronics and
   Telecommunication
Rungta College of Engineering and
   Technology
Bhilai, India

Sivaji Bandyopadhyay
Department of Computer Science and
   Engineering
National Institute of Technology
Silchar, India

Zahra Nasr Esfahani
Biomedical Engineering Department,
   Faculty of Engineering
University of Isfahan
Isfahan, Iran

Negar Maleki Far
Biomedical Engineering Department,
   Faculty of Engineering
University of Isfahan
Isfahan, Iran

Farzaneh Fazilati
Biomedical Engineering Department,
   Faculty of Engineering
University of Isfahan
Isfahan, Iran

Zahra Ghanbari
Faculty of Biomedical Engineering
Amirkabir University of Technology
Tehran, Iran

Monali Gulhane
Department of Computer Science and
   Engineering
Koneru Lakshmaiah Education
   Foundation
Guntur, India

Lalita Gupta
Department of Electronics and
   Communication Engineering
Maulana Azad National Institute of
   Technology
Bhopal, India

Aayesha Hakim
Department of Electronics Engineering
Veermata Jijabai Technological Institute
Mumbai, India

T. K. Muhamed Jishad
Department of Electrical Engineering
National Institute of Technology Calicut
Kozhikode, India

Mislav Jordanić
Biomedical Engineering Research
Centre (CREB), Automatic Control
Department (ESAII)
Universitat Politècnica de
   Catalunya – BarcelonaTech
Barcelona, Spain

Rabiah Abdul Kadir
Institute of IR4.0
Universiti Kebangsaan Malaysia
Bangi, Malaysia

Smith K. Khare
Department of Electronics and
Communication Engineering
PDPM-Indian Institute of Information
Technology, Design and
Manufacturing
Jabalpur, India

Abdullah Faiz Ur Rahman Khilji
Department of Computer Science and
Engineering
National Institute of Technology
Silchar, India

Sahinur Rahman Laskar
Department of Computer Science and
Engineering
National Institute of Technology
Silchar, India

C. H. Laxmi Bala
Department of Computer Science
IIIT Nuziveedu
Nuziveedu, India

Joan Francesc Alonso López
Biomedical Engineering Research
Centre (CREB), Automatic Control
Department (ESAII)
Universitat Politècnica de
Catalunya – BarcelonaTech
Barcelona, Spain

Maya Silvi Lydia
Faculty of Computer Science and
Information Technology
Universitas Sumatera Utara
Medan, Indonesia

Gaurav Makwana
Department of Electronics and
Communication Engineering
Maulana Azad National Institute of
Technology
Bhopal, India

M. B. Malarvili
Universiti Teknologi Malaysia (UTM)
Johor Bahru, Malaysia

Marjan Mansourian
Department of Biostatistics and
Epidemiology, School of Public
Health
Isfahan University of Medical Sciences
Isfahan, Iran

Hamid Reza Marateb
Biomedical Engineering Department,
Faculty of Engineering
University of Isfahan
Isfahan, Iran

K. Mohanasundaram
AP, School of Business,
Alliance University
Bangalore, India

Mohammad Reza Mohebbian
Department of Electrical and Computer
Engineering
University of Saskatchewan
Saskatoon, Canada

Mohammad Hassan Moradi
Faculty of Biomedical Engineering
Amirkabir University of Technology
Tehran, Iran

Alexie Mushikiwabeza
Universiti Teknologi Malaysia (UTM)
Johor Bahru, Malaysia
and
University of Rwanda(UR)/Center
    of Excellence in Biomedical
    Engineering and E-Health(CEBE)
Kigali, Rwanda

Y. V. Narayana
Department of Electronics and
    Communication Engineering
Tenali Engineering College
Guntur, India

Farzad Ziaie Nezhad
Biomedical Engineering Department,
    Faculty of Engineering
University of Isfahan
Isfahan, Iran

Marjan Nosouhi
Biomedical Engineering Department,
    Faculty of Engineering
University of Isfahan
Isfahan, Iran

Partha Pakray
Department of Computer Science and
    Engineering
National Institute of Technology
Silchar, India

Allam Jaya Prakash
IEEE
Piscataway, New Jersey

Mohsen Rastegari
Biomedical Engineering Department,
    Faculty of Engineering
University of Isfahan
Isfahan, Iran

Ravi
Department of Electronics and
    Communication Engineering
Delhi Technological University (DTU)
New Delhi, India

T. Sajana
Department of Computer Science and
    Engineering
Koneru Lakshmaiah Education
    Foundation
Guntur, India

M. Sanjay
Department of Electrical Engineering
National Institute of Technology Calicut
Kozhikode, India

Saunak Samantray
IEEE
Piscataway, New Jersey

Abdulhamit Subasi
Faculty of Medicine
University of Turku
Turku, Finland
and
Department of Computer Science
Effat University
Jeddah, Saudi Arabia

Sachin Taran
Department of Electronics and
    Communication Engineering
Delhi Technological University (DTU)
New Delhi, India

K. A. Venkatesh
Department of Mathematics, Computer
    Science and Engineering
Myanmar Institute of Information
    Technology (MIIT)
Mandalay, Myanmar

Miguel Ángel Mañanas Villanueva
Biomedical Engineering Research
    Centre (CREB), Automatic Control
    Department (ESAII)
Universitat Politècnica de
    Catalunya-BarcelonaTech
Barcelona, Spain

Khan A. Wahid
Department of Electrical and Computer
    Engineering
University of Saskatchewan
Saskatoon, Canada

Ram Narayan Yadav
Department of Electronics and
    Communication Engineering
Maulana Azad National Institute of
    Technology
Bhopal, India

Fatemeh Yusefi
Biomedical Engineering Department,
Faculty of Engineering
University of Isfahan
Isfahan, Iran

1 Classification of Alertness and Drowsiness States Using the Complex Wavelet Transform-Based Approach for EEG Records

Sachin Taran and Ravi
Delhi Technological University (DTU)

Smith K. Khare
PDPM-Indian Institute of Information
Technology, Design and Manufacturing

Varun Bajaj
PDPM-Indian Institute of Information
Technology, Design and Manufacturing

G. R. Sinha
Myanmar Institute of Information Technology (MIIT)

CONTENTS

1.1 INTRODUCTION

Drowsiness is an intermediate stage of awakened and asleep stages, which considerably affects the alertness and decision-making ability of the brain. This is a potential problem in drivers that leads to road accidents. As per a survey conducted by National Highway Safety, 56,000–100,000 crashes and accidents happen due to drowsiness affecting the drivers, which subsequently causes more than 1500 casualties and 71,000 injuries annually [1]. There is an estimation of loss of economy due to these causalities, which is around 230 billion USD as per the report of Federal Highway Administration [2]. Drowsiness also negatively affects all types of real-time human operations like medical and power plant services [3]. As the traffic load is increasing day by day, drowsiness becomes a serious problem, because it attracts the researchers to find the solution. So various monitoring techniques have been proposed by researchers, like optical and sensor-based monitoring.

Reliable monitoring of drowsiness is the main aim of the development of a new driver assistance system that can avert a number of accidents. The development of different systems is based on a number of factors such as behavior of driver and performance of the system. However, the factors or parameters vary due to different types of vehicle and dynamic driving conditions [4,5]. In another method, a camera is placed to capture the upper body movement like head and eye movement [6,7]. However, this technique fetches good results, but it compromises with driver privacy. Nearly all drivers dislike a continuous monitoring by the camera. So, it is not feasible to commercially promote this technique. In another approach, biomedical signals are used for drowsiness detection [8–12].

Biomedical signals are very helpful for extracting the information to measure the response of a person especially when they are in a drowsiness state (DS). In the biomedical approach, electrocardiograph (ECG) [10] and electrooculography [11] are used by a number of researchers, and electroencephalography (EEG) is reported as the most commonly used biomedical modality for the determination of the brain agitation or similar states [13–17]. EEG is also a standard technique in sleep studies. The other advantages of EEG are noninvasive, high temporal resolution, and low cost [18–22]. The major issue in biomedical signal monitoring is the placing of sensors and cables on the body but this issue can be easily resolved by placing a wireless sensor in a nonintrusive system [12].

Brain electrical activity plays a vital role in the identification of different states like awakens and asleep stages. In a different brain state, the electrical activity wave pattern reflects unique characteristics that can be intensively studied. So EEG gets more attention from the researchers in comparison with other techniques. In Ref. [9], the eight-channel EEG, ECG, Electromyography (EMG), and Electrooculography (EOG) data were recorded, and different types of entropies with relative band ratios of EEG are used to analyze drowsiness. The loss of alertness in drivers is correlated with the power spectrum of the alpha band in EEG [23]. For the monitoring of the driver, in this experiment, they used a total of 34 EEG/ECG/EOG electrodes. Most of the earlier mentioned techniques used multiple leads of the EEG channel, which is uncomforting for the drivers. In Ref. [24], a single EEG channel for monitoring a DS based on a comparative estimation of the relative band of EEG is proposed. Apart from the aforementioned electrode-based technique, a wireless EEG-based interface is also proposed to improve the comfort of the driver [12].

An EEG-based self-organized neural fuzzy system is suggested to predict and monitor the drowsiness of the driver [25]. In Ref. [26], the Hermite function is used as a basis function for the decomposition of the drowsiness EEG signal. In another method, RR-time series features and EEG features were combined using deep neural networks for the classification of sleep stages [27]. For increasing the comfort and real-time monitoring of driver state, a Bluetooth-enabled fully wearable EEG system is introduced [28]. In this method, the smartwatch is used for the evaluation of the proposed model in a real-time way. A multimodal approach uses both video information and biomedical signal in order to determine the state of drowsiness [29]. The data-driven method of empirical mode decomposition (EMD) with the Hilbert transform is used to analyze the DS [30]. The variational nonlinear chirp-mode decomposition-based features are given for the detection of drowsiness [31]. The tunable-Q wavelet transform (TQWT) and optimized TQWT-based methods are introduced for the classification of alertness state (AS) and DS [32,33]. In EEG records, automatic drowsiness detection is performed by spectral, wavelet, and multimodal analyses (MA) [34,35]. The different machine learning algorithms are compared for the detection of drowsiness [36].

The methods in the literature are limited by their effectiveness in terms of accuracy due to several drawbacks. The fast Fourier transform-based methods have a problem of localization in time frequency [37]. The autoregressive model is noise-prone. Wavelets and TQWT-based methods require a choice of wavelets and tuning parameters [38–41], while EMD is purely mathematical [42]. It is hard to explore a specific basis in Hermite function-based method for the EEG signal. Hence, in this chapter, a dual-tree complex wavelet transform (DTCWT)-based approach is suggested for the classification of AS and DS.

1.2 METHODOLOGY

Figure 1.1 shows the signal flow of EEG signal in the proposed method. The various subsections of the proposed methodology are dataset, DTCWT, feature extraction, and machine learning algorithms.

1.2.1 DATASET

In this work, the EEG record of 16 male subjects from the MIT-BIH Polysomnographic database is used [43,44]. The mean weight and age of all patients are 119 and 43, respectively. In this database, single-channel EEG data are recorded (used channel C3-O1, C4-A1, and O2-A1) with the sampling frequency of 250 Hz. The EEG epochs are labeled as "awake stage" and "Stage I" for alertness and drowsiness, respectively. The total available epochs corresponding to alertness and drowsiness are 667 and 488, respectively. An example of AS and DS classes of EEG signals is shown in Figure 1.2.

1.2.2 DUAL-TREE COMPLEX WAVELET TRANSFORM

The DTCWT was suggested by Kingsbury in 1998 that aims at addressing the limitations of discrete wavelet transform (DWT) such as shift variant behavior and low directional selectivity. It has all the properties of shift-invariant and directional selectivity for two and higher dimensions [45]. It is a redundant transform with a

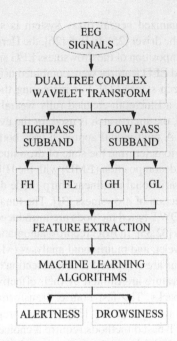

FIGURE 1.1   DTCWT-based method for alertness and drowsiness states classification.

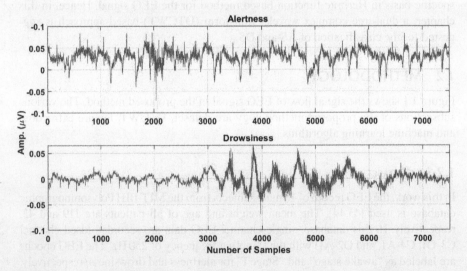

FIGURE 1.2   AS and DS classes of EEG signals.

redundancy factor of 2^d for a d-dimensional signal. In DTCWT architecture, two parallel DWTs are employed using the low-pass and high-pass filters, which are called real and imaginary trees. In the dual-tree approach, two real DWT methods are employed: the first DWT gives the real part, and the second gives the imaginary part. For these two real wavelet transforms, two different sets of filters are employed with each satisfying a perfect reconstruction condition [45,46].

The wavelet function $\psi(t)$ of DTCWT is a complex, which can be represented as [46]

$$\psi(t) = \psi_h(t) + i\psi_g(t)$$

where $\psi_h(t)$ and $\psi_g(t)$ are the two real wavelets, which are basically the Hilbert transform pair. A finite energy EEG signal s(t) can be represent by the wavelets and scaling function as [46],

$$s(t) = \sum_{k=-\infty}^{\infty} p(k)\varnothing(t-k) + \sum_{j=0}^{\infty}\sum_{k=-\infty}^{\infty} d(j,k)2^{\frac{j}{2}}\psi\left(2^j t - k\right).$$

where $p(k)$ and $d(j,k)$ are the scaling and wavelet coefficients, respectively (Figures 1.3 and 1.4).

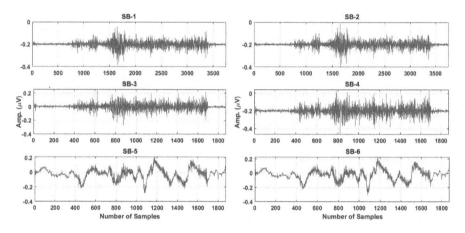

FIGURE 1.3 DTCWT provided bandlimited components of AS class EEG signals.

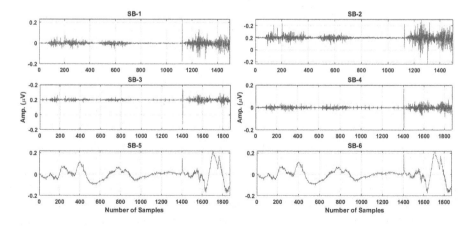

FIGURE 1.4 DTCWT provided bandlimited components of DS class EEG signals.

1.2.3 FEATURE EXTRACTION

The different time-domain measures are used for assessing the characteristics of DTCWT provided bandlimited components. The measures are Higuchi fractal derivative, Hurst exponent, temporal moments (third, fourth, fifth), trimean, minimum value, maximum value, integrated EEG, mean, log-energy, interquartile range, quantiles (first, second, third), and Hjorth parameters (activity, mobility, complexity) [47–53].

1.2.4 CLASSIFICATION

The purpose of this work lies in classifying the DS and AS [54,55]. The numerous studies suggest a large number of clarifiers used for the classification task but few methods only dominate in the classification of EEG signals and its applications. Based on an extensive study of usability, performance, and robustness, we investigated a number of classifiers and used few of them in the current classification work. The support vector machine (SVM), decision trees, and k-nearest neighbor (k-NN) have been tested. We also studied some variants of decision trees [56] such as medium, fine, and coarse trees. Discriminant analysis [1,55] was also evaluated by exploiting its variants, namely, quadratic and linear. Among variants of k-NN [57], fine, medium, coarse, cubic, cosine, and weighted methods were studied. The cubic, linear, quadratic, and Gaussian models of SVM [58–60] were investigated. The boosted and bagged trees of ensemble method were studied with an emphasis on subspace and RUSBoosted trees. The performance of the proposed classifier was evaluated with respect to all these major classifiers.

1.3 RESULTS AND DISCUSSION

The suggested framework aims to develop an efficient DCTWT-based drowsiness detection. At first, EEG signals of DS and AS are decomposed into three-level low-pass and high-pass sub-bands by taking J as 2. The low-pass and high-pass sub-bands so obtained are later employed for feature extraction. Several features are evaluated from the sub-bands, and most significant features are selected for the classification of AS and DS. In this method, 18 features are selected using the Kruskal–Wallis test. The probability of chi is checked for each sub-band and feature. The values of chi are found to be well below the standard one, i.e., 0.05. These features are Higuchi fractal derivative, Hurst exponent, TM3, TM4, TM5, trimean, minima, maxima, integrated EEG, mean, log energy, interquartile range, quantile 1, quantile 2, quantile 3, Hjorth activity, Hjorth mobility, and Hjorth complexity, respectively. The feature matrix so formed is applied to different linear as well as nonlinear classification algorithms. The training and testing done by ten cross-validation method. A complete input matrix is divided into ten random disjoint sets, of which nine sets are considered for training and the remaining one is utilized for testing. It is noteworthy to mention that the input matrix for AS is 667×18 (signals × features) and for the DS, it is 488×18 (signals × features), respectively. Seven types of classification algorithms along with different kernels have been used for the separation of AS and DS. Decision tree (fine, coarse, and medium), linear discriminant analysis (LDA), logistic regression (LR), Naïve Bayes (Gaussian and kernel), SVM (linear, fine Gaussian, medium Gaussian, and coarse Gaussian), k-NN (fine, medium, coarse, cosine, cubic, and weighted), and ensemble (bagged, boosted, and RUSBoosted) classifiers are employed. A common experimental platform is maintained, and uniform hyperparameters are selected for each classifier. The number of maximum splits is maintained to be 100, 4, and 20 for a fine, coarse, and medium trees. Gini's diversity index split criteria are used, and surrogate decision splits are kept off. Default parameters are set for LDA, LR, and Naïve Bayes. The automatic kernel scale is used for the linear SVM, while the manual kernel-scale model is used in the case of fine Gaussian, medium Gaussian, and coarse Gaussian SVM with a kernel-scale factor of 0.9, 3.6, and 14, respectively. The box constraint level is kept at 1 for all the kernels of SVM. The number of neighbors for fine kernel is 1 and for the coarse kernel, it is 100, while that for cosine, cubic, medium, and the weighted kernel is 10. Euclidean distance metric with equal distance weight is used. The maximum number of splits for boosted and RUSBoosted tree is 20, while for the bagged tree, the maximum number of splits is kept to be 1154. The number of learners is selected to be 20 and the learning rate is set at 0.1, respectively.

Table 1.1 shows the accuracy comparison of different classifiers for six sub-bands. The highest classification accuracy achieved with the fine and medium kernel of a decision tree is 80.6% and 83.5% in SB-6, while for the coarse kernel, it is 82.7% in SB-4.

TABLE 1.1
Accuracy Comparison with Different Machine Learning Algorithms

Decision Tree	SB-1	SB-2	SB-3	SB-4	SB-5	SB-6
Fine tree	73	69.6	72.1	79.7	73.6	80.6
Medium tree	75.9	72.5	77.3	82.9	79.2	83.5
Coarse tree	75	72.2	77.6	82.7	79	82.1
LDA	71.9	72.2	73.8	82.6	73.2	83.3
LR	73.2	73.2	75.7	82.9	74.5	83.4
Naïve Bayes						
Gaussian	57.2	58.1	60.3	65.2	60	**65.3**
Kernel	66.4	66.7	**70**	68.8	69.5	68.4
SVM						
Linear	72.6	72.4	76.3	84	76.9	**84.2**
Fine Gaussian	76.8	76.2	74.5	83.9	75	**84.2**
Medium Gaussian	76.6	75.4	77.6	84.1	76.7	**84.7**
Coarse Gaussian	71.5	71.8	75.4	82.4	76.1	**82.7**
KNN						
Fine	70.6	70	71	**81.4**	71.8	**81.4**
Medium	77	71.3	76.2	82.7	75.7	**83.2**
Coarse	72.8	74.3	74.7	80.3	73.9	**80.4**
Cosine	76.5	76.6	74.5	**83.1**	74.5	82.6
Cubic	75.7	75.8	74	**83.4**	74.7	82.8
Weighted	77.7	75.8	75.4	83.9	75.1	**84.2**
Ensemble						
Boosted tree	77.8	76.6	78.7	82.4	79.2	**85.1**
Bagged tree	76.4	76.6	77.7	83.9	78.4	**84.5**
RUSBoosted tree	78.01	76.97	79.48	84.59	73.59	**86.41**

The least performer SB is SB-2 for the fine, medium, and coarse kernel with an accuracy (ACC) of 69.6%, 72.5%, and 72.2%, respectively. The maximum ACC obtained with LDA and LR is 83.3% and 83.4% in SB-6, while the minimum is 71.9% and 73.2% in SB-1. The Gaussian and kernel-based Naïve Bayes classifiers provide a maximum ACC of 65.3% and 70% in SB-6 and SB-3, while the lowest in SB-1 with 57.2% and 66.4%. SVM classifier proved to be best in SB-6 with an ACC of 84.2%, 84.2%, 84.7%, and 82.7% using the linear, fine, medium, and coarse Gaussian kernels. The minimum ACC with the linear, fine, medium, and coarse Gaussian kernel is 72.4%, 74.5%, 75.4%, and 71.5% in SB-2, SB-3, SB-2, and SB-1, respectively.

The maximum ACC with a fine kernel of KNN is 81.4% in SB-4 and SB-6, while for the medium, coarse, and weighted kernel, it is 83.2%, 80.4%, and 84.2% in SB-6 and for cosine and cubic kernel of KNN an ACC is 83.1% and 83.4%, respectively, which shows its best performance in SB-4. The ensemble variants of boosted, bagged, and RUSBoosted tree are the highest in SB-6 with an ACC of 85.1%, 84.5%, and 86.41%, while the least-performing SB are SB-2, SB-1, and SB-5 having an ACC of 76.6%, 76.4%, and 73.59%, respectively. As seen from the table, the highest ACC is achieved using the RUSBoosted tree. Therefore, to get a detailed insight into the proposed method, sensitivity (SEN), specificity (SFE), and precision (PRE) are evaluated. The bar representation of classification of ACC for each variant of a machine learning algorithm for six sub-bands is shown in Figure 1.5. Figure 1.5a shows the classification rate using a decision tree; (b) and (c) represent the classification rate using LDA, LR, and NB. Classification rates for SVM, KNN, and ensemble classifiers are represented by (d), (e), and (f), respectively.

Table 1.2 highlights the values of performance parameters obtained for six sub-bands. As evident from the table, ACC and SEN are the highest in SB-6 with values of 86.41% and 89.97%, while the lowest in SB-5 with values of 73.59% and 77.10%, respectively. The values of SFE and PRE are highest for SB-4 with values of 82.02% and 86.96%, while the least in SB-5 with 68.79% and 77.21%. The bar representation of performance parameters for six sub-bands is shown in Figure 1.6. The class-wise accuracy for the AS and DS for each sub-band is shown in Table 1.3. SB-6 provides the highest classification rate for AS and DS with an 86.06% correct classification rate for AS and 86.89% classification rate for DS. The least classification rate for the AS is 77.21%, and for the DS, it is 68.65% in SB-5, respectively.

Finally, the area under the curve (AUC) and receiver-operating characteristics (ROC) are shown in Figure 1.7. Figures 1.7a–f denotes the AUC for SB-1-SB-6. The AUC obtained for SB-1 is 84% with the current classifier identified at TFR and FPR of (0.83, 0.28), and for SB-2, the AUC is 84% at TPR and FPR of (0.8, 0.22). The AUC obtained for SB-3 and SB-4 is 86% and 89% at a TFR and FPR of (0.82, 0.25) and (0.88, 0.25), whereas the least AUC is obtained for SB-5 with 83% at TPR and FPR of (0.81, 0.29), and the highest AUC of 92% is obtained for SB-6 at a TPR and FPR of (0.86, 0.16), respectively.

To prove the effectiveness of the proposed method over the existing state-of-the-art models, the classification of ACC is compared as shown in Table 1.4.

In Refs. [34] and [36], Correa et al. used power spectral density (PSD) and MA methods for the extraction of different spectral and statistical moments. These spectral and statistical features are classified with artificial neural networks (ANNs).

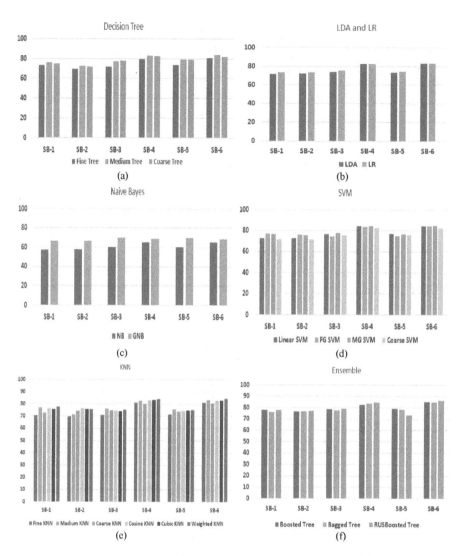

FIGURE 1.5 Bar representation of classification accuracy for (a) DT, (b) LDA and LR, (c) NB, (d) SVM, (e) KNN, and (f) ensemble.

TABLE 1.2
Performance Parameters Using RUSBoosted Classifier

SB	SB-1	SB-2	SB-3	SB-4	SB-5	SB-6
ACC	78.01	76.97	79.48	84.59	73.59	**86.41**
SEN	79.88	79.36	83.49	86.44	77.10	**89.97**
SFE	75.22	73.52	74.46	**82.02**	68.79	82.01
PRE	82.76	81.26	80.36	**86.96**	77.21	86.06

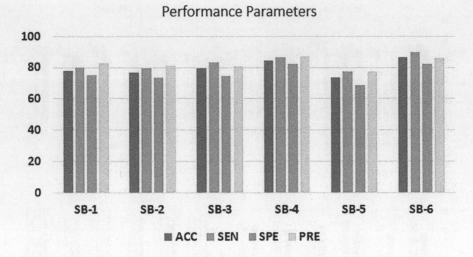

FIGURE 1.6   Bar representation of performance parameters using RUSBoosted classifier.

TABLE 1.3
Percentage Confusion Matrix Using RUSBoosted Classifier

SB	SB-1	SB-2	SB-3	SB-4	SB-5	SB-6
AS-True	82.76	81.26	80.36	86.96	77.21	86.06
AS-False	17.24	18.74	19.64	13.04	22.79	13.94
DS-False	28.48	28.89	21.72	18.65	31.35	13.11
DS-True	71.52	71.11	78.28	81.35	68.65	86.89

The method in Ref. [34] managed to attain an ACC of 84.1%, whereas the method in Ref. [36] reached an ACC of 85.66%. Belakhdar et al. [36] extracted the frequency domain features decomposed with fast Fourier transform (FFT). The feature matrix of these frequency domain features is classified with ANN with an ACC of 84.75%. Tripathy and Acharya [27] used bandpass filtering (BPF) to separate different rhythms of EEG signals. The features extracted from the rhythms are later classified by ANN to claim an ACC of 85.51%. On the other hand, the proposed method uses DTCWT to decompose the EEG signal into multiple low-pass and high-pass sub-bands. Eighteen features are extracted and classified using different variants of machine learning algorithms. The ensemble RUSBoosted classifier proved to be best among all with an ACC of 86.41%, which is higher than all the methods used in the discussion for comparison.

1.4   CONCLUSION

This chapter explores the effectiveness of complex wavelet transform for the classification of AS and DS of EEG records. The EEG signals are analyzed using DTCWT into finite frequency range components. Further, the components are used for the extraction

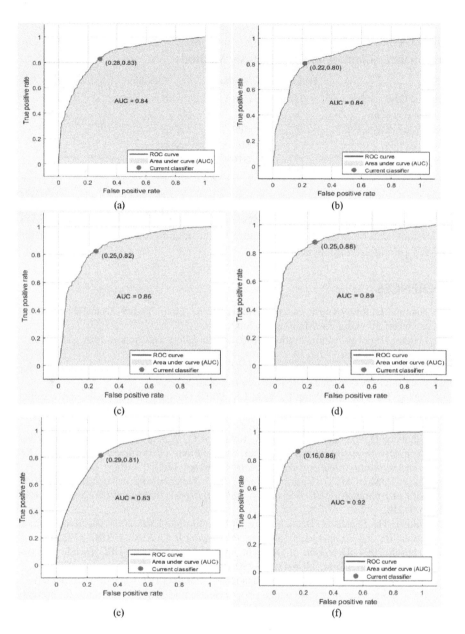

FIGURE 1.7 ROC and AUC obtained using RUSBoosted classifier for (a) SB-1, (b) SB-2, (c) SB-3, (d) SB-4, (e) SB-5, and (f) SB-6.

of several time-domain measures. The performance of classification utilizing the measure is evaluated over the different machine learning algorithms. The RUSBoosted classifier produced the best performance of 86.41% for the proposed method as compared to other tested classifiers. The proposed method also provides the best classification performance as compared to other existing methods. This can benefit in the

TABLE 1.4

Performance Summary of Same Dataset Methods

Authors	Features	Method	Classifier	Accuracy (%)
Correa and Leber [34]	12	PSD	ANN	84.1
Belakhdar et al. [36]	9	FFT	ANN	84.75
Tripathy and Acharya [27]	10	BPF	ANN	85.51
Correa et al. [35]	7	MA	ANN	85.66
Proposed	18	DTCWT	RUSBoosted tree	86.41

design of automated drowsiness detection system. The method can also be applied to the classification of other physiological signals. The classification of AS and DS EEG signals can be improved in the future by introducing the new measures for the DTCWT-provided components.

REFERENCES

1. National-Highway-Traffic-Safety-Administration, 2011. FARS Data-Tables. http://wwwfars.nhtsa.dot.gov/Main/reportslinks.aspx.
2. Federal-Highway-Administration, 2010. Office Safety Integration. http://safety.fhwa.dot.gov/facts stats/.
3. Baker K, Olson J and Morisseau D, 1994. Work practices, fatigue, and nuclear power plant safety performance. *Human Factors*, 36(2), 244–257.
4. Chang TH, Hsu CS, Wang C, Yang LK, 2008. Onboard measurement and warning module for irregular vehicle behavior. *IEEE Transactions on Intelligent Transportation Systems*, 9(3), 501–513.
5. Sandberg D, Åkerstedt T, Anund A, Kecklund G, Wahde M, 2011. Detecting driver sleepiness using optimized nonlinear combinations of sleepiness indicators. *IEEE Transactions on Intelligent Transportation Systems*, 12(1), 97–108.
6. Smith P, Shah M, Da-Vitoria-Lobo N, 2003. Determining driver visual attention with one camera. *IEEE Transactions on Intelligent Transportation Systems*, 4(4), 205–218.
7. Dikkers HJ, Spaans M, Dactu D, Novak M, Rothkrantz LJM, 2004. Facial recognition system for driver vigilance monitoring. *Proceeding IEEE SMC*, 4, 3787–3792.
8. Faber J, 2004. Detection of different levels of vigilance by EEG pseudo spectra. *NeuralNetwork World*, 14(3–4):285–290.
9. Papadelis C, Chen Z, Kourtidou-Papadeli C, Bamidis PD, Chouvarda I, Bekiari E, 2007. Monitoring sleepiness with on-board electrophysiological recordings for preventing sleep deprived traffic accidents. *Clinical Neurophysiology*, 9(118), 1906–1922.
10. Tasaki M, Sakai M, Watanabe M, Wang H, Wei D, 2010. Evaluation of drowsiness during driving using electrocardiogram: A driving simulation study. *IEEE International 351 Conference on CIT*, Bradford, UK 1480, 1485.
11. Fabbri M, Provini F, Magosso E, Zaniboni A, Bisulli A, Plazzi G, Ursino M, 2009. Detection of sleep onset by analysis of slow eye movements: A preliminary study of MSLT recordings. *Sleep Medicine*, 10, 637–640.
12. Lin C-T, Che-Jui C, Bor-Shyh L, Shao-Hang H, Chih-Feng C, I-Jan W, 2010. A real-time wireless brain–computer interface system for drowsiness detection. *IEEE Transactions on Biomedical Circuits and Systems*, 4(4), 214–122.

13. Demir F, Bajaj V, Ince MC, Taran S, Şengur A, 2019. Surface EMG signals and deep transfer learning-based physical action classification. *Neural Computing and Applications*, 31(12), 8455–8462.
14. Taran S, Bajaj V, Sharma D, Siuly S, Sengur A, 2018. Features based on analytic IMF for classifying motor imagery EEG signals in BCI applications. *Measurement*, 116, 68–76.
15. Chaudhary S, Taran, S, Bajaj V, Sengur A, 2019. Convolutional neural network based approach towards motor imagery tasks EEG signals classification. *IEEE Sensors Journal*, 19(12), 4494–4500.
16. Crespel A, Gélisse P, Bureau M, Genton P, 2005. *Atlas of Electroencephalography.* Paris: Eurotext.
17. Chaudhary S, Taran S, Bajaj V, Siuly S, 2020. A flexible analytic wavelet transform based approach for motor-imagery tasks classification in BCI applications. *Computer Methods and Programs in Biomedicine*, 187, 105325.
18. Bajaj V, Taran S, Tanyildizi E, Sengur A, 2019. Robust approach based on convolutional neural networks for identification of focal EEG signals. *IEEE Sensors Letters*, 3(5), 1–4.
19. Krishna AH, Sri AB, Priyanka KYVS, Taran S, Bajaj V, 2018. Emotion classification using EEG signals based on tunable-Q wavelet transform. *IET Science, Measurement and Technology*, 13(3), 375–380.
20. Khare SK, Bajaj V, 2020. Time-frequency representation and convolutional neural network-based emotion recognition. *IEEE Transactions on Neural Networks and Learning Systems*, 1–9. doi: 10.1109/TNNLS.2020.3008938.
21. Taran S, Bajaj V, 2018. Clustering variational mode decomposition for identification of focal EEG signals. *IEEE Sensors Letters*, 2(4), 1–4.
22. Khare SK, Bajaj V, Siuly S, Sinha GR, 2020. Classification of schizophrenia patients through empirical wavelet transformation using electroencephalogram signals. *Modelling and Analysis of Active Biopotential Signals in Healthcare*, 1, 1.1–5.26.
23. Nikhil RP, Chuang CY, Ko LW, Chao CF, Jung TP, Liang SF, Lin CT, 2008. EEG-based subject- and session-independent drowsiness detection: an unsupervised approach. *EURASIP Journal on Advances in Signal Processing*, 2008, 1–11.
24. Picot A, Charbonnier S, Caplier A, 2009. Monitoring drowsiness on-line using a single encephalographic channel. In: de Mello CAB, editor, *Biomedical Engineering*. Croatia: In Tech, pp. 145–164.
25. Lin FC, Ko LW, Chuang CH, Su, TP, Lin CT, 2012. Generalized EEG-based drowsiness prediction system by using a self-organizing neural fuzzy system. *IEEE Transactions on Circuits and Systems I: Regular Papers*, 59(9), 2044–2055.
26. Taran S, Bajaj V, 2018. Drowsiness detection using adaptive Hermite decomposition and extreme learning machine for electroencephalogram signals. *IEEE Sensors Journal*, 18(21), 8855–8862.
27. Tripathy RK, Acharya UR, 2018. Use of features from RR-time series and EEG signals for automated classification of sleep stages in deep neural network framework. *Biocybernetics and Biomedical Engineering*, 38(4), 890–902.
28. Li G, Lee BL, Chung WY, 2015. Smartwatch-based wearable EEG system for driver drowsiness detection. *IEEE Sensors Journal*, 15(12), 7169–7180.
29. Anitha C, 2019. Detection and analysis of drowsiness in human beings using multi-modal signals. In: *Digital Business*. Cham: Springer, pp. 157–174.
30. Taran S, Bajaj V, 2018, October. Drowsiness detection using instantaneous frequency based rhythms separation for EEG signals. *In 2018 Conference on Information and Communication Technology (CICT)*, Jabalpur, India (pp. 1–6). IEEE.
31. Khare SK, Bajaj V, Sinha GR, 2020. Automatic drowsiness detection based on variational nonlinear chirp mode decomposition using electroencephalogram signals. *Modelling and Analysis of Active Biopotential Signals in Healthcare*, 1, 5.1–5.25.

32. Bajaj V, Taran S, Khare SK, Sengur A, 2020. Feature extraction method for classification of alertness and drowsiness states EEG signals. *Applied Acoustics*, 163, 107224.

33. Khare SK, Bajaj V, 2020. Optimized tunable Q wavelet transform based drowsiness detection from electroencephalogram signals. *IRBM*, 53(12), 163.

34. Correa AG, Leber EL, 2010 August. An automatic detector of drowsiness based on spectral analysis and wavelet decomposition of EEG records. *In 2010 Annual International Conference of the IEEE Engineering in Medicine and Biology,* Buenos Aires, Argentina (pp. 1405–1408). IEEE.

35. Correa AG, Orosco L, Laciar E, 2014. Automatic detection of drowsiness in EEG records based on multimodal analysis. *Medical Engineering and Physics*, 36(2), 244–249.

36. Belakhdar I, Kaaniche W, Djmel R, Ouni B, 2016, March. A comparison between ANN and SVM classifier for drowsiness detection based on single EEG channel. *In 2016 2nd International Conference on Advanced Technologies for Signal and Image Processing (ATSIP)*, Monastir, Tunisia (pp. 443–446). IEEE.

37. Taran S, Bajaj V, 2019. Sleep apnea detection using artificial bee colony optimize hermite basis functions for EEG signals. *IEEE Transactions on Instrumentation and Measurement*, 69(2), 608–616.

38. Khare SK, Bajaj V, Sinha, GR, 2020. Adaptive tunable Q wavelet transform based emotion identification. *IEEE Transactions on Instrumentation and Measurement*, 69(12), 9609–9617.

39. Taran S, Sharma PC, Bajaj V, 2020. Automatic sleep stages classification using optimize flexible analytic wavelet transform. *Knowledge-Based Systems*, 192, 105367.

40. Khare SK, Bajaj V, 2020. Constrained based tunable Q wavelet transform for efficient decomposition of EEG signals. *Applied Acoustics*, 163, 107234.

41. Sravani C, Bajaj V, Taran S, Sengur A, 2020. Flexible analytic wavelet transform based features for physical action identification using sEMG signals. *IRBM*, 41(1), 18–22.

42. Khare SK, Bajaj V, 2020. A facile and flexible motor imagery classification using electroencephalogram signals. *Computer Methods and Programs in Biomedicine*, 197, 105722.

43. Goldberger AL, Amaral LA, Glass L, Hausdorff JM, Ivanov PC, Mark RG, Mietus JE, Moody GB, Peng CK, Stanley HE, 2000. PhysioBank, PhysioToolkit, and PhysioNet: components of a new research resource for complex physiologic signals. *Circulation*, 101(23), e215–e220.

44. Ichimaru Y, Moody GB, 1999. Development of the polysomnographic database on CD-ROM. *Psychiatry and Clinical Neurosciences*, 53(2), 175–177.

45. Kingsbury NG, 1998, August. The dual-tree complex wavelet transform: A new technique for shift invariance and directional filters. *In IEEE Digital Signal Processing Workshop, South Lake* Tahoe, CA (vol. 86, pp. 120–131). Citeseer.

46. Selesnick IW, Baraniuk RG, Kingsbury NC, 2005. The dual-tree complex wavelet transform. *IEEE Signal Processing Magazine*, 22(6), 123–151.

47. Taran S, Bajaj V, 2019. Emotion recognition from single-channel EEG signals using a two-stage correlation and instantaneous frequency-based filtering method. *Computer Methods and Programs in Biomedicine*, 173, 157–165.

48. Taran S, Bajaj V, Sharma D, 2017, August. TEO separated AM-FM components for identification of apnea EEG signals. *In 2017 IEEE 2nd International Conference on Signal and Image Processing (ICSIP)*, Singapore (pp. 391–395). IEEE.

49. Taran S, Bajaj V, Siuly S, 2017. An optimum allocation sampling-based feature extraction scheme for distinguishing seizure and seizure-free EEG signals. *Health Information Science and Systems*, 5(1), 7.

50. Chada S, Taran S, Bajaj V, 2020. An efficient approach for physical actions classification using surface EMG signals. *Health Information Science and Systems*, 8(1), 3.

51. Taran S, Bajaj V, 2017. Rhythm-based identification of alcohol EEG signals. *IET Science, Measurement and Technology*, 12(3), 343–349.

52. Taran S, Bajaj V, Sharma D, 2017. Robust Hermite decomposition algorithm for classification of sleep apnea EEG signals. *Electronics Letters*, 53(17), 1182–1184.

53. Taran S, Bajaj V, 2019. Motor imagery tasks-based EEG signals classification using tunable-Q wavelet transform. *Neural Computing and Applications*, 31(11), 6925–6932.

54. Lotte F. et al., 2018. A review of classification algorithms for EEG-based brain: Computer interfaces to cite this version : A review of classification algorithms for EEG-based brain-computer interfaces. *Human Brain Mapping*, 38(11), 270–278, doi: 10.1002/hbm.23730.

55. Subasi A, Erçelebi E, 2005. Classification of EEG signals using neural network and logistic regression. *Computer Methods and Programs in Biomedicine*, 78(2), 87–99, doi: 10.1016/j.cmpb.2004.10.009.

56. Wang T, Li Z, Yan Y, Chen H, 2007. A survey of fuzzy decision tree classifier methodology. *Advances in Soft Computing*, 40(3), 959–968, doi: 10.1007/978-3-540-71441-5_104.

57. Jabbar MA, Deekshatulu BL, Chandra P, 2013. Classification of heart disease using K- nearest neighbor and genetic algorithm. *Procedia Technology*, 10, 85–94, doi: 10.1016/j.protcy.2013.12.340.

58. SVMS.org, 2010. Introduction to support vector machines, vol. 2011, January 1st, p. 1, [Online]. Available: http://www.svms.org/introduction.html.

59. Guenther N, Schonlau M, 2016. Support vector machines. *Stata Journal*, 16(4), 917–937, doi: 10.1177/1536867x1601600407.

60. Bennett KP, Campbell C, 2000. Support vector machines: Hype or hallelujah? *SIGKDD Explorations Newsletter*, 2(2), 1–13, doi: 10.1145/380995.380999.

2 Stochastic Event Synchrony Based on a Modified Sparse Bump Modeling: Application to PTSD EEG Signals

Zahra Ghanbari and Mohammad Hassan Moradi
Amirkabir University of Technology

CONTENTS

2.1 INTRODUCTION

Coordination of coupled systems is defined as synchrony [1]. Synchronization concept in chaotic systems has found diverse applications in different fields [2]. Synchrony has found application in studying EEG signals for detecting disorders [3,4], or their effects [5–8]; it is also used in cognitive [9,10] and sleep studies [11], and so on.

Sparse bump modeling, bump modeling in brief, is an approach of extracting oscillatory burst from biomedical signals, especially EEG. It is a 2D extension of the initial version, which was introduced in 2007 [12]. Extracted bursts are organized activities, most likely representative of local synchronizations [13]. Oscillatory activity can be divided into burst and background activity. Background part of EEG signal is composed of regular waves. In contrast, bursts are considered as transient events that have higher amplitudes. Bursts are considered to have a particular functional role, which is different with the activity of background EEG. Encountering a stimulus, there can be two patterns in EEG in response to it: synchronization and desynchronization associated with events. These phenomena are interpreted as the organization of the spontaneous oscillations of brain [14,15]. Bump modeling was first applied to local field potential (LFP) signals [16]. Then, it was used for the early detection of Alzheimer's disease [17]. Bump modeling has also been used to study sleep EEG [18,19].

Stochastic event synchrony (SES) is proposed as a synchrony measure for computing synchrony between two sparse binary strings. SES benefits from making inference in a probabilistic model [20]. Its 2D extension is proposed based on the sparse bump modeling [21]. SES has been applied to EEG signals for the purpose of early diagnosis of Alzheimer's disease [22].

In this chapter, we propose a modified version of sparse bump modeling. Our modified version makes benefits from a powerful transform called the second-order synchrosqueezed wavelet transform, which provides time-frequency maps with more resolutions. It was used on EEG signals in our previous works [23,24]. Then, SES based on this modified sparse bump modeling will be calculated. Our proposed method, which will be called the modified SES, is applied to EEG signals. Signals are collected from participants with posttraumatic stress disorder (PTSD), as well as two control groups. As combat-related PTSD is studied in our work, in addition to healthy controls, trauma-exposed non-PTSD participants are considered as the second control group. Signals are recorded in resting-state eyes closed and resting-state eyes open.

The rest of this chapter is organized as follows: Section 2.2 is dedicated to sparse bump modeling. Section 2.3 provides information about SES. Section 2.4 is focused on synchrosqueezing wavelet transform. Section 2.5 is a brief explanation of our proposed method based on the concepts introduced in previous sections. Section 2.6 is dedicated to data. Section 2.7 reports the results. Finally, Section 2.8 is specialized to conclusion.

2.2 SPARSE BUMP MODELING

This section is dedicated to bump modeling. Bump modeling can be summarized as the following steps [16]:

    i. Transferring signal to time-frequency domain using the wavelet transform
    ii. z-score normalizing the plane
    iii. Describing the map as a set of windows
    iv. Adapting parametric functions within these windows, according to the decreasing order of energy values.

The complex Morlet wavelet is used to generate the time-frequency map corresponding to signal X:

$$C_x(t,s) = \int X(\tau) v^* \left(\frac{\tau - t}{s} \right) d\tau \qquad (2.1)$$

Time-frequency planes are normalized to a reference signal, R. R is the wavelet transformed into C_r. The average amplitudes and standard deviations corresponding to C_r are computed as $M_f = [\mu_1, ..., \mu_F]$, and $S_f = [\sigma_1, ..., \sigma_F]$, associated with each frequency, F. After normalization, the z-scored map is calculated as:

$$Z_x(f,t) = \frac{C_x(f,t) - \mu_f}{\sigma_f} \qquad (2.2)$$

Considering the positive z-score values, if the signal in hand is associated with a stimulation, these oscillatory peaks are the most likely components of them. On the other hand, if the signal is corresponding to the resting state, these oscillations represent the organized oscillatory bursts. It should be mentioned that although z-score values are in $\mathbb{R}$, only values in $\mathbb{R}^+$ are acceptable as input values in bump modeling. For modeling positive z-score values, negative components of the plane are rejected according to a threshold. The obtained map after thresholding, Z_x^Φ, is as follows:

$$Z_x^\Phi = 0.5 \left[(Z_x + \Phi) + |(Z_x + \Phi)| \right] \qquad (2.3)$$

where Φ is the z-score offset, and usually, we have $\Phi \in [0,3]$.

Likewise, the above state, considering negative z-score values, for the signal associated with a stimulus, these oscillatory peaks are the most likely components of negative z-score values. On the contrary, for resting-state signals, these oscillations represent a local disorganization of oscillatory activity. These values in $\mathbb{R}^-$ are extracted into Z_x^-, as a thresholded plane:

$$Z_x^- = -0.5 \left[|(Z_x)| - (Z_x) \right] \qquad (2.4)$$

Z_x is expressed based on a set of windows, $\omega(s, \tau)$, where s and τ stand for the scale and step in time-frequency plane, respectively. Moreover, dimensions of ω are height and width, H and W, respectively, which are specified based on the time-frequency resolution of the central frequency of the window.

After normalizing the time-frequency plane, half-ellipsoid functions are used to model it, which are defined as:

$$\Psi(A, h, \omega, f, t, y, x) = 1 + \frac{(x-t)^2}{\omega^2} - \frac{(y-f)^2}{h^2} \qquad (2.5)$$

Consequently, bumps are calculated as:

$$\zeta(A,h,\omega,\,f,t,y,x) = \begin{cases} 0 & \Psi < \lambda \\ A\sqrt{\Psi} & \text{otherwise} \end{cases} \tag{2.6}$$

Therefore, error of adaptation that should be minimized is:

$$\text{Er}(A,h,\omega,f,t,y,x) = \sum_{x=1}^{W}\sum_{y=1}^{H}\omega_{y,x}(s,\tau) - \zeta(A,h,\omega,f,t,y,x)^2 \tag{2.7}$$

If $\Psi(A,h,\omega,f,t,y,x) < \lambda$ holds, derivatives of bumps are zero. On the contrary, for $\Psi(A,h,\omega,f,t,y,x) \geq \lambda$, derivatives of Er with respect to A, h, ω, f, and t are equal to $-2\sqrt{E}\sqrt{\Psi}$, $-2\sqrt{E}\dfrac{A(y-f)^2}{h^3\sqrt{\Psi}}$, $-2\sqrt{E}\dfrac{A(x-t)^2}{\omega^3\sqrt{\Psi}}$, $-2\sqrt{E}\dfrac{A(y-f)}{h^2\sqrt{\Psi}}$, $-2\sqrt{E}\dfrac{A(x-t)}{l^2\sqrt{\Psi}}$, respectively.

In order to improve adaptation, parameters are optimized stepwise based on a priority, which is related to the order of their derivatives [16]', i.e., h, ω, f, t, and A, respectively.

In bump modeling, it is important to properly choose the window, and the function will be estimated within its boundaries. In the first version of bump modeling, this window has been proposed as [16]:

$$\Omega = \arg\max_{s,\tau} \sum_{y}\sum_{x}\omega_{y,x}(s,\tau) \tag{2.8}$$

However, in the latest version, a method based on matching pursuit [25] is used. In this approach, content of windows are matched with the initial bump function $\zeta_{s,\tau}(y,x) = \zeta(A_{s,\tau}\,,\,h_{s,\tau},\,l_{s,\tau},\,f_{s,\tau},\,y,x)$, where $x_{s,\tau} = h_{s,\tau} = \dfrac{L}{2}$, $f_{s,\tau} = h_{s,\tau} = H/2$; moreover, the highest peak in the window is:

$$A_{s,\tau} = \max_{x,y}(\omega_{y,x}(s,\tau)) \tag{2.9}$$

2.3 STOCHASTIC EVENT SYNCHRONY

This section is dedicated to SES. In the first part, one-dimensional point processes are discussed. The next part will explain the multidimensional extension of SES.

2.3.1 ONE-DIMENSIONAL POINT PROCESSES

In the first step, one-dimensional point processes S and S' are considered. They can be also mentioned as event strings. Our goal is to compute the synchrony between S and S'. Two event strings are called synchronous or locked if they are identical except in three items: a shift in time, small deviations in the times of event occurrence (timing jitter), and omitting or inserting a few events. To speak more accurately, two event strings are synchronous if the timing jitter is small compared with the average interevent time. Moreover, the number of deleted or inserted events is a small proportion of the total number of events. Based on the above definition, SES is introduced as a measure of synchrony between two event strings. In one-dimensional case, SES is proposed as a triplet, namely, δ_t, s_t, ρ_{sp}, where δ_t represents the shift in time, s_t is the timing jitter, and ρ_{sp} stands for the percentage of the spurious events, which indicates events in one string which cannot be paired with an event in the other one. ρ_{sp} is defined as follows [21]:

$$\rho_{sp} = \frac{n_{sp} + n'_{sp}}{n + n'} \tag{2.10}$$

where n_{sp} and n'_{sp} represent the total number of spurious events in S and S', respectively. n and n' are the total number of events in S and S', respectively. Characterizing synchrony with three parameters provides us with the opportunity of distinguishing different synchrony forms. These parameters are calculated by making an inference in a probabilistic model. To explain this model, assume a symmetric process of generating S and S'. First of all, an event string, V, of length l is generated so that the events V_k are uniformly distributed in the $[0, T_0]$ interval and are mutually independent. Z and Z' are obtained as a result of applying delay values of $-\delta_t/2$ and $\delta_t/2$ to V, respectively. Moreover, a slight perturbation is applied to the obtained events occurrence times. The timing jitter has a $s_t/2$ variance. Finally, S and S' are generated from Z and Z' via omitting some events, which correspond to each pair of (Z_k, Z'_k), where Z_k or Z'_k is omitted with the probability of p_s. Consequently, the following statistical model will be achieved:

$$p(s, s', b, b', v, \delta_t, s_t, l) = p(s \mid b, v, \delta_t, s_t) p(s' \mid b', v, \delta_t, s_t)$$

$$\times p(b, b' \mid l) p(v \mid l) p(l) p(\delta_t) p(s_t) \tag{2.11}$$

where b and b' represents the binary strings illustrating the events of S and S' are spurious as follows:

$$\begin{cases} B_k = 1, & S_k \text{ is spurious} \\ B_k = 0, & \text{otherwise} \end{cases} \tag{2.12}$$

We have the similar equation for B'_k. There is a geometric a priori associated with the length l:

$$p(l) = (1 - \lambda) \lambda^l \tag{2.13}$$

where $\lambda \in (0,1)$, and $p(v \mid l) = T_0^{-l}$. For the b and b', a priori is as follows:

$$p(b, b'|l) = (1 - p_s)^{n+n'} p_s^{2l-n-n'} = (1 - p_s)^{n+n'} p_s^{n_{\text{sp}}^{\text{total}}} \qquad (2.14)$$

where $n_{\text{sp}}^{\text{total}}$ represents the total number of spurious events in S and S', and we have:

$$n_{\text{sp}}^{\text{total}} = n_{\text{sp}} + n_{\text{sp}}' = 2l - n - n' \qquad (2.15)$$

where $n_{\text{sp}} = \sum_{k=1}^{n} b_k = l - n'$, and similarly, $n_{\text{sp}}' = \sum_{k=1}^{n'} b_k = l - n$.

Conditional distribution in string event S is as follows:

$$p(s \mid b, v, \delta_t, s_t) = \prod_{k=1}^{n} \left(\mathbb{N} \left(s_k - v_{ik}; -\frac{\delta_t}{2}, \frac{s_t}{2} \right) \right)^{1-b_k} \qquad (2.16)$$

where V_{ik} stands for the event in V, which is associated with S_k, and $\mathbb{N}(s; m, x)$ represents a univariate Gaussian distribution with mean m and variance x. In a similar manner for S', we have:

$$p(s' \mid b', v, \delta_t, s_t) = \prod_{k=1}^{n'} \left(\mathbb{N} \left(s_k' - v_{i'k}; \frac{\delta_t}{2}, \frac{s_t}{2} \right) \right)^{1-b_k} \qquad (2.17)$$

While encoding the prior information about δ_t and s_t is not essential, the following priors can be considered:

$$p(\delta_t) = 1, \; p(s_t) = 1 \qquad (2.18)$$

Therefore, by marginalizing (2.11) with respect to v, it can be written as follows:

$$p(s, s', b, b', \delta_t, s_t, l) = \int p(s, s', b, b', v, \delta_t, s_t, l) \, dv \propto \beta^{n_{\text{sp}}^{\text{total}}} \prod_{k=1}^{n_{\text{nonsp}}} \mathbb{N}(s_{jk}' - s_{jk}; \delta_t, s_t) \qquad (2.19)$$

where $n_{\text{nonsp}} = n + n' - l$ represents the number of non-spurious event pairs, (s_{jk}, s_{jk}') is the pair of non-spurious events, and $\beta = p_s \sqrt{\lambda / T_0}$. As B and B' can be calculated from J and J', the above model will be denoted as $p(s, s', j, j', \delta_t, s_t)$ in the following. l is omitted according to $l = n + n' - n_{\text{nonsp}}$, which indicates that having s, s', b, and b' and given n, n', and n_{nonsp}, l is determined. Moreover, there is no need to specify λ, T_0, and p_s individually since these parameters appear via β, which is used to control the number of spurious events.

Our goal is to calculate δ_t and s_t for event strings S and S'. ρ_{sp} is computed as follows:

$$\rho_{sp} = \frac{\sum_{k=1}^{n} b_k + \sum_{k=1}^{n'} b'_K}{n+n'} \tag{2.20}$$

Among different solutions for the above inference problem, a cyclic maximization is used. First of the two initial values are chosen, i.e., $\hat{\delta}_t^{(0)}$ and $\hat{s}_t^{(0)}$. Second, the update rules are applied until convergence, or alternatively, the determined time is elapsed. The rules are as follows:

$$\left(\hat{j}^{(i+1)}, \hat{j}'^{(i+1)}\right) = \underset{b,b'}{\arg\max}\, p\left(s,s',j,j',\hat{\delta}_t^{(i)},\hat{s}_t^{(i)}\right) \tag{2.21}$$

$$\left(\hat{\delta}_t^{(i+1)}, \hat{s}_t^{(i+1)} = \underset{\delta_t,\, s_t}{\arg\max}\, p\left(s,s',\hat{j}^{(i+1)},\hat{j}'^{(i+1)},\delta_t,s_t\right)\right) \tag{2.22}$$

The first update can be performed through applying the max-product algorithm to an appropriate factor graph or via applying Viterbi algorithm to a proper trellis [26]. This is a dynamic time warping-like procedure [27]. The second update results in the empirical mean and variance values over non-spurious events.

2.3.2 MULTIDIMENSIONAL POINT PROCESSES

Although the algorithm that will be explained in this section can be applied to multidimensional point processes in general, multidimensional point processes in the time-frequency plane will be considered here.

Assume two continuous time signals, for example, different channels of EEG signals. First, the time-frequency representation associated with each signal is approximated as a summation of half-ellipsoid basis functions, called "bumps." A bump is determined by five parameters: time, frequency, width, height, and amplitude, which are denoted as $X, \mathcal{F}, \Delta X, \Delta \mathcal{F}, \mathcal{M}$. Consequently, bump models are obtained as:

$$Y = \left(\left(X_1,\mathcal{F}_1,\Delta X_1,\Delta\mathcal{F}_1,\mathcal{M}_1\right),\ldots,\left(X_n,\mathcal{F}_n,\Delta X_n,\Delta\mathcal{F}_n,\mathcal{M}_n\right)\right) \tag{2.23}$$

$$Y' = \left(\left(X'_1,\mathcal{F}'_1,\Delta X'_1,\Delta\mathcal{F}'_1,\mathcal{M}'_1\right),\ldots,\left(X'_{n'},\mathcal{F}'_{n'},\Delta X'_{n'},\Delta\mathcal{F}'_{n'},\mathcal{M}'_{n'}\right)\right) \tag{2.24}$$

which are actually 5D point processes demonstrating the most prominent oscillatory activity.

In the following, the multidimensional extension of SES is discussed. Similar to the one-dimensional case, a bump may appear in a time-frequency plane, while it does not exist in the other one, or it may appear in both time-frequency planes, but at slightly different positions. In the first case, bumps are called spurious bumps, and in the latter one, they are called non-spurious. In the multidimensional case, the interdependence of two models, Y and Y', are modeled using five parameters, including δ_t, s_t, and ρ_{sp}, which are common with the one-dimensional case, in addition to δ_f and s_f. δ_f represents the average frequency offset between pairwise non-spurious bumps, and s_f stands for the variance of the frequency offset between pairwise non-spurious bumps. The alignment of Y and Y' is performed based on the five parameters in addition to the inference algorithm explained before, which will be denoted as $\theta = (\delta_t, s_t, \delta_f, s_f)$. Equation (2.19) can be extended to the time-frequency domain:

$$p(y, y', j, j', \theta) \propto \beta^{n_{\text{sp}}^{\text{total}}} \prod_{k=1}^{n_{\text{nonsp}}} \mathbb{N}\left(\frac{x'_{k'} - x_k}{\Delta x_k + \Delta x'_{k'}}; \delta_t, s_t \right)$$

$$\times \mathbb{N}\left(\frac{f'_{k'} - f_k}{\Delta f_k + \Delta f'_{k'}}; \delta_f, s_f \right) p(\delta_t) p(s_t) p(\delta_f) \qquad (2.25)$$

where $x'_{k'} - x_k$ is the timing offset and $f'_{k'} - f_k$ is the frequency offset, which are normalized to the values of bumps width and height, respectively. In one-dimensional case, J, J', and θ can be calculated using the cyclic maximization. On the contrary, in the multidimensional case, this cannot be used. Permutation of events should be allowed. Therefore, j_k and $j'_{k'}$ are not necessarily increased monotonically. Consequently, the state space increases very fast, which causes the Viterbi or the max-product algorithms to be impractical.

This issue is addressed through applying the max-product algorithm to a cyclic graph of studying system. Although this will result in a suboptimal procedure of achieving pairwise alignments of bumps, it is practical. Therefore, a new representation of the above model is proposed in Ref. [28]. Corresponding to each pair of events, Y_k and $Y'_{k'}$, a binary variable $C_{kk'}$ is defined as follows:

$$C_{kk'} = \begin{cases} 1 & \text{if } Y_k \text{ and } Y'_{k'} \text{ compose a pair of none-spurious event} \\ 0 & \text{otherwise} \end{cases} \qquad (2.26)$$

The following constrains are held, due to the fact that each event in Y is corresponding to the maximum one event in Y':

$$\sum_{k'=1}^{n'} C_{1k'} = \wp_1 \in \{0,1\}, \sum_{k'=1}^{n'} C_{2k'} = \wp_2 \in \{0,1\}, ..., \sum_{k'=1}^{n'} C_{nk'} = \wp_n \in \{0,1\} \quad (2.27)$$

Likewise, each event in Y' is corresponding to one event in Y at most and can be described by a set of constrains, similarly. We can denote the relationship between sequences $\wp$ and $\wp'$ and sequences B and B' as $B_k = 1 - \wp_k$ and $B'_k = 1 - \wp'_k$, respectively. The whole model can be written as:

$$p(y,y',b,b',c,\theta) \propto \prod_{k=1}^{n} \left(\beta\delta[b_k - 1] + \delta[b_k]\right) \prod_{k'=1}^{n'} \left(\beta\delta[b'_k - 1] + \delta[b']\right)$$

$$\prod_{k'=1}^{n'} \left(\mathbb{N}\left(\frac{x'_{k'} - x_k}{\Delta x_k + \Delta x'_{k'}}; \delta_t, s_t\right) \mathbb{N}\left(\frac{f'_{k'} - f_k}{\Delta f_k + \Delta f'_{k'}}; \delta_f, s_f\right)\right)^{c_{kk'}}$$

$$\times p(\delta_t) p(s_t) p(\delta_f)$$

$$\prod_{k=1}^{n}\left(\delta\left[b_k + \sum_{k'}^{n'} c_{kk'} - 1\right]\right)\prod_{k'}^{n'}\left(\delta\left[b'_{k'} + \sum_{k=1}^{n} c_{kk'} - 1\right]\right) \tag{2.28}$$

As there is no need of prior knowledge about δ_t and δ_f, priors are considered as: $p(\delta_t) = p(\delta_f) = 1$. However, for s_t and s_f, there is prior information. In better words, having smaller values for s_f compared to s_t is rational since a bump that appears in a time-frequency plane is expected to be found in the approximately same frequency in the other plane, but in more different time (with more timing offset). The afore-mentioned prior knowledge is encoded via conjugate prior values for s_t and s_f, which are scaled inverse chi-square distributions. The parameter $\theta = (\delta_t, s_t, \delta_f, s_f)$ and the alignment $C = (C_{11}, C_{12}, \ldots, C_{nn'})$ are determined by maximum a posteriori estimation:

$$\left(\hat{c}, \hat{\theta}\right) = \arg\max_{c,\theta} p(y, y', c, \theta) \tag{2.29}$$

where $p(y, y', c, \theta)$ can be calculated from the following marginalizing over b and b':

$$p(y, y', c, \theta) \propto \prod_{k=1}^{n}\left(\beta\delta\left[\sum_{k'=1}^{n'} c_{kk'}\right] + \delta\left[\sum_{k'=1}^{n'} c_{kk'} - 1\right]\right)$$

$$\cdot \prod_{k'=1}^{n'}\left(\beta\delta\left[\sum_{k'=1}^{n} c_{kk'}\right] + \delta\left[\sum_{k'=1}^{n} c_{kk'} - 1\right]\right) \tag{2.30}$$

$$\cdot \prod_{k=1}^{n}\prod_{k'=1}^{n'}\left(\mathbb{N}\left(\frac{x'_{k'} - x_k}{\Delta x_k + \Delta x'_{k'}}; \delta_t, s_t\right) \mathbb{N}\left(\frac{f'_{k'} - f_k}{\Delta f_k + \Delta f'_{k'}}; \delta_f, s_f\right)\right)^{c_{kk'}} p(\delta_t) p(s_t)$$

Estimation of $\hat{\rho}_{\text{sp}}$ from $\hat{c}$ is as follows:

$$\hat{\rho}_{\text{sp}} = \frac{\sum_{k=1}^{n} \hat{b}_k + \sum_{k=1}^{n'} \hat{b}'_k}{n + n'} = \frac{n + n' - 2\sum_{k=1}^{n}\sum_{k'=1}^{n'} \hat{c}_{kk'}}{n + n'} \tag{2.31}$$

A maximum posteriori estimate for Equation (2.29) does not make sense. Therefore, it is calculated using the cyclic maximization based on the following update rules, until convergence or lapsing the defined time:

$$\hat{c}^{(i+1)} = \arg\max_{c} p\left(y, y', c, \hat{\theta}^{(i)}\right) \tag{2.32}$$

$$\hat{\theta}^{(i+1)} = \arg\max_{\theta} p\left(y, y', \hat{c}^{(i+1)}, \theta\right) \tag{2.33}$$

Initial values are considered as: $\hat{\delta}_t^{(0)} = \delta_f^{(0)}$, $\hat{s}_t^{(0)} = \hat{s}_{0,t}$, and $\hat{s}_f^{(0)} = s_{0,f}$. $\hat{\theta}^{(i+1)}$ can be obtained in a closed form. $\hat{\delta}_t^{(i+1)}$ is the sample mean of the timing offset, which is calculated over all non-spurious event pairs. Similarly, $\hat{\delta}_f^{(i+1)}$, $\hat{s}_t^{(i+1)}$, and $\hat{s}_f^{(i+1)}$ are computed.

In contrast, the optimal pairwise alignment C is not straightforward. It is solved via applying the max-product algorithm to the cyclic factor graph associated with the model (2.30). Inference algorithm is as follows:

$\mu\uparrow(c_{kk'})$ and $\mu\uparrow'(c_{kk'})$ are the upward propagated messages. On the other hand, $\mu\downarrow(c_{kk'})$ and $\mu\downarrow'(c_{kk'})$ are the downward propagated messages. First, we have the following initialization:

$$\mu\uparrow(c_{kk'}) = \mu\uparrow'(c_{kk'}) \propto \left(\mathbb{N}\left(\frac{x'_{k'} - x_k}{\Delta x_k + \Delta x'_{k'}}; \delta_t, s_t \right) \mathbb{N}\left(\frac{f'_{k'} - f_k}{\Delta f_k + \Delta f'_{k'}}; \delta_f, s_f \right) \right)^{c_{kk'}} \tag{2.34}$$

When messages $\mu\uparrow(c_{kk'})$ and $\mu\uparrow'(c_{kk'})$, $k = 1, 2, \ldots, n; k' = 1, 2, \ldots, n'$, are the initialized messages, $\mu\downarrow(c_{kk'})$ and $\mu\downarrow'(c_{kk'})$ are alternatively updated through:

$$\begin{pmatrix} \mu\downarrow(c_{kk'} = 0) \\ \mu\downarrow(c_{kk'} = 1) \end{pmatrix} \propto \begin{pmatrix} \max(\beta, \max_{l' \neq k'} \mu\uparrow(c_{kl'} = 1) / \left(\mu\uparrow(c_{kl'} = 0) \right) \\ 1 \end{pmatrix} \tag{2.35}$$

$$\begin{pmatrix} \mu\downarrow'(c_{kk'} = 0) \\ \mu\downarrow'(c_{kk'} = 1) \end{pmatrix} \propto \begin{pmatrix} \max(\beta, \max_{l \neq k} \mu\uparrow'(c_{lk'} = 1) / \left(\mu\uparrow(c_{lk'} = 0) \right) \\ 1 \end{pmatrix} \tag{2.36}$$

Moreover, $\mu\uparrow(c_{kk'})$ and $\mu\uparrow'(c_{kk'})$ are alternatively updated through:

$$\mu\uparrow(c_{kk'}) \propto \mu\downarrow'(c_{kk'}) \left(\mathbb{N}\left(\frac{x'_{k'} - x_k}{\Delta x_k + \Delta x'_{k'}}; \delta_t, s_t \right) \mathbb{N}\left(\frac{f'_{k'} - f_k}{\Delta f_k + \Delta f'}; \delta_f, s_f \right) \right)^{c_{kk'}} \tag{2.37}$$

$$\mu\uparrow'(c_{kk'}) \propto \mu\downarrow(c_{kk'}) \left(\mathbb{N}\left(\frac{x'_{k'} - x_k}{\Delta x_k + \Delta x'_{k'}}; \delta_t, s_t \right) \mathbb{N}\left(\frac{f'_{k'} - f_k}{\Delta f_k + \Delta f'_{k'}}; \delta_f, s_f \right) \right)^{c_{kk}} \tag{2.38}$$

The above updates are done until convergence, or elapsing time. It should be mentioned that despite the fact that there is no guarantee for convergence of cyclic graph, [28] expresses that the four above updates always converge to a fixed point, based on their experiments.

After a brief introduction to bump modeling and SES, in the following section the concept of synchrosqueezed wavelet transform is explained.

2.4 SYNCHROSQUEEZED WAVELET TRANSFORM

Synchrosqueezed wavelet transform is reviewed in this section. The first part is dedicated to the original 1D form, and the second part of this section is focused on the 2D version.

2.4.1 THE ORIGINAL FORM

Synchrosqueezed wavelet transform was proposed by Daubechies et al. [29]. Synchrosqueezing is a special case of reallocation methods. Assume a time-frequency plane, $R(t, \omega)$. A reallocation approach is used to sharpen it by allocating points of this plane to another point set, $R(t', \omega')$. The mentioned allocation is defined as the local behavior of $R(t, \omega)$ around (t, ω). Assuming a signal, $x(t)$, a continuous wavelet transform is:

$$W_x = \int x(t) \frac{1}{\sqrt{a}} \overline{\Psi\left(\frac{t-b}{a}\right)} dt \qquad (2.39)$$

where φ represents a wavelet chosen approximately. Reallocating $W_x(a, b)$ yields a concentrated time-frequency plane. Instantaneous frequency lines will be extracted as a result. The wavelet-based synchrosqueezing transform (WSST) was primarily applied to auditory signals [30]. Signals containing constituent components with time-varying oscillatory characteristics are illustrated as:

$$x(t) = \sum_{k=1}^{K} x_k(t) \qquad (2.40)$$

where $x_k(t)$ denotes a Fourier-like oscillatory component. $x_k(t)$ represents a mono-component asymptotic $AM - FM$ signal, and we have

$$x_k(t) = A_k(t) \cos(\phi_k(t)) \qquad (2.41)$$

where $A_k(t)$ stands for the instantaneous amplitude, and $\phi_k(t)$ represents the instantaneous phase. The instantaneous frequency is obtained as follows:

$$f_{\text{inst}}(t) = d\phi_k(t) / dt \qquad (2.42)$$

Calculating WSST can be summarized in the following steps:

i. Complex continuous wavelet transform is computed according to Equation (2.39). φ as the mother wavelet should satisfy the condition for any ε as follows:

$$\int_0^\infty \frac{1}{\varepsilon} |\Psi(\varepsilon)|^2 \, d\varepsilon < \infty \tag{2.43}$$

For $\varepsilon < 0$, if the condition $\hat{\psi}(\varepsilon)$ holds, $\psi(\varepsilon)$ is analytic. Then, $\psi(t)$ is written as:

$$\Psi(t) = \psi(t) e^{j\omega_0 t} \tag{2.44}$$

where $\psi(t)$ is the window function, and ω_0 stands for the central angular frequency of the wavelet. Wavelet ridge is defined as:

$$P = \left\{ (a_r, b) \in \mathbb{R}^2; M_x(a_r, b) \right\} = \max \left(\left| W_x(a_i, b) \right| \right) \tag{2.45}$$

where (a_r, b) represents the ridge points at b, and $\left| W_x(a_i, b) \right|$ the wavelet coefficient modulus. As in actual applications the analytic form of the signal is not usually available, the extracted ridge is used to obtain the instantaneous frequency of the signal:

$$f_{\text{inst}}(t) = \frac{\omega_0}{2\pi a_r(t)} \tag{2.46}$$

The instantaneous amplitude is computed as follows:

$$A_{\text{inst}}(t) \approx \frac{2 \left| W_x(a_r(t), t \right|}{\sqrt{a_r(t)} \left| \hat{\psi}(0) \right|} \tag{2.47}$$

where $\hat{\psi}(0)$ represents the Fourier transform of $\psi(t)$ at $\omega = 0$. WSST is proposed to overcome the problem of blurring in time-frequency planes. Denoting $f_{\text{inst}}(t)$ by f and $A_{\text{inst}}(t)$ by A, we will have:

$$W_x(a, b) = \frac{1}{\sqrt{a}} \int s(\varepsilon) \hat{\Psi}(a\varepsilon) e^{ib\varepsilon} d\varepsilon = \frac{A}{2\sqrt{a}} \hat{\psi}(af) e^{ibf} \tag{2.48}$$

$\hat{\Psi}(\varepsilon)$ is concentrated around $\varepsilon = f_0$. As a result, $W_x(a,b)$ will be concentrated around $a = \omega_0 / \omega$. In contrast, $W_x(a,b)$ usually spreads out over a region around $a = f_0 / f$. Regardless of values, $W_x(a, b)$ represents an oscillatory behavior with original frequency, f, at point b.

ii. For each (a,b), a primary estimation of the phase transform is obtained as follows:

$$f_{\text{inst}}(a,b) = \frac{i}{W_x(a,b)} \, \partial W_x(a,b) / \partial b \tag{2.49}$$

where $W_x(a,b) \neq 0$. This formulation will suppress the influence of wavelet on W_x.

iii. Reassignment of $W_x(a,b)$ yields WSST, which is formulated as follows:

$$X_{s,\bar{\varepsilon}}^{\alpha}(b,\omega) = \int_{A_{\bar{\varepsilon},x}(b)} \frac{1}{\alpha} W_x(a,b) g\left(\frac{\omega - \omega_x(a,b)}{\alpha}\right) a^{\frac{3}{2}} da \tag{2.50}$$

where g denotes the smoothing function, α represents the accuracy value, and $\bar{\varepsilon}$ is the threshold. Moreover, $A_{\bar{\varepsilon},x}(b) = \left\{a; |W_x(a,b)| > \bar{\varepsilon}\right\}$ frequency, f, and scaling variables, a, b, are the discrete values. Consequently, $W_x(a,b)$ is merely computed in discrete values. $(b,a) \rightarrow (b, f_{\text{inst}}(a,b))$ describes a mapping from the time domain to the time-frequency domain. Due to the above mapping, synchrosqueezing transform, $X(f, b)$, is only calculated at the center of the successive bins:

$$\left[f_l - 0.5(f_l - f_{l-1}), f_l + 0.5(f_l - f_{l-1}) \right] \tag{2.51}$$

Thus, synchrosqueezing transform is defined as follows [29]:

$$X(f_l,b) = \frac{1}{\Delta f} \sum_{a_k : |f_x(a_k,\, b) - f_l| \leq \frac{\Delta f}{2}} W_x(a_k,b) a_k^{-\frac{3}{2}} (\Delta a)_k \tag{2.52}$$

According to the following equation, it has been proved that WSST has the perfect concentration property:

$$\int_0^{\infty} W_x(a, b) a^{-\frac{3}{2}} da = \frac{1}{2\pi} \int_{-\infty}^{\infty} \int_0^{\infty} \hat{s}(\varepsilon) \overline{\hat{\Psi}(a\varepsilon)} e^{ib\varepsilon} / a \, da \, d\varepsilon$$

$$= \frac{1}{2\pi} \int_0^{\infty} \int_0^{\infty} \hat{s}(\varepsilon) \overline{\hat{\Psi}(a\varepsilon)} e^{ib\varepsilon} / a \, da \, d\varepsilon \tag{2.53}$$

2.4.2 Second-Order Wavelet-Based SST (WSST2)

Although WSST is a powerful method for boosting the time-frequency representation, it is only applicable to a class of multicomponent signals constituting slightly perturbed purely harmonic modes. In order to address the issue, an extension of WSST is proposed. This improved version is called as the second-order wavelet-based synchrosqueezing transform (WSST2). WSST2 benefits from a more accurate instantaneous frequency estimation [31]. As the first step, defining a second-order local modulation operator is needed. Then, it will be used to compute the instantaneous frequency estimate. The modulation operator is associated with the first-order derivatives (with respect to t) of the reassignment operators.

Corresponding to a given signal $x \in L^\infty(\mathbb{R})$, for any (t,a), the complex reassignment operators are defined as:

$$\tilde{\omega}_f(t,a) = \frac{1}{2\pi i} \frac{\partial_t W_x^\Psi(t,a)}{W_x^\Psi(t,a)} \tag{2.54}$$

$$\tilde{\tau}_x(t,a) = \frac{\int_{\mathbb{R}} \tau x(\tau) a^{-1} \overline{\Psi\left(\frac{\tau-t}{a}\right)} d\tau}{W_x^\Psi(t,a)} = t + a \frac{W_x^{t\Psi}(t,a)}{W_x^\Psi(t,a)} \tag{2.55}$$

where $W_x^\Psi(t,a) \neq 0$. Operators defined in a manner that $t\Psi \in L^1(\mathbb{R})$ and $\Psi' \in L^1(\mathbb{R})$ hold. Consequently, the second-order local complex modulation operator is defined as:

$$\tilde{q}_{t,x}(t,a) = \frac{\partial_t \tilde{\omega}_x(t,a)}{\partial_t \tilde{\tau}_x(t,a)}, \quad \partial_t \tilde{\tau}_x(t,a) \neq 0 \tag{2.56}$$

One can define another second-order local modulation operator based on partial derivatives with respect to a:

$$\tilde{q}_{a,x}(t,a) = \frac{\partial_a \tilde{\omega}_x(t,a)}{\partial_a \tilde{\tau}_x(t,a)} \tag{2.57}$$

Since $\tilde{q}_{a,x}(t,a)$ and $\tilde{q}_{t,x}(t,a)$ have the same properties, the improved instantaneous frequency estimation corresponding to the time-frequency representation of $x \in L^\infty(\mathbb{R})$ is

$$\text{Re}\left\{\tilde{\omega}_x^{[2]}(t,a)\right\} \tag{2.58}$$

where $\text{Re}\{.\}$ denotes the real part, where

$$\tilde{\omega}_x^{[2]}(t,a) = \begin{cases} \tilde{\omega}_x(t,a) + \tilde{q}_{t,x}(t,a)\left(t - \tilde{\tau}_x(t,a)\right) & \text{if } \partial_t \tilde{\tau}_x(t,a) \neq 0 \\ \tilde{\omega}_x(t,a) & \text{otherwise} \end{cases} \tag{2.59}$$

For a Gaussian modulated linear chirp x, as $x(t) = A(t)e^{i2\pi\varphi(t)}$, where $\log A(t)$ and $\varphi(t)$ are both quadratic, it is proven that $\mathrm{Re}\{\tilde{\omega}_x^{[2]}(t,a)\}$ is an exact estimate of $\dfrac{d\varphi(t)}{dt}$ [31]. Instantaneous frequency corresponding to a more general mode can be estimated with Gaussian amplitude using $\mathrm{Re}\{\tilde{\omega}_x^{[2]}(t,a)\}$. In this case, the estimation error just involves the phase derivatives of orders more than three.

Given a $x \in L^\infty(\mathbb{R})$, $\tilde{\omega}_x(t,a)$ can be computed using just five continuous wavelet transform (CWT)s:

$$\tilde{\omega}_x(t,a) = -\frac{1}{i2\pi a}\frac{W_x^{\Psi'}(t,a)}{W_x^{\Psi}(t,a)} \tag{2.60}$$

For $\tilde{q}_{t,f}(t,a)$, it can be written in a similar manner:

$$\tilde{q}_{t,f}(t,a) = \frac{1}{i2\pi a^2}\frac{W_x^{\Psi''}(t,a)W_x^{\Psi}(t,a) - W_x^{\Psi'}(t,a)^2}{W_x^{t\Psi}(t,a)W_x^{\Psi'}(t,a) - W_x^{t\Psi'}(t,a)W_x^{\Psi}(t,a)} \tag{2.61}$$

where $t \to W^{\Psi'}, W^{t\Psi}, W^{\Psi''}, W^{t\Psi'}$ are CWTs associated with x obtained in $L^1(\mathbb{R})$, based on wavelets $\Psi', t\Psi, \Psi'', t\Psi'$, respectively. The above formulation can be easily derived from:

$$\partial_t^p W_x^{\Psi}(t,a) = (-a)^{-p} W_x^{\Psi^p}(t,a) \tag{2.62}$$

The second-order WSST, WSST2, is calculated by replacing $\tilde{\omega}_x(t,a)$ with $\hat{\omega}_x^{[2]}(t,a)$ in Ref. [32]:

$$S_{2,W_x^{\Psi}}^{\gamma}(t,\omega) := \int_{|W_x^{\Psi}(t,a)|>\gamma} W_x^{\Psi}(t,a)\delta\left(\omega - \hat{\omega}_x^{[2]}(t,a)\right)da\,/\,a \tag{2.63}$$

Finally, s_k is retrieved via replacing $S_{2,W_x^{\Psi}}^{\gamma}(t,\omega)$ in Ref. [33].

2.4.2.1 Numerical Implementation of WSST2

This section is dedicated to the numerical implementation of WSST2. Consume signal x which is uniformly discretized in $[0,1]$, at values $t_m = m/n$ with $m = 0,\ldots,n-1$, and $n = 2^L, L \in \mathbb{N}$. First, W_x^{Ψ} is discretized at points $\left(\dfrac{m}{n},a_j\right)$, where $a_j = \dfrac{1}{n}2^{\frac{j}{n_v}}$, $j = 0, \ldots, Ln_v$. n_v represents the voice number. n_v is a user-defined controlling parameter, which controls the scale number. Proposed values for n_v are 32 and 64 [34]. Discrete wavelet transform (DWT) in Fourier domain associated with x is defined as follows:

$$W_x^{\Psi}\left(t_m, a_j\right) \approx W_{d,x}^{\hat{\Psi}}\left(m, j\right) := \left(\mathcal{F}_d^{-1}\left(\left(\mathcal{F}_d\left(x\right) \odot \overline{\hat{\Psi}_j}\right)\right)\right)_m \qquad (2.64)$$

where $\mathcal{F}_d\left(x\right)$ represents the standard discrete Fourier transform (DFT) of signal x, and $\mathcal{F}_d^{-1}\left(x\right)$ stands for the inverse of $\mathrm{F}_d\left(x\right)$, i.e., iDFT. $\hat{\Psi}_{j,q} = \hat{\Psi}\left(a_j q\right), q = 0, 1, \ldots, n-1$. $\odot$ represents the elementwise multiplication. The complex estimate of the second-order modulation operator, $\tilde{q}_{t,x}$, is computed as [35]:

$$\tilde{q}_{d,t,x}\left(m, j\right) = \frac{i2\pi\left(W_{d,x}^{\hat{\Psi}}\left(m, j\right) W_{d,x}^{\varepsilon^2\hat{\Psi}}\left(m, j\right) - \left(W_{d,x}^{\varepsilon\hat{\Psi}}\left(m, j\right)\right)^2\right)}{a_j^2\left[\left(W_{d,x}^{\hat{\Psi}}\left(m, j\right)\right)^2 + W_{d,x}^{\hat{\Psi}}\left(m, j\right) W_{d,x}^{\varepsilon\hat{\Psi}'}\left(m, j\right) - W_{d,x}^{\hat{\Psi}'}\left(m, j\right) W_{d,x}^{\varepsilon\hat{\Psi}}\left(m, j\right)\right]}$$

$$(2.65)$$

where $W_{d,x}^{\varepsilon^2\hat{\Psi}}$, $W_{d,x}^{\varepsilon\hat{\Psi}}$, $W_{d,x}^{\varepsilon\hat{\Psi}'}$, and $W_{d,x}^{\hat{\Psi}'}$ stand for DWTs of x, which are calculated based on wavelets $\varepsilon \to \varepsilon^2\hat{\psi}$, $\varepsilon \to \varepsilon\hat{\psi}$, $\varepsilon \to \varepsilon\hat{\psi}'$, and $\varepsilon \to \hat{\psi}'$, respectively.

Using definitions as follows:

$$\tilde{\omega}_{d,x}\left(m, j\right) = \frac{W_{d,x}^{\varepsilon\hat{\Psi}}\left(m, j\right)}{a_j W_{d,x}^{\hat{\Psi}}\left(m, j\right)} \qquad (2.66)$$

$$\tilde{\tau}_{d,x}\left(m, j\right) = t + \frac{a_j}{i2\pi} \frac{W_{d,x}^{\hat{\Psi}'}\left(m, j\right)}{W_{d,x}^{\hat{\Psi}}\left(m, j\right)} \qquad (2.67)$$

A discrete form of a second-order complex instantaneous frequency estimate corresponding to x is quantified based on the following formulation:

$$\tilde{\omega}_{d,x}^{[2]}\left(m, j\right) = \begin{cases} \tilde{\omega}_{d,x}\left(m, j\right) + \tilde{q}_{d,t,x}\left(m, j\right)\left(t - \tilde{\tau}_{d,x}\left(m, j\right)\right) & \text{if } \partial_t \tilde{\tau}_{d,x}\left(m, j\right) \neq 0 \\ \tilde{\omega}_{d,x}\left(m, j\right) & \text{otherwise} \end{cases}$$

$$(2.68)$$

where $\partial_t \tilde{\tau}_{d,x}\left(m, j\right)$ is equal to

$$\frac{W_{d,x}^{\hat{\Psi}}\left(m, j\right) W_{d,x}^{\varepsilon\hat{\Psi}'}\left(m, j\right) - W_{d,x}^{\hat{\Psi}'}\left(m, j\right) W_{d,x}^{\varepsilon\hat{\Psi}}\left(m, j\right) + \left(W_{d,x}^{\hat{\Psi}}\left(m, j\right)\right)^2}{W_{d,x}^{\hat{\Psi}}\left(m, j\right)^2} \qquad (2.69)$$

The discrete instantaneous frequency estimate is obtained as the real part of $\tilde{\omega}_{d,x}^{[2]}\left(m, j\right)$ written as:

$$\hat{\omega}_{d,x}^{[2]}\left(m, j\right) = \mathrm{Re}\left\{\tilde{\omega}_{d,x}^{[2]}\left(m, j\right)\right\} \qquad (2.70)$$

2.4.2.2 Quantifying WSST2

This section is dedicated to calculating WSST2. First of all, the frequency domain should be spelt. It should be emphasized that each scale a_j is equal to the inverse of frequency; $f_j = \left(a_j\right)^{-1} = 2^{-\frac{j}{n_v}} n$. Frequency bins corresponding to the wavelet representation are as follows:

$$W_j = \left[0.5\left(f_{j+1} + f_j \right), 0.5\left(f_j + f_{j-1} \right) \right] \qquad (2.71)$$

where f_{Ln_v} is equal to 0 and f_{-1} approaches to ∞, and $0 \leq j \leq L_{n_v} - 1$. Therefore, the second-order synchrosqueezing operator is calculated as:

$$S_{d,2,x}^{\gamma}\left(\frac{m}{n} f_j \right) = \sum_{\mathbb{G}_d(j)} W_{d,x}^{\hat{\Psi}}(m,l) \frac{\log 2}{n_v} \qquad (2.72)$$

where, $\mathbb{G}_d\left(j \right) = \left\{ 0 \leq l \leq L_{n_v} \text{ s.t. } \hat{\omega}_{d,x}^{[2]}(m,l) \in W_j \text{ and } \left| W_{d,x}^{\hat{\Psi}}(m,l) \right| > \gamma \right\}$

For each t_m, the following equation holds:

$$f_k\left(\frac{m}{n} \right) \approx \frac{1}{C_{d,\psi,k}'} \sum_{l \in Y_k(m)} S_{d,2,x}^{\gamma}(m,\omega_l) \qquad (2.73)$$

where f_k represents a mode, $C_{d,\psi,k}'$ is a discrete approximation of $C_{\psi,k}'$, and $Y_k(m)$ stands for a set of measures corresponding to a narrow frequency band around the ridge curve of the kth mode. Method proposed in Refs. [35,36] is applied to extract this ridge.

It should be mentioned that the accuracy of the set $Y_k(m)$ is dependent on the frequency of the kth mode.

2.5 STOCHASTIC EVENT SYNCHRONY BASED ON A MODIFIED BUMP MODELING

Related concepts, including bump modeling, one- and multidimensional SES, are explained in the previous sections. Moreover, WSST and WSST2 are illustrated. Bump modeling benefits from pruning the time-frequency plane using half-ellipsoid functions. In the original sparse bump modeling, wavelet transform with complex Morlet is used. Our contribution is introducing a modified version of sparse bump modeling using WSST2 instead of original wavelet transform. WSST2 can provide the time-frequency planes with higher resolutions. SES works based on bump modeling. So, it is supposed that a new version of bump modeling affects SES as well.

2.6 DATA DESCRIPTION AND PREPROCESSING

Data is described in this section. Resting-state EEG signals are recorded in two states: resting-state eyes closed and eyes open. Our study was performed with 45 right-handed Iranian men participants. Fifteen of them were diagnosed with chronic PTSD, due to participating in Iran-Iraq war (1980–1988). Fifteen other participants have been exposed to the same trauma but diagnosed as non-PTSD. Fifteen other participants were non-trauma-exposed healthy controls. The aforementioned groups, i.e., combat-related PTSD, combat trauma-exposed-non-PTSD, and healthy controls, will be mentioned as PTSD, non-PTSD, and control, respectively.

Exclusion criteria were current symptoms or history of a diagnosed neurological disorder, psychosis, and current substance abuse or dependence. Participants of the non-PTSD and control groups are recruited from the local community. They were not taking medications with potential psychoactive influences. In addition, the control group participants did not express any history of a major trauma, i.e., a serious disease, surgery or physical injury, car accident, sexual assault, and also combat experience.

PTSD or non-PTSD labels were assigned based on the examinations made by a psychiatrist and a psychologist. All of the participants were asked to answer the structured clinical interview for Diagnostic and Statistical Manual of Mental Disorder, 5th edition (DSM[5th]). Additionally, they were asked to answer the questionnaire of PTSD checklist (PCL) as well as depression, anxiety stress scales (DASS-21) questionnaire. PCL questionnaire monitors and measures trauma-related symptoms severity in the previous week. DASS-21 questionnaire monitors anxiety, stress, and depressive severity in the previous 2 weeks. Clinical and demographic information of participants is summarized in Table 2.1.

Our study has been confirmed via ethical approvals from Amirkabir University of Technology, and Sadr Hospital, Tehran, Iran. First, participants were totally informed about the study procedure. Then, they were asked to sign a written consent form. At the end of the session, participants received a financial compensation for their participation.

EEG data was recorded in one session, in a quiet room with normal lightening. Data was recorded using g.Tec system with active electrodes. Signals are recorded in 16 channels, and 1200 Hz is used as the sampling frequency. Recording sites are

TABLE 2.1
Clinical and Demographical Information of Participants [37,38]

Characteristics	Control Mean ± std	Non-PTSD Mean ± std	PTSD Mean ± std
Age (years)	50.867 ± 3.870	53.1 ± 2.0	52.8 ± 4.3
Marital state (1: married, 0: divorced)	1 ± 0	1 ± 0	0.8 ± 0.4
Number of children	1.8 ± 0.8	2.4 ± 0.7	2.5 ± 1.0
Education (years)	11.9 ± 2.5	14.8 ± 1.9	9.0 ± 3.5
Age of trauma	-	17.6 ± 1.2	18.6 ± 3.2
Depression and anxiety	9.1± ± 3.8	8.7 ± 6.8	50.0 ± 6.9
PTSD	-	7.7 ± 4.0	55.5 ± 12.7

considered as AF3, AF4, F3, Pz, F4, F8, FC3, FC4, T8, CP3, P5, P6, PO3, PO4, O1, and O2. Fz and left ear were used as ground and reference, respectively. A band-pass filter with $0.1-$ Hz and $100-$ Hz cutoff frequencies was applied during data acquisition. Furthermore, a $50-$ Hz notch filter was used. Impedance of electrodes has been kept below 5 KΩ during recording data. EEG signals were recorded for 5 minutes in resting-state eyes closed. Similarly, EEG signals were recorded for 5 minutes in resting-state eyes open.

To avoid the potential bias caused by considering the left ear as the reference, proper re-referencing is applied. First of all, gross movement artifacts are removed. At the next step, independent component analysis (ICA) is applied to de-noise signals. The next step is filtering signals into delta band $(1-4\ Hz)$, theta band $(4-8\ Hz)$, alpha band $(8-13\ Hz)$, beta band $(13-25\ Hz)$, and gamma band $(25-40\ Hz)$. Filtered signals are segmented into epochs of 1000 ms length. The same is done for the whole band signal. For more details, see [37,38].

2.7   RESULTS

In the present section, the results of applying the proposed method to PTSD EEG data are reported. Moreover, original SES is applied to data, and the results are compared. As expressed before, data is recorded in the resting-state eyes open and the resting-state eyes closed. Signals are studied in the whole band, as well as five sub-bands for these two states. Results are reported using statistical analysis, i.e., analysis of variance (ANOVA) at 99% confidence level. As mentioned, SES provides five output parameters: time delay, timing jitter variance, ratio of spurious events, variance of the frequency offset (frequency jitter), and the average interevent frequency offset in the time-frequency map. Investigating these features reveals that ρ_{sp} which shows the ratio of spurious events, has the best performance among five features provided either by the modified SES or either by the original one. Therefore, in this part, the results are reported based on this feature.

First, we use ρ_{sp} to study whether three groups are separable or not. Results are summarized in Table 2.2 for the resting-state eyes closed, and similarly in Table 2.3, corresponding to the resting-state eyes open.

In the next part, pairwise separability of groups is studied based on ρ_{sp}. Results for the resting-state eyes closed and open are summarized in Tables 2.4 and 2.5, respectively.

TABLE 2.2

p-Values Associated with Separability of Three Groups in Resting-State Eyes Closed, Based on SES and Modified SES

Classes	Delta	Theta	Alpha	Beta	Gamma	Whole Band
Modified SES	0.001	0.318	0.061	0.095	0.257	< 0.0001
SES	0.181	0.706	0.094	0.619	1	0.001

TABLE 2.3
***p*-Values Associated with Separability of Three Groups in Resting-State Eyes Open, Based on SES and Modified SES**

Classes	Delta	Theta	Alpha	Beta	Gamma	Whole Band
Modified SES	0.010	0.174	0.072	< 0.0001	0.047	< 0.0001
SES	0.319	1	0.571	0.054	0.202	< 0.0001

TABLE 2.4
***p*-Values Associated with Separability of Pairwise Groups in Resting-State Eyes Closed, Based on SES and Modified SES**

Classes	Method	Delta	Theta	Alpha	Beta	Gamma	Whole Band
Control-non-PTSD	Modified SES	< 0.0001	0.012	< 0.0001	0.006	0.417	< 0.0001
	SES	0.008	0.959	0.390	0.368	1	0.002
Control-PTSD	Modified SES	< 0.0001	0.004	0.176	< 0.001	0.304	0.003
	SES	< 0.0001	0.567	1	0.002	1	0.014
Non-PTSD-PTSD	Modified SES	< 0.0001	1	0.219	0.016	0.169	< 0.0001
	SES	0.013	1	0.617	0.852	0.007	

TABLE 2.5
***p*-Values Associated with Separability of Pairwise Groups in Resting-State Eyes Open, Based on SES and Modified SES**

Classes	Method	Delta	Theta	Alpha	Beta	Gamma	Whole Band
Control-non-PTSD	Modified SES	< 0.0001	0.245	0.002	< 0.0001	0.142	< 0.0001
	SES	< 0.0001	0.306	0.111	0.005	0.185	0.382
Control-PTSD	Modified SES	0.004	< 0.0001	< 0.0001	< 0.0001	0.01	< 0.0001
	SES	0.130	1	0.106	0.009	0.016	< 0.0001
Non-PTSD-PTSD	Modified SES	0.357	0.001	< 0.0001	< 0.0001	< 0.0001	< 0.0001
	SES	1	0.150	0.281	0.270	0.216	< 0.0001

2.8 CONCLUSION

In this chapter, a modified version of bump modeling is proposed. This modified version is generated making benefits from the second-order syncrosqueezed wavelet transform (WSST2). WSST, as an EMD-like method, is inherently a combination of wavelet transform and reallocation methods. Therefore, it provides the time-frequency maps with higher resolution and less blurring compared to wavelet transform. WSST2 is a modified version of WSST. Bump modeling approximates the corresponding time-frequency map of the signal with a sum of half-ellipsoid basis functions, called bumps. Our contribution is using WSST2 for introducing a modified version of bump modeling. SES is a synchrony measure based on bump modeling. It describes the alignment of two point processes based on five parameters: time delay, variance of timing jitter, ratio of spurious events, the variance of the frequency offset (frequency jitter), and the average frequency offset between events. Using the proposed modified bump modeling, we achieve a modified SES as well. The modified SES is applied to EEG signals of the resting-state eyes closed and open, recorded from participants with combat-related PTSD and two control groups, namely, trauma-exposed non-PTSD and healthy controls. Five aforementioned features are extracted using the modified SES and the original one. Results are assessed using ANOVA at 99% confidence level. Our results reveal the superiority of ρ_{sp}, the ratio spurious events, compared to other extracted features. Moreover, the results indicate that the modified SES has a better performance compared to the original one.

ACKNOWLEDGMENT

The authors would like to express their appreciation to Professor F. B. Vialatte, Professor J. Dauwels, Professor M. B. Shamsollahi, Professor A. R. Moradi, and Dr. J. Mirzaei for their kind advices.

REFERENCES

1. Ghanbari, Z., and M. H. Moradi. 2020b. "FSIFT-PLV: An emerging phase synchrony index." *Biomedical Signal Processing and Control* 57:101764. doi: 10.1016/j.bspc.2019.101764.
2. Mormann, F., K. Lehnertz, P. David, and C. E. Elger. 2000. "Mean phase coherence as a measure for phase synchronization and its application to the EEG of epilepsy patients." *Physica D: Nonlinear Phenomena* 144(3):358–369. doi: 10.1016/S0167-2789(00)00087-7.
3. Omidvarnia, A., G. Azemi, P. B. Colditz, and B. Boashash. 2013. "A time–frequency based approach for generalized phase synchrony assessment in nonstationary multivariate signals." *Digital Signal Processing* 23(3):780–790. doi: 10.1016/j.dsp.2013.01.002.
4. Shahsavari Baboukani, P., G. Azemi, B. Boashash, P. Colditz, and A. Omidvarnia. 2019. "A novel multivariate phase synchrony measure: Application to multichannel newborn EEG analysis." *Digital Signal Processing* 84:59–68 doi: 10.1016/j.dsp.2018.08.019.
5. Kawano, T., N. Hattori, Y. Uno, K. Kitajo, M. Hatakenaka, H. Yagura, H. Fujimoto, T. Yoshioka, M. Nagasako, H. Otomune, and I. Miyai. 2017. "Large-scale phase synchrony reflects clinical status after stroke: An EEG study." *Neurorehabilitation and Neural Repair* 31(6):561–570. doi: 10.1177/1545968317697031.
6. Bajaj, V., K. Rai, A. Kumar, D. Sharma, and G. K. Singh. 2017. "Rhythm-based features for classification of focal and non-focal EEG signals." *IET Signal Processing* 11(6):743–748.
7. Taran, S., and V. Bajaj. 2018b. "Rhythm-based identification of alcohol EEG signals." *IET Science, Measurement and Technology* 12(3):343–349.

8. Taran, S., and V. Bajaj. 2018a. "Drowsiness detection using instantaneous frequency based rhythms separation for EEG signals." *2018 Conference on Information and Communication Technology (CICT)*, Jabalpur, India, 26–28 October 2018.

9. Fell, J., and N. Axmacher. 2011. "The role of phase synchronization in memory processes." *Nature Reviews Neuroscience* 12:105. doi: 10.1038/nrn2979.

10. Glennon, M., M. A. Keane, M. A. Elliott, and P. Sauseng. 2016. "Distributed cortical phase synchronization in the EEG reveals parallel attention and working memory processes involved in the attentional blink." *Cerebral Cortex* 26(5):2035–2045. doi: 10.1093/cercor/bhv023.

11. Mezeiová, K., and M. Paluš. 2012. "Comparison of coherence and phase synchronization of the human sleep electroencephalogram." *Clinical Neurophysiology* 123(9):1821–1830. doi: doi: 10.1016/j.clinph.2012.01.016.

12. Dubois, R., P. Maison-Blanche, B. Quenet, and G. Dreyfus. 2007. "Automatic ECG wave extraction in long-term recordings using Gaussian mesa function models and nonlinear probability estimators." *Computer Methods Programs Biomed* 88(3):217–233. doi: 10.1016/j.cmpb.2007.09.005.

13. Vialatte, F.-B., J. Dauwels, J. Solé-Casals, M. Maurice, and A. Cichocki. 2009. "Improved sparse bump modeling for electrophysiological data." *Advances in Neuro-Information Processing*, Berlin, Heidelberg.

14. Başar, E., A. Gönder, and P. Ungan. 1976. "Important relation between EEG and brain evoked potentials." *Biological Cybernetics* 25(1):27–40. doi: 10.1007/BF00337046.

15. Basar, E., T. Demiralp, M. Schürmann, C. Basar-Eroglu, and A. Ademoglu. 1999. "Oscillatory brain dynamics, wavelet analysis, and cognition." *Brain and Language* 66(1):146–183. doi: 10.1006/brln.1998.2029.

16. Vialatte, F. B., C. Martin, R. Dubois, J. Haddad, B. Quenet, R. Gervais, and G. Dreyfus. 2007. "A machine learning approach to the analysis of time–frequency maps, and its application to neural dynamics." *Neural Networks* 20(2):194–209. doi: 10.1016/j.neunet.2006.09.013.

17. Vialatte, F., A. Cichocki, G. Dreyfus, T. Musha, T. M. Rutkowski, and R. Gervais. 2005. "Blind source separation and sparse bump modelling of time frequency representation of EEG signals: New tools for early detection of Alzheimer's disease." *2005 IEEE Workshop on Machine Learning for Signal Processing*, Mystic, Connecticut, USA, 28–30 September 2005.

18. Ghanbari, Z., M. Najafi, and M. B. Shamsollahi. 2011. "Sleep spindles analysis using sparse bump modeling." *2011 1st Middle East Conference on Biomedical Engineering*, Sharjah, United Arab Emirates, 21–24 February 2011.

19. Najafi, M., Z. Ghanbari, B. Molaee-Ardekani, M. Shamsollahi, and T. Penzel. 2011. "Sleep spindle detection in sleep EEG signal using sparse bump modeling." *2011 1st Middle East Conference on Biomedical Engineering*, Sharjah, United Arab Emirates, 21–24 February 2011.

20. Dauwels, J., F. Vialatte, and A. Cichocki. 2007. "A novel measure for synchrony and its application to neural signals." *2007 IEEE International Conference on Acoustics, Speech and Signal Processing – ICASSP'07*, Honolulu, Hawaii, 15–20 April 2007.

21. Dauwels, J., F. Vialatte, T. Rutkowski, and A. Cichocki. 2007. "Measuring neural synchrony by message passing." *Proceedings of the 20th International Conference on Neural Information Processing Systems,* Vancouver, British Columbia, Canada.

22. Dauwels, J., F. Vialatte, C. Latchoumane, J. Jeong, and A. Cichocki. 2009. "EEG synchrony analysis for early diagnosis of Alzheimer's disease: A study with several synchrony measures and EEG data sets." *2009 Annual International Conference of the IEEE Engineering in Medicine and Biology Society*, Hilton Minneapolis, Minnesota, USA, 3–6 September 2009.

23. Ghanbari, Z., and M. H. Moradi. 2016. "Synchrosqueezing transform: Application in the analysis of the K-complex pattern." *2016 23rd Iranian Conference on Biomedical Engineering and 2016 1st International Iranian Conference on Biomedical Engineering (ICBME)*, Tehran, Iran, 24–25 November 2016.

24. Ghanbari, Z., and M. H. Moradi. 2017. "K-complex detection based on synchrosqueezing transform." *AUT Journal of Electrical Engineering* 49(2):214–222. doi: 10.22060/eej.2017.12577.5096.

25. Mallat, S. G., and Z. Zhifeng. 1993. "Matching pursuits with time-frequency dictionaries." *IEEE Transactions on Signal Processing* 41(12):3397–3415. doi: 10.1109/78.258082.

26. Loeliger, H. 2004. "An introduction to factor graphs." *IEEE Signal Processing Magazine* 21(1):28–41. doi: 10.1109/MSP.2004.1267047.

27. Myers, C. S., and L. R. Rabiner. 1981. "A comparative study of several dynamic time-warping algorithms for connected-word recognition." *The Bell System Technical Journal* 60(7):1389–1409. doi: 10.1002/j.1538–7305.1981.tb00272.x.

28. Dauwels, Justin, F. Vialatte, T. Rutkowski, and A. Cichocki. 2007. "Measuring neural synchrony by message passing." *Proceedings of the 20th International Conference on Neural Information Processing Systems*, Vancouver, British Columbia, Canada.

29. Daubechies, I., J. Lu, and H.-T. Wu. 2011. "Synchrosqueezed wavelet transforms: An empirical mode decomposition-like tool." *Applied and Computational Harmonic Analysis* 30(2):243–261. doi: 10.1016/j.acha.2010.08.002.

30. Daubechies, I., and S. H. Maes. 2017. "A nonlinear squeezing of the continuous wavelet transform based on auditory nerve models."

31. Oberlin, T., and S. Meignen. 2017. "The second-order wavelet synchrosqueezing transform." *2017 IEEE International Conference on Acoustics, Speech and Signal Processing (ICASSP)*, New Orleans, USA, pp. 3994–3998.

32. Meignen, S., T. Oberlin, and S. McLaughlin. 2012. "A New algorithm for multicomponent signals analysis based on synchrosqueezing: With an application to signal sampling and denoising." *IEEE Transactions on Signal Processing* 60(11):5787–5798.

33. P. Flandrin, *Time-Frequency/Time-Scale Analysis*, vol. 10. Cambridge, MA: Academic Press, 1998.

34. Pham, D.-H., and S. Meignen. 2017. Second-order synchrosqueezing transform: The wavelet case, comparisons and applications.

35. Carmona, R. A., W. L. Hwang, and B. Torresani. 1997. "Characterization of signals by the ridges of their wavelet transforms." *IEEE Transactions on Signal Processing* 45(10):2586–2590.

36. Pham, D., and S. Meignen. 2017a. "High-order synchrosqueezing transform for multicomponent signals analysis: With an application to gravitational-wave signal." *IEEE Transactions on Signal Processing* 65(12):3168–3178.

37. Ghanbari, Z., M. H. Moradi, A. Moradi, and J. Mirzaei. 2020. "Resting state functional connectivity in PTSD veterans: An EEG study." *Journal of Medical and Biological Engineering.* doi: 10.1007/s40846-020-00534-7.

38. Ghanbari, Z., and M. H. Moradi. 2020. Fuzzy scale invariant feature transform phase locking value and its application to PTSD EEG data. In Bajaj, V., and G. R. Sinha (eds), *Modelling and Analysis of Active Biopotential Signals in Healthcare,* vol. 1. Bristol: IOP Publishing, pp. 2–25.

3 HealFavor: A Chatbot Application in Healthcare

Abdullah Faiz Ur Rahman Khilji,
Sahinur Rahman Laskar, and Partha Pakray
National Institute of Technology Silchar

Rabiah Abdul Kadir
Institute of IR4.0, Universiti Kebangsaan Malaysia

Maya Silvi Lydia
Universitas Sumatera Utara

Sivaji Bandyopadhyay
National Institute of Technology Silchar

CONTENTS

3.1 INTRODUCTION

Applications that greatly help in improving the user experience are adopted well by users; hence, they are used in a wide variety of products ranging from travel, food, e-commerce, and other such domains [1]. But such a system has not been developed to an advanced stage for the healthcare domain.

It is well known that healthcare-based artificial intelligence (AI) research draws great costs [2] and attention from the international community [3]. Any significant advancement for an automated prognosis after diagnosing the patient would be of immense use. Improved quality and wide-scale accessibility are very important in today's age of disparity between consumer affordability and healthcare costs. There is also the need for reducing the communication gap between a physician and the patient. Since our current tools and methods proscribe the given prolonged exercise, innovation is necessary. Such a singular application to improve the prevailing situation is a chatbot.

A chatbot is a bilateral application that works in both directions in order to simulate user interactions based on the given input by the user [4]. Since such a system learns the patterns and knowledge from past conversations, it is able to handle new situations based on the given experience. But for such a system to make correct predictions, we need a sufficient dataset. The dataset thus involved must cover the majority of the use cases the patient might experience. Also, it must be well reviewed by a medical expert. Such a system would help in greatly scaling up the medical accessibility over large sections of deprived society. Since for our system, suitable data was not available, we have developed a suitable dataset that caters to the frequently occurring medical conditions and have developed a prototype system to utilize the same and offer personalized suggestions to the patients.

Initial sections of this chapter discuss the related words and the strategy we have used to prepare the dataset. We have also discussed the framework we have used in order to test the dataset with a machine learning (ML) model. We have also extended it to employ other accessibility techniques like machine translation (MT) to cater to people from different linguistic backgrounds.

3.1.1 HEALTHCARE

The field of healthcare is a knowledge-driven domain [5], which consists of research articles, experimental findings, a large number of narratives, and notes. It contains both structural and unstructured data [6], making it difficult for expert systems to decipher the hidden knowledge beneath this trove of information [7]. Thus, to access this meaningful information, we require advanced text processing techniques [8]. Various structural and nonstructural information has been usefully extracted using natural language processing (NLP) techniques, some of which would be discussed in the subsequent sections.

3.1.2 NLP IN HEALTHCARE

The language we used to communicate is a combination of various characters that combine to form a well-defined structure [9]. There are sets of rules that help us in deciphering the structure of this mode of communication [10]. Thus, the natural language is the language used for communication by humans that includes but is not limited to English, Hindi, and Assamese. All over the world, there are many such diverse yet structured modes of communication [11]. There are also diverse set of definitions for processing the natural language, which are known as computation linguistics when processed by a computer [12]. NLP can also be termed as defining computationally usable algorithms or techniques for analyzing the text the human speaks so that a machine could use human-like abilities for a range of tasks [13]. These tasks more often than not involve multidisciplinary research majorly in the domain of human–computer interaction [5]. NLP is also related to logic, philosophy, cognitive, and science [14]. NLP consists of two principal tasks, namely, natural language understanding (NLU) and natural language generation (NLG) [15].

As we can understand from the name, NLU can be conceived as a process by which the intention of the text is comprehended by a machine [16]. Thus, NLU processes the text to a machine-understandable form [17]. On the other hand, NLG refers to the process of constructing the natural text to satisfy the constraints of the given task [18].

Subsequent sections are divided as follows. Section 3.2 discusses the related work pertaining to the chat-based system. Section 3.3 discusses the data preparation methodology and the processing techniques utilized to make the data in a machine-understandable form. Section 3.4 discusses the prototype system architecture. Section 3.5 discusses the MT concepts in detail. Section 3.6 discusses the evaluation techniques used. Finally, we conclude in Section 3.7 and have listed the future works in Section 3.8.

3.2 THEORETICAL BACKGROUND

3.2.1 CHATBOTS IN HEALTHCARE

Contemporary chatbot-based applications involved the usage of rudimentary abilities to mimic user-based conversations. One such contemporary application was ELIZA [19]; here, the model used some generic statements which were barely minimum for the interaction to proceed with the system. Further, the initial study gave birth to ALICE [20]; here, artificial intelligence markup language (AIML) was leveraged based on different patterns that were stored in the knowledge base for similar purposes. Cleverbot [21] and Jabberwacky [22] are some contemporary implementations that understand the stored scripted patterns. The work based on Ref. [23] uses graphs that are finite states in memory to customize and cater to each user's interaction separately.

Apart from this, there also exist developer-friendly systems like Google's Dialogflow[1] and IBM's Watson,[2] which allow users to build their own systems by configuring intents and entities [24].

Recent years have witnessed quite many systems that have been designed keeping in mind the medical requirements [25]. Chatbots have tried to replace generally physician or a nurse. There exists an agent named Molly, a nurse who has been found to improve the overall efficiency of monitoring the patient by 20%. DeepMind health AI [26] facilitates a quick diagnosis based on the patient. The HealthTap [27] startup has been focused to connect doctors with patients to facilitate a better interaction, resulting in the overall well-being of the patient. Applications like these also help in improving the overall mental health of a patient [28]. This involves extracting intents from the text using NLU with recognizing emotions, thus allowing an automated psychiatric counseling. This has also helped in intervening in the alcohol consumption activities of young adults [28]. Wysa is a chatbot-based mind coach that allows users to achieve their mental goals through a series of steps [29] involving text-based messaging.

There are some applications where a natural setup is made wherein the user interacts with the system and the log files are collected and analyzed [30,31]. Work has also been done to analyze the log files of a banking chatbot [32]. Also, various patterns had been discovered due to the intercoder reliability metric. A complete data-driven analysis for graph-driven chatbots is performed by Ref. [33]. Research has also shown that people have responded socially even when the user was aware that his/her answers are responded to by a machine [34,35], giving an impetus to the chatbot evolution. It is also shown that informal speech increases the human nature in a chatbot [36]. Thus, the adoption of conversational human voice reduces the chances of having unsocial feeling such a chat-based online service. Some researchers have studied the effect of chatbot on the well-being of the youth by cognitive behavior therapy [37]. Some chat-based applications have also shown to help adolescents with issues like drug and alcohol [38]. Morgan et al. in their work [39] developed a chat-based system for children to consult regarding their legal health benefits that they are entitled to. In the growing virtual world, chatbots have also shown an increased usage in the online education sector as shown by Ref. [40]. Many other domains have also employed the use of question answering-based chatbot systems [41].

With the increasing use of mobile device users, there is a demand of chatbot system that supports a multilingual facility. Most of the chatbot systems assist on a single language (English), and a very limited research work has been done on the multilingual chatbot [42]. In Ref. [42], they used Google translate [43] to inject a multilingual facility in the chatbot. Multilingual chatbot requires the translation ability that can allow communication across the world to yield online interaction, thereby improving customer service, hence having great business potential. To develop a multilingual chatbot, MT plays a vital role by allowing an automatic translation facility in the chatbot. The available online translation service like Google covers 103 languages throughout the globe but still needs to cover many low-resource languages especially found in the Asian country like India. In order to provide a multiple-language

---

[1] https://dialogflow.com.
[2] https://www.ibm.com/watson.

translation in the chatbot, there are lots of research scopes though the work done is limited. Moreover, healthcare chatbot demands a multilingual facility [42] that can be benefited for the common people who need quick health-related supports in their native languages and are not comfortable in English conversation.

3.2.2 RESEARCH GAPS

There is a lack of research in healthcare chatbot system that can provide a quick assistance for the health-related issues to a large number of people among various backgrounds. As discussed in Section 3.2.1, there are various applications that cater to the various needs of different users, but a general purpose-based system is required so that a large proportion of the patients which display common symptoms are taken care of. This would definitely allow the healthcare institutions to spend more time with those patients who are more critical and require a serious attention.

Moreover, a diverse country like India has people belonging to various linguistic backgrounds, and thus, the current healthcare system faces some limitations in diagnosing people from the large underdeveloped section of the society due to various technical and financial limitations. Much of this could be solved in the today's virtual age by implementing such a system and incorporating a multilingual model based on the MT algorithms.

In the subsequent sections, we have discussed how such a system can practically be achieved and implemented at a large scale.

3.2.3 PRIVACY CONCERNS

The healthcare field is very sensitive [44] regarding the patient's data, and hence, the dataset generated needs to be taken care with the utmost security. All care is taken so that the user's data is backed up at secure locations, as various researchers [45] have shown that customers are concerned about disclosing their personal information to online services like chatbot. Several researchers have also linked privacy concerns with information disclosure [46–48].

3.3 DATA PREPARATION METHODOLOGY

For preparing a suitable data corpus for the prototype system of HealFavor [49], the authors have undertaken the following steps to obtain a suitable data for the system. Since the concerned domain is healthcare, steps have been taken to ensure the authenticity of the data.

3.3.1 DATA SOURCES AND OBSERVATIONS

For curating the dataset for the system, we have used the resources available on Drugs.com[3] to extract credible information. WebMD[4] was also used to collect information related to different ailments and medicine [50,51]. Stanford Medicine[5] was

---

[3] https://www.drugs.com/.
[4] https://www.webmd.com/.
[5] http://med.stanford.edu/.

used to gain some insights regarding different ailments. To make our dataset authentic and up to date with various health guidelines and measures, we consulted healthcare physicians to gain an insight into commonly occurring diseases. We were able to then differentiate between commonly occurring symptoms (like headache, stomach pain, abdominal pain) and commonly occurring skin diseases. The physicians also briefed us to carefully differentiate between diarrhea and loose motions, cold, and cough. We have also included precautionary measures and diagnosis related to highly contagious diseases like coronavirus disease (COVID-19) [52]. This addition has been made to cater to the growing worldwide pandemic.

3.3.2 QUALITY AND FILTERING

The dataset that we proposed underwent levels of filters to ensure a uniformity throughout. Since there are large number of diseases that may have overlapping symptoms, these need to be taken care of. There are various naming conventions and commonly spoken words related to a particular ailment. Thus, large common decision boundaries require to be formulated, in order to obtain suitable classification sets of the sickness the patient suffers. Typically, it is the case that oneself is not well aware of the definition and in practice the meaning of a particular symptom, and hence needs to be supported through guide questions where the ailing person interacts and makes the system and himself clear with what he is suffering with. Hence, the system needs to guide the patient with questions for a robust diagnosis.

Since the knowledge base of such a system is quite small in comparison with the expertise of physicians, it is impractical to answer all sets of questions. Hence, questions related to critical or chronic conditions are immediately flagged and reported so that further diagnosis can be made by an expert as these may require a complex use case which the system cannot handle at this stage. The conditions involving chronic ailments and diseases affecting the vital organs are recognized through keywords that red flag the patient and initiate a chain process where the user is advised to immediately consult a specialist. On recommendation of health experts, we have improvised and developed a curated list of such terms that are deemed as critical. Since we might have left out some terms which might be related to one of the words in the critical list, we use high-dimensional word vectors as discussed in Section 3.4 to overcome this issue.

3.3.3 PRE-PROCESSING

To make the data suitable for training our system, following the work of Ref. [49], we have undertaken some pre-processing steps. The data is pre-processed using common techniques as shown in Ref. [53]. These methods include tokenization, removal of stop words, and stemming. The given text is segmented or tokenized with the help of various delimiting characters like punctuations or spaces. After tokenizing, the stop words are removed. Also, all the text is converted to lowercase.

Apart from the heuristics-based pre-processing techniques, other advanced pre-processing techniques were explored as well. Lemmatization is a technique that replaces the word with its base form (Ref. [54]); in comparison with stemming, a word stem is not produced. Lemmatization here offers to replace the word with its basic form. Since the pre-processing has to be performed on a chat-based dataset, many short forms of the words also exist: for example, "gud" instead of "good," "2mrv" instead of "tomorrow," "gr8" instead of "great." The dataset was thus subjected to phonetic-based microtext normalization [55].

3.3.4 DATA REPRESENTATION

For the deep learning model in Ref. [20] for the chat-based system, there are two principal components. One is for the NLU model, which is used to classify the intent of the patient, and this component is formulated as a multiclassification problem. The second component is formulated as a sequence prediction task, wherein each intent is a part of the sequence, and based on these intents, the next action of the patient is predicted. Inherently, the NLU model predicts the intents based on the predefined list of intents, and the sequence model predicts the output so that the conversation goes on.

For the former component, we require sentences that have the same intent and occur in separate conversations. This problem is hence proposed as a multiclassification problem. Also, it should be sufficiently large corpus to cover the different commonly occurring intents. This intent is sent to the sequence prediction module whose task is to decipher the disease based on the list of given symptoms. For the latter component of the model, we surveyed on the use cases the patient might undertake and prepared a list of intents. This dataset was iteratively improved as the system was being used, thereby reinforcing the confidence of the prediction model for both multiclassification and sequence prediction tasks.

The sequence prediction task is also a challenging component, which can be broken into two main components, viz., booking-related and health-related. As the booking-related path has a more definite structure, it has a more predictable form, and hence, definite structures are present for the same in the dataset. The health-related component in the dataset has a more complex structure. A sample data representation is depicted in Figure 3.1. In this example, we have shown that two separate paths are taken into consideration based on the input by the patient. If the question is health-related, it proceeds on one path, and if the question is related to booking, it proceeds on a separate path. Even while the question is related to health, there are different branches with which the system proceeds. At the end, if the patient is satisfied, the system ends the conversation gracefully. The system would mainly cater to health-based and appointment booking questions. An exhaustive dataset would be prepared considering both the avenues. For paths that are unpredictable, sets of sentence sequences need to be prepared that are complete in nature. Moreover, the domain is healthcare and the problem has to be undertaken systematically for a stable system.

The data for the system is split into quadruples and stored in the textual form for processing and analysis. These four sections in text files contain the exhaustive data for both the tasks discussed above. The remaining two sections provide for the domain and the system actions. The concerned data in our work is available at the HealFavor repository[6] at GitHub. The data at this repository contains different fold-

---

[6] https://github.com/cnlp-nits/HealFavor.

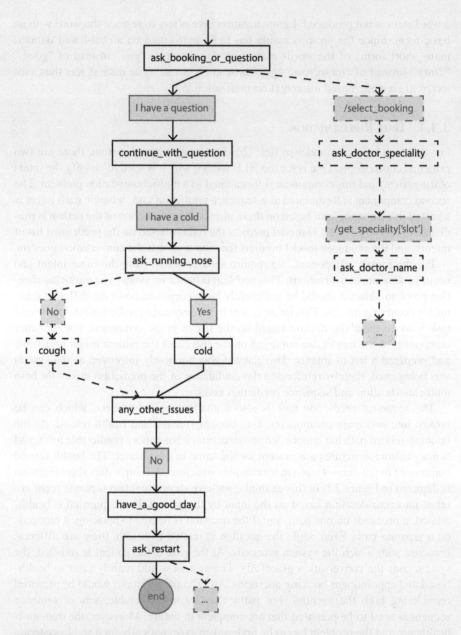

FIGURE 3.1   Sample data representation.

ers, viz., actions, domains, nlu, and stories. The actions folder contains the files which will be used in the system to generate a reply for the patient. The domains folder contains a list of all the intents, entities, slots, and templates used to decipher the data inputted by the patient to the system, whereas nlu contains the probable list of statements corresponding to each intent. And, stories are the sample sequences which the system can be run. It uses this data to train the deep learning model. It is to be noted that the data provided at this repository is a sample data and not the complete version of the dataset used to train the model.

3.4 PROTOTYPE SYSTEM ARCHITECTURE

To facilitate easier access of healthcare information across the masses following Ref. [49], we have designed and developed our own chatbot system that utilizes the Rasa [56] framework. For this system, we have used the recurrent embedding dialogue policy [57] that has used Facebook artificial intelligence research's (FAIR) algorithm [58]. This policy uses a general-purpose embedding based on deep learning to solve a diverse nature of tasks. To compare the present state with the probable states, we have used the cosine approach for similarity calculations.

Khilji et al. [49] have used bag-of-words (BoW) representation [59] to leverage the feature vectors which were then streamed into the embedding layer. The output is also fed to layers of dense neurons whose resultant helps us in predicting the best probable output of the system in response to the patient's input. The recurrent network [60] obtains the patient message and the past system output to calculate attention [61]. Since at every time stamp future inputs are not known, the interpolation gate is kept empty. The recurrent layer receives outputs from both the attention and embedding neurons, which are then fed into a separate embedding layer. The dialogue generation uses the output of the final layer along with the vectors of the previous system-generated response to answer the patient. The long short-term memory (LSTM) states are randomly multiplied with attention probabilities element-wise to accommodate random steps. Figure 3.2 summarizes the architecture for our system. The following loss function is used to calculate the similarity at recurrent time stamps.

$$x = \mu_- + \max_d \left(\mathrm{csim}(c, d_-) \right) \tag{3.1}$$

$$y = \mu_+ - \mathrm{csim}(c, d_+) \tag{3.2}$$

$$l(t) = \mathrm{maximum}(0, x) + \mathrm{maximum}(0, y) \tag{3.3}$$

In Equations (3.1) and (3.2), the cosine similarity evaluated between the embedding generated c and the target embedding d_+ is denoted by "csim." In order to achieve higher correct similarity predictions and higher loss in case of incorrect predictions, the loss is calculated by negative sampling of incorrect actions d_- x and y are shown in Equations (3.1) and (3.2), respectively.

3.4.1 SYSTEM DESCRIPTION: LEVERAGING THE DEEP LEARNING MODEL

The ML model is trained on the prepared dataset by Ref. [49] for the proposed system. The prototype system is discussed in Section 3.4.2.

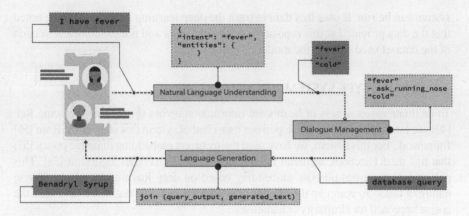

FIGURE 3.2 Prototype system architecture.

3.4.2 HEALFAVOR: OUR PROTOTYPE SYSTEM

In order to allow for a complete utilization of our proposed dataset, we would be requiring a system. This will allow for a better evaluation of our model and would enable us to check the robustness of the system.

Figure 3.3 depicts a working example of the system that uses our underlying dataset to book an appointment with the patients' requirement while the dataset is fetched from the relevant hospital's database.

The system also interacts with a real-time Mongo database (MongoDB), which is used to store the user data to analyze and provide better services to the patient. It also allows managing various small- and large-scale outbreaks, which could be better handled if managed by a central entity.

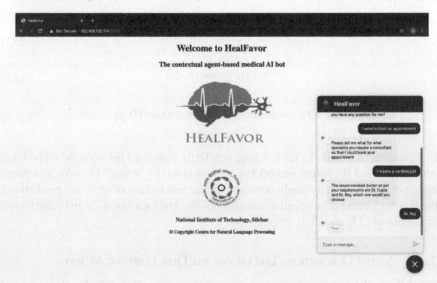

FIGURE 3.3 Prototype system working demo.

We have also worked upon to display some capabilities, advantages, and working of a MT system with such a healthcare-based model. The MT model enables people from various linguistic groups to access the system. The translation is performed in the backend, and the model receives the English text for processing.

3.5 MACHINE TRANSLATION

3.5.1 MACHINE TRANSLATION AND ITS APPROACHES

NLP is a limb of (AI, which aids the computer to understand the human spoken languages also known as natural languages. MT is the subfield of NLP that covers language-ungraspable issues via the automatic translation of a sentence from one natural language to another. Broadly, there are two types of approaches in MT: rule-based and corpus-based, as shown in Figure 3.4. Also, Figure 3.5 depicts periods of the MT approaches.

3.5.2 RULE-BASED MACHINE TRANSLATION

Rule-based MT [62], also known as knowledge-based or classical MT, is based on the set of rules of the linguistic information about the source and target languages, which is further categorized into three types, viz., dictionary-based MT (direct systems), transfer-based MT, and interlingual MT, as shown in Figure 3.6, by the famous Vauquois triangle [63].

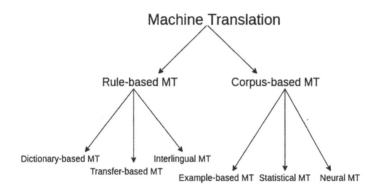

FIGURE 3.4 Machine translation approaches.

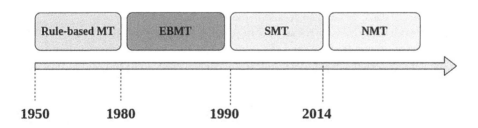

FIGURE 3.5 Periods of the MT approaches.

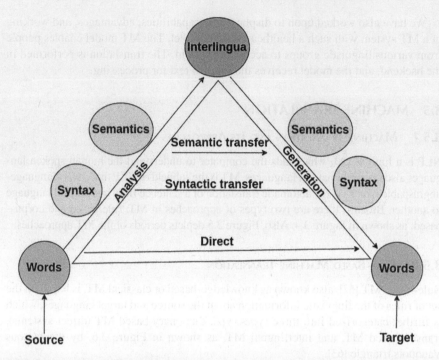

FIGURE 3.6 The Vauquois triangle.

In this triangle, there are two stages: the analysis stage, as we move towards the top, and the generation stage, as we move away from the top. The analysis stage requires more linguistic knowledge to obtain more information about the source language text, which increases the difficulty level towards the tip of the triangle. The obtained information or features from the analysis stage are used in the generation stage to generate the target language text.

At the bottom of the Vauquois triangle, the direct translation is the most-naive type of translation, wherein simply, each source word is translated into the target word. The first stage of direct translation is the morphological analysis of the source sentence, and then using a bilingual lexicon, a word-by-word translation is achieved. After this, based on the set of rules, output translation is reordered. Lastly, with the help of a morphological generator, the resulting translation is corrected. This correction is with respect to the correct variant of a morphological word, including number, gender, tense, case, verbal aspect, and name.

The next level is transfer-based MT, in which there are two types of translations. One is the synthetic transfer, and the other one is the semantic transfer approach. In the synthetic transfer approach, the source sentence is analyzed on the basis of syntax, and the output syntactic structure is mapped to a new syntactic structure in the target language by the set of rules. And the actual source words are translated into the target words, following the direct translation approach. The merits of the synthetic transfer approach are to generate the target sentences by taking care of the word order, syntax, and grammar. Similarly, the semantic transfer approach is based on mapping the semantic structure of the source sentence to the corresponding target sentence using the predefined example semantic rules in addition to the synthetic analysis.

The advantage of the semantic transfer approach is to generate target sentences by encountering the word ambiguity problem. Both the direct and transfer approaches solely rely on a predefined set of rules that map words, syntax, and semantic roles from one language to another language. But for multiple languages, there is a need of reconstruction of rules for each language pair. To resolve this issue, at the top of the Vauquois triangle, an alternative approach of interlingual MT is introduced. Here, source language text is analyzed and transformed into an abstract language-independent representation (interlingua) and then target language is generated from the interlingua representation. Interlingual MT is effective when the source and target languages are very different from each other, like Arabic and English. The major drawbacks of such rule-based approaches are the unavailability of good dictionaries and the expensive procedure required to construct new dictionaries, and also the time-consuming method to define a set of rules for each and every language. Therefore, corpus-based MT, also known as data-driven MT, is introduced.

3.5.3 CORPUS-BASED MACHINE TRANSLATION

The corpus-based MT relies on a parallel corpus of the source and target language, and eliminates the need for linguistic knowledge to create a set of rules. It is a language-independent MT, where any source language can be translated to any target language. Example-based machine translation (EBMT) [64,65] is the naive approach in the corpus-based MT. The key idea behind the EBMT is the analogy (text similarity) concept. In EBMT, similar sentences from the parallel corpus which contain similar phrases to the source language sentence are selected, and then, the selected sentences are used to translate the phrases of the source sentence to the target sentence, and lastly, these phrases are combined to generate the target language sentence. The main drawback of EBMT is that in real-time scenarios, we cannot cover various types of sentences by examples only. To encounter this issue, statistical machine translation (SMT) is introduced [66,67].

In this approach, statistical models are generated whose parameters are estimated or learned from the analysis of parallel corpus. The obtained statistical model is used to predict the target sentence for the given source sentence. There are various types of SMT, namely, word-based translation, phrase-based translation, syntax-based translation, and hierarchical phrase-based translation, but the most acceptable and widely used approach is phrase-based SMT. Prior to neural MT, phrase-based SMT [68] achieves the state-of-the-art approach. SMT considers the translation task as a probabilistic task by predicting the best translation for a given source sentence. The abstract pictorial representation of SMT is shown in Figure 3.7. To evaluate the resultant conditional probability of the source sentence, given the target sentence and probability of the target sentence in SMT the translation model (TM) and language model (LM) are used [69]. The decoder is responsible for searching the best translation using the beam search strategy.

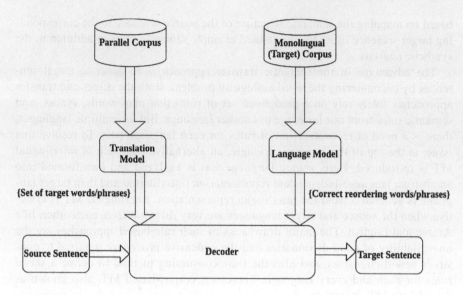

FIGURE 3.7   Abstract diagram of SMT.

Despite the advantages of phrase-based SMT over EBMT and rule-based MT, the followings are the demerits which lead to the development of neural machine translation (NMT) [70–72].

- **End-to-end problem**: SMT lacks an end-to-end solution; it is based on a combination of different modules like TM and LM. The problem is that if any of the module is updated, then the entire translation system needs to be updated. Also, the various components increase the system complexity.
- **Long-term dependency issue and lack of context-analyzing ability**: The TM and n-gram LM could not capture all the important information for very long-type sentences since it is not feasible in practice.
- SMT is not able to learn translation of words and phrases which are not available in the corpus; this issue is known as out of vocabulary problem. Also, visualization of word embedding is not feasible in SMT.
- SMT lacks generalization since continuous space representation is not possible. Such representation (a vector form of real numbers) is very effective to capture various properties of languages.
- SMT is limited to text-based MT, and the multimodal-based system is not feasible in such a system which is very effective in the case of low-resource language translation.

The NMT has the ability to learn a model in an end-to-end fashion, by mapping the source and the corresponding target sentence. The trained NMT model can translate from source language to any target language. In an NMT model, the primary unit is an artificial neuron that resembles a biological neuron. The main issue of SMT is that it creates a model context by considering the collection of phrases with a limited size, since data sparsity lowers the quality if phrase size increases. Also, in the feed-forward neural network-based NMT, the phrase pairs score is calculated by assuming the fixed length of the phrases. However, in a real-time scenario, the source and target phrase length of the translation are not fixed. To deal with variable-length phrases, recurrent neural network (RNN)-based NMT [70,72] can be used by processing each word in a sequence of any arbitrary length through continuous space representations. These representations help in learning long-distance relationship in between words in a particular sequence. Also, RNN maintains and updates a memory called as a state while processing each word in a sequence. Moreover, to improve learning for the long-term features, RNN adopts an LSTM [73] or gated recurrent unit (GRU) [74] for encoding and decoding. The NMT system has two major ingredients: the encoder and decoder. The encoder is responsible to compress the entire source sentence into a context vector, and the decoder is used to decode the target sentence from the context vector. NMT-based systems achieves the state-of-the-art performance in both high- and low-resource language translations [75–79].

3.5.4 MACHINE TRANSLATION IN HEALFAVOR APPLICATION

MT plays an important role in the HealFavor application, by allowing the multilingual chat facility through the automatic translation. In this application, the user can ask health-related issues in their own language and chatbot can assist the user in the same language. Therefore, this application can be helpful for the local population who are not comfortable covering in English. In a multilingual country like India, where multiple languages exist, multilingual-based healthcare chatbot can help the local community by providing a quick assistance in their own languages on health-related problems. Figure 3.8 depicts one example where a user asks a health-related question in Hindi and gets the response in Hindi. Here, the MT module acts as an interface to translate Hindi to English for the processing in HealFavor application, and gets the text input in English. It then translates English to Hindi to communicate with the user. Thus, the MT system acts as an intermediary between the user and the deep learning model. The user does not see the text in English but all processing is performed in the default language of the model. Moreover, this will allow for greater coverage of the issues of diverse linguistic groups across all sections of the society, and will prevent the language as being a barrier from real applications that could benefit humankind.

FIGURE 3.8   Machine translation in HealFavor.

3.6 EVALUATION

In order to examine the robustness of the proposed dataset, Rasa framework is used. This scalable framework allows the model to be represented as a well-built system that utilizes the dataset for allowing for a greater degree of response and suggestions. The framework allows for a smooth implementation of the inherent ML models and allows it to be used with a reinforced approach to improve the model, together with allowing an integrated approach with the system.

To check the adaptability and hear the feedback of actual users, we have also conducted a user experience survey, details of which are thoroughly discussed in Section 3.6.1.

3.6.1 USER EXPERIENCE SURVEY

The notion of user experience survey has been widely accepted in the human–computer interaction domain [1,80] even though it has no clear-defining boundaries.

There is no standard evaluation metric available to measure the performance of the healthcare chatbot system [81]. Therefore, to determine an approximate system's performance, we have chosen a survey of user experience. The purpose of which is to exhaustively evaluate our system before it enters the phase of large-scale testing. A total of four evaluators were considered for this task who tested 200 questions on the system. The answers were evaluated on the basis of one to three scale, as shown in Table 3.1.

3.6.2 ACCURACY

The accuracy can be measured using Equation (3.4) of the proposed previous work [49]. Here, n_t is the sum of total scores with respect to each question, which results in 292, and n_m is obtained by multiplying the total number of questions with the highest score (3) that gives 600. Thus, the accuracy gives 48.66% from the four human evaluators (HEs), as shown in Table 3.2.

$$\text{Accuracy} = \frac{n_t}{n_m}$$

$$(3.4)$$

TABLE 3.1
Score Description

Score	Description
1	Out of domain
2	Not correct
3	Correct

TABLE 3.2
Human Evaluation Result

HE-1	HE-2	HE-3	HE-4	n_t	Accuracy
73	69	76	74	292	48.66%

For our work, four HEs (HE-1, HE-2, HE-3, and HE-4) were considered. The evaluators used the system and rated the output of the system as described in Table 3.1. The data given by these evaluators was used to give the final result.

3.7 CONCLUSION

In our work, we have discussed various ways the growing technology could benefit healthcare by providing basic facilities to remote locations. It could greatly benefit mankind in several ways. One such work has been thoroughly discussed in this chapter. We have also presented a suitable dataset for a chat-based healthcare system. A prototype of which has already been made for evaluation and to check its feasibility for a large-scale implementation.

3.8 FUTURE WORKS

In our work, we have made an in-depth survey of the various chatbot applications in healthcare and also discussed in detail the working of our chat-based healthcare application named as HealFavor. The dataset has been increased from Ref. [49] to incorporate the safety measures required during a pandemic, especially for COVID-19. The evaluation result using the prototype system has also been presented. Improvement to the dataset to include more ailments is required. Various precautionary measures by health experts are also to be included to improve people's life. Subsequent works would also include multilingual as well as multimodal functionalities to the system.

ACKNOWLEDGMENTS

This work was supported by the Department of Science & Technology (DST) and Science and Engineering Research Board (SERB), Government of India and the Research Project Grant No. CRD/2018/000041. The authors would like to thank Center for Natural Language Processing (CNLP) and Department of Computer Science and Engineering at National Institute of Technology, Silchar, for providing the requisite support and infrastructure to execute this work.

REFERENCES

1. Sauro J, Lewis JR. *Quantifying the User Experience: Practical Statistics for User Research.* Amsterdam /Waltham, MA: Elsevier/Morgan Kaufmann; 2012.
2. Dranove D, Forman C, Goldfarb A, et al. The trillion dollar conundrum: Complementarities and health information technology. *American Economic Journal: Economic Policy* 2014;6:239–270.
3. Puaschunder JM. The legal and international situation of AI, robotics and big data with attention to healthcare. Report on Behalf of the European Parliament European Liberal Forum, New York; 2019.
4. Yan M, Castro P, Cheng P, et al. Building a chatbot with serverless computing. *Proceedings of the 1st International Workshop on Mashups of Things and APIs - MOTA'16*, Trento, Italy: ACM Press; 2016, pp. 1–4.
5. Ondo N., Olaleke JO. A systematic review of natural language processing in healthcare. *IJITCS* 2015;7:44–50.

6. Jiang F, Jiang Y, Zhi H, et al. Artificial intelligence in healthcare: Past, present and future. *Stroke and Vascular Neurology* 2017;2:230–243.
7. Luo L, Li L, Hu J, et al. A hybrid solution for extracting structured medical information from unstructured data in medical records via a double-reading/entry system. *BMC Medical Informatics and Decision Making* 2016;16:1–14.
8. Murdoch TB, Detsky AS. The inevitable application of big data to health care. *JAMA* 2013;309:1351–1352.
9. Popescu M, Ionescu RT. The story of the characters, the DNA and the native language. Proceedings of the Eighth Workshop on Innovative Use of NLP for Building Educational Applications, Atlanta, GA; 2013, pp. 270–278.
10. Nadkarni PM, Ohno-Machado L, Chapman WW. Natural language processing: An introduction. *Journal of the American Medical Informatics Association* 2011;18:544–551.
11. Henderson JC, Brill E. Exploiting diversity in natural language processing: Combining parsers. *1999 Joint SIGDAT Conference on Empirical Methods in Natural Language Processing and Very Large Corpora*, Colloege Park, MD; 1999.
12. Cohen KB, Hunter L. Natural language processing and systems biology. In: Artificial Intelligence Methods and Tools for Systems Biology. Dordrecht, The Netherlands: Springer; 2004, pp. 147–173.
13. Drake M. *Encyclopedia of Library and Information Science*, Second Edition. Boca Raton, FL: CRC Press; 2003.
14. Copestake A. *Natural Language Processing: Part 1 of Lecture Notes*. Cambridge: Ann Copestake Lecture Note Series; 2003.
15. Regina B, Michael C. *Natural Language Processing: Background and Overview*. Cambridge: Barzilay and Collins Lecture Note Series; 2005.
16. Allen J. *Natural Language Understanding*. London: Pearson; 1995.
17. Saetre R. GeneTUC: Automatic information extraction from biomedical texts. *Proceedings of Computer Science Graduate Students Conference, Norwegian University of Science and Technology (NTNU)*, Trondheim, Norway; 2004.
18. Siddharthan A, Reiter E, Dale R. *Building Natural Language Generation Systems*. Cambridge: Cambridge University Press; 2000, 234 p. *Natural Language Engineering* 2001;7:271–274.
19. Weizenbaum J. ELIZA: A computer program for the study of natural language communication between man and machine. *Commun ACM* 1966;9:36–45.
20. Wallace R. The elements of AIML style. Alice AI Foundation; 2003, 139.
21. Cleverbot.com - a clever bot [Internet]. [cited 2020 June 20]. Available from: https://www.cleverbot.com/.
22. Jabberwacky - live chat bot [Internet]. [cited 2020 March 5]. Available from: http://www.jabberwacky.com/.
23. Divya S, Indumathi V, Ishwarya S, et al. A self-diagnosis medical chatbot using artificial intelligence. *Journal of Web Development and Web Designing* 2018;3:1–7.
24. Janarthanam S. *Hands-on Chatbots and Conversational UI Development: Build Chatbots and Voice User Interfaces with Chatfuel, Dialogflow, Microsoft Bot Framework, Twilio, and Alexa Skills*. Birmingham: Packt Publishing Ltd; 2017.
25. Sherer SA. Patients are not simply health it users or consumers: The case for "e Healthicant" applications. *Communications of the Association for Information Systems* 2014;34:17.
26. Winter JS, Davidson E. Big data governance of personal health information and challenges to contextual integrity. *The Information Society* 2019;35:36–51.
27. Chung K, Park RC. Chatbot-based healthcare service with a knowledge base for cloud computing. *Cluster Computing* 2019;22:1925–1937.
28. Oh KJ, Lee D, Ko B, et al. A chatbot for psychiatric counseling in mental healthcare service based on emotional dialogue analysis and sentence generation. *Proceedings - 18th IEEE International Conference on Mobile Data Management, MDM 2017*, Daejeon; 2017.

29. Inkster B, Sàrda S, Subramanian V. An empathy-driven, conversational artificial intelligence agent (Wysa) for digital mental well-being: Real-world data evaluation mixed-methods study. *JMIR Mhealth Uhealth* 2018;6:e12106.

30. Huang J, Zhou M, Yang D. Extracting chatbot knowledge from online discussion forums. IJCAI, Hyderabad, India; 2007, pp. 423–428.

31. Rivolli A, Amaral C, Guardão L, et al. KnowBots: Discovering relevant patterns in chatbot dialogues. International Conference on Discovery Science, Split, Croatia, Springer; 2019. pp. 481–492.

32. Li C-H, Chen K, Chang Y-J. When there is no progress with a task-oriented chatbot: A conversation analysis. Proceedings of the 21st International Conference on Human-Computer Interaction with Mobile Devices and Services, Taipei, Taiwan; 2019, pp. 1–6.

33. Jalota R, Trivedi P, Maheshwari G, et al. An approach for ex-post-facto analysis of knowledge graph-driven chatbots: The DBpedia chatbot. International Workshop on Chatbot Research and Design, Amsterdam, Springer; 2019, pp. 19–33.

34. Nass C, Moon Y. Machines and mindlessness: Social responses to computers. *Journal of Social Issues* 2000;56:81–103.

35. Reeves B, Nass CI. *The Media Equation: How People Treat Computers, Television, and New Media Like Real People and Places.* Cambridge: Cambridge University Press; 1996.

36. Araujo T. Living up to the chatbot hype: The influence of anthropomorphic design cues and communicative agency framing on conversational agent and company perceptions. *Computers in Human Behavior* 2018;85:183–189.

37. Fitzpatrick KK, Darcy A, Vierhile M. Delivering cognitive behavior therapy to young adults with symptoms of depression and anxiety using a fully automated conversational agent (Woebot): A randomized controlled trial. *JMIR Mental Health* 2017;4:e19.

38. Crutzen R, Peters G-JY, Portugal SD, et al. An artificially intelligent chat agent that answers adolescents' questions related to sex, drugs, and alcohol: An exploratory study. *Journal of Adolescent Health* 2011;48:514–519.

39. Morgan J, Paiement A, Seisenberger M, et al. A chatbot framework for the children's legal centre. *The 31st International Conference on Legal Knowledge and Information Systems (JURIX),* Groningen; 2018.

40. Tegos S, Demetriadis S, Psathas G, et al. A configurable agent to advance peers' productive dialogue in MOOCs. *International Workshop on Chatbot Research and Design,* Amsterdam. Springer; 2019, pp. 245–259.

41. Khilji AFUR, Manna R, Laskar SR, et al. Question classification and answer extraction for developing a cooking QA system. *Computación y Sistemas* 2020;24(2):921–927.

42. Vanjani M, Aiken M, Park M. Chatbots for multilingual conversations. *Journal of Management Science and Business Intelligence* 2019;4:19–24.

43. Patil S, Davies P. Use of google translate in medical communication: Evaluation of accuracy. *BMJ* 2014;349:g7392.

44. Venkatasubramanian K, Gupta SK. Security solutions for pervasive healthcare. *Security in Distributed, Grid, Mobile, and Pervasive Computing* 2007;349:1–6.

45. Følstad A, Nordheim CB, Bjørkli CA. What makes users trust a chatbot for customer service? An exploratory interview study. *International Conference on Internet Science,* St. Petersburg, Russia, Springer; 2018, pp. 194–208.

46. Barnes SB. A privacy paradox: Social networking in the United States. *First Monday* 2006;11(9):11–15.

47. Baruh L, Secinti E, Cemalcilar Z. Online privacy concerns and privacy management: A meta-analytical review. *Journal of Communication* 2017;67:26–53.

48. Dinev T, Hart P. An extended privacy calculus model for E-commerce transactions. *Information Systems Research* 2006;17:61–80.

49. Khilji AFUR, Laskar SR, Pakray P, et al. HealFavor: Dataset and a prototype system for healthcare chatbot. *2020 International Conference on Data Science, Artificial Intelligence, and Business Analytics (DATABIA),* Medan, Indonesia, 2020, pp. 1–4.

50. Kuhn M, Letunic I, Jensen LJ, et al. The SIDER database of drugs and side effects. *Nucleic Acids Research* 2016;44:D1075–D1079.
51. Tran N. WebMD: Transforming the US health care industry. Technical report; 2000.
52. Mehta P, McAuley DF, Brown M, et al. COVID-19: Consider Cytokine Storm Syndromes and Immunosuppression. *Lancet (London, England).* 2020;395:1033.
53. Uysal AK, Gunal S. The impact of preprocessing on text classification. *Information Processing and Management* 2014;50:104–112.
54. Plisson J, Lavrac N, Mladenic D. *A Rule Based Approach to Word Lemmatization. Proceedings of the 7th International Multi-Conference Information Society IS-2004,* Ljubljana, Slovenia; 2004, 4.
55. Satapathy R, Guerreiro C, Chaturvedi I, et al. Phonetic-based microtext normalization for twitter sentiment analysis. *2017 IEEE International Conference on Data Mining Workshops (ICDMW),* New Orleans, LA, 2017, pp. 407–413.
56. Bocklisch T, Faulkner J, Pawlowski N, et al. Rasa: Open source language understanding and dialogue management. arXiv:171205181 cs. 2017.
57. Vlasov V, Drissner-Schmid A, Nichol A. Few-shot generalization across dialogue tasks. arXiv:181111707 cs. 2018.
58. Wu LY, Fisch A, Chopra S, et al. StarSpace: Embed all the things! *Thirty-Second AAAI Conference on Artificial Intelligence,* New Orleans, LA, 2018.
59. Zhang Y, Jin R, Zhou Z-H. Understanding bag-of-words model: A statistical framework. *International Journal of Machine Learning and Cybernetics* 2010;1:43–52.
60. Mikolov T, Kombrink S, Burget L, et al. Extensions of recurrent neural network language model. *2011 IEEE International Conference on Acoustics, Speech and Signal Processing (ICASSP),* Prague, Czech Republic, IEEE; 2011, pp. 5528–5531.
61. Vaswani A, Shazeer N, Parmar N, et al. Attention is all you need. *Advances in Neural Information Processing Systems* 2017;30:5998–6008.
62. Eisele A, Federmann C, Saint-Amand H, et al. Using Moses to integrate multiple rule-based machine translation engines into a hybrid system. Proceedings of the Third Workshop on Statistical Machine Translation, Columbus, OH; 2008, pp. 179–182.
63. Vauquois B. A survey of formal grammars and algorithms for recognition and transformation in mechanical translation. IFIP Congress, Edinburgh; 1968, pp. 1114–1122.
64. Nagao M. A framework of a mechanical translation between Japanese and English by analogy principle. In Elithorn A, Banerji R (eds), *Artificial and Human Intelligence (Edited Review Papers Presented at the International NATO Symposium on Artificial and Human Intelligence),* North-Holland, Amsterdam, The Netherlands; 1984, 173–180; repr. in Nirenburg, et al. pp. 351–354.
65. Somers H. Example-based machine translation. *Machine Translation* 1999;14:113–157.
66. Brown PF, Cocke J, Della Pietra SA, et al. A statistical approach to machine translation. *Computational Linguistics* 1990;16:79–85.
67. Koehn P. *Statistical Machine Translation.* Cambridge: Cambridge University Press; 2009.
68. Koehn P, Och FJ, Marcu D. *Statistical Phrase-Based Translation.* California: University of Southern California Marina Del Rey Information Sciences Inst; 2003.
69. Och FJ, Tillmann C, Ney H. Improved alignment models for statistical machine translation. *1999 Joint SIGDAT Conference on Empirical Methods in Natural Language Processing and Very Large Corpora,* College Park, MD, 1999.
70. Cho K, Van Merriënboer B, Gulcehre C, et al. Learning phrase representations using RNN encoder-decoder for statistical machine translation. arXiv preprint arXiv:14061078, 2014.
71. Devlin J, Zbib R, Huang Z, et al. Fast and robust neural network joint models for statistical machine translation. *Proceedings of the 52nd Annual Meeting of the Association for Computational Linguistics (Volume 1: Long Papers),* Baltimore, MD; 2014, pp. 1370–1380.

72. Sutskever I, Vinyals O, Le QV. Sequence to sequence learning with neural networks. *Advances in Neural Information Processing Systems* 2014;27:3104–3112.
73. Hochreiter S, Schmidhuber J. Long short-term memory. *Neural Computation* 1997;9:1735–1780.
74. Cho K, Van Merriënboer B, Bahdanau D, et al. On the properties of neural machine translation: Encoder-decoder approaches. arXiv preprint arXiv:14091259. 2014;
75. Pathak A, Pakray P. Neural machine translation for Indian languages. *Journal of Intelligent Systems* 2019;28:465–477.
76. Pathak A, Pakray P, Bentham J. English–Mizo machine translation using neural and statistical approaches. *Neural Computing and Applications* 2019;31:7615–7631.
77. Laskar SR, Dutta A, Pakray P, et al. Neural machine translation: English to Hindi. *2019 IEEE Conference on Information and Communication Technology*, Azerbaijan, Baku. IEEE; 2019, pp. 1–6.
78. Laskar SR, Pakray P, Bandyopadhyay S. Neural machine translation: Hindi-Nepali. *Proceedings of the Fourth Conference on Machine Translation (Volume 3: Shared Task Papers, Day 2)*, Florence, Italy; 2019, pp. 202–207.
79. Laskar SR, Singh RP, Pakray P, et al. English to Hindi multi-modal neural machine translation and Hindi image captioning. *Proceedings of the 6th Workshop on Asian Translation*, Hong Kong, China; 2019, p. 62–67.
80. Law EL-C, Roto V, Hassenzahl M, et al. Understanding, scoping and defining user experience: A survey approach. *Proceedings of the SIGCHI Conference on Human Factors in Computing Systems*, Boston, MA; 2009, pp. 719–728.
81. Shawar BA, Atwell E. Different measurement metrics to evaluate a chatbot system. *Proceedings of the Workshop on Bridging the Gap: Academic and Industrial Research in Dialog Technologies*, Rochester, NY, 2007, pp. 89–96.

4 Diagnosis of Neuromuscular Disorders Using Machine Learning Techniques

Abdulhamit Subasi
University of Turku
Effat University

CONTENTS

4.1   INTRODUCTION

The neuromuscular system is composed of two subsystems: nervous system and skeletal muscular system. Several nerves or muscle fibers rarely far removed from the symptoms might cause neuromuscular disorders. Neuromuscular disorders occurring in the neuromuscular junctions, the nervous system, and the muscle fibers range from an insignificant loss of strength to amputation due to muscle or neuron death. Hence, the precise determination of the disease is essential to focus on the treatment. Neuromuscular disorders are diagnosed by means of electromyography (EMG), and the type of etiology, pathology, and location are examined by means of the EMG signals. Although the EMG supports neurologists in their diagnosis, in a more complicated case, additional treatments such as medical imaging techniques (including ultrasound or MRI and muscle biopsies) are essential. The identification of neuromuscular disorders is usually carried out by qualified and expert neurologists who examine the EMG waveforms using muscle acoustics and needle conduction studies. Complications became apparent when qualified and expert neurologists do not exist to handle the demand of patients and, henceforth, it is becoming crucial to build a computer-aided diagnosis (CAD) system based on the EMG signal analysis. These types of CAD frameworks will support neurologists for the diagnosis of neuromuscular disorders without using advanced medical imaging methods (MRI and ultrasound) and muscle biopsies [1–6].

Neuromuscular diseases change the morphology and physiology of motor units (MUs) of the muscle. The cause of the myopathic disease is the atrophy or death of motor fibers; on the other hand, the cause of the neuropathic disease is the damage or death of motor neurons. The firing patterns of the MUs and the shape or characteristics of the motor unit action potentials (MUAPs) are affected by the results of these changes. Hence, investigating the MUAPs formed by the MUs of a muscle assists neurologists to discover the category of disease that might be affected by the muscle [2,7–9].

The crucial phase for the recognition of MUAPs and the analysis of neuromuscular disorders utilizing EMG signals is the feature extraction. The goal behind the EMG signal processing is to discover the distinctive and informative features from the MUAP waveform, which may designate a particular muscle action. Informative and distinctive features can describe neuromuscular diseases, nerve health, and muscle strength [2,3,10]. The performance of EMG signal recognition might be affected by the success of the feature extraction techniques, the quality of the acquired signals, and classifier type. As a result, the performance of the EMG signal recognition scheme can be enhanced by resolving difficulties met with these characteristics. In recent times, different feature extraction approaches were proposed to improve EMG signals for recognition. Conventional approaches extract efficient features in time [10–12], frequency [1,7,13], or time-frequency domain [2,14].

Nevertheless, the frequency and time-frequency domain features are vulnerable to the changes of EMG morphology between different patients, and the time-domain features are sensitive to the noise artifacts. Thus, a combination of different types of features or more complex methods [1] were proposed to solve these problems. On the other hand, wavelet transform (WT) approach characterizes the signal by means of various scales and translations [1,13]. The discrete wavelet transform (DWT) is utilized to decompose a signal into different sub-bands to prevent the effect of noise, and features based on sub-band components are extracted to yield a clear frequency distribution. The wavelet packet decomposition (WPD) employs both the high- and low-frequency components [15–17] to achieve a better frequency resolution of the decomposed signal. The tunable-Q-factor wavelet transform (TQWT) is thought as a robust transform in order to analyze the oscillatory signals [18]. To investigate the desired signal with TQWT, its input parameters are tuned. The number of oscillations of the wavelet can be controlled by Q-factor [19]. The dual-tree complex wavelet transform (DT-CWT) contains two values that have real values. The DWT has a shift-variant feature because of the decimation procedure. Therefore, different set of wavelet coefficients can be produced because of the minor changes in the input signal. To get rid of this problem, Kingsbury [20] presented the DT-CWT, which has an approximate shift-invariant property and a better resolution as compared to the DWT. The DT-CWT employs two equivalent decimated filter-bank trees with coefficients that have real values created at each tree [21].

The main aim of this chapter is to compare the wavelet-based feature extraction methods to describe the MUAP patterns to analyze the neuromuscular disease. The extracted features from wavelet sub-bands achieve important variations among the healthy, ALS (amyotrophic lateral sclerosis), and myopathic subjects, and as a consequence, they produce efficient features for EMG signal classification. The extracted features from wavelet sub-bands will be used as an input data for the classifiers that classify EMG signals to diagnose neuromuscular disorders [1,2,13]. Accordingly, the EMG signals should be analyzed systematically for an efficient automatic EMG signal recognition. Thus, numerous CAD systems have been realized for the identification of neuromuscular diseases by using EMG signals [1,2,7,13]. In this chapter, wavelet-based signal decomposition techniques are compared to diagnose neuromuscular disorders by utilizing EMG signals. The total classification accuracy, area under ROC (receiver-operating characteristic) curve (AUC), F-measure, and kappa statistic are used to assess the performance of the machine learning techniques for EMG signal classification.

4.2 LITERATURE REVIEW

The shapes and sounds in EMG signals include a valuable basis of information for the identification, handling, and control of neuromuscular diseases. By using pattern recognition methods, these constraints may be evaluated by an expert, qualitatively or quantitatively. Because of the advantages of quantitative EMG process, the development of reliable automatic MUAP classifiers has been discovered, and many schemes have now been introduced for this aim, but the performance of current approaches is not sufficiently enough to be employed in clinical environments. Kamali et al. [22]

suggested a new classification strategy focused on the hybrid serial/parallel architecture ensemble of support vector machines (SVMs) to decide the class label (normal, neuropathic, or myopathic) for a given MUAP. Using an EMG signal decomposition method, the developed framework employs time-domain and time-frequency domain features of MUAPs derived from an EMG signal. Different classification methods have been studied, including single and multiple classifiers with many features subsets. In addition, the multiclassifier using multiple feature sets to combine base classifiers revealed an average performance of 97% accuracy.

Kamali et al. [23] proposed the duration and recurrence characteristics of MUAPs concentrated from EMG indicator and provide a discriminative majority of data for neuromuscular problem analysis and medication. Such effects on approved programmed analytics systems using MUAP offers, however, would not yet be convincing. The primary goals in outlining a characterization system for MUAP are to achieve secondary arrangement precision that should be used within the context of clinical preference. To this end, a better classifier may be suggested in this study on moving forward with the execution of MUAP arrangement done by estimating the population name of a MUAP given.

In combination with four separate features, Mishra et al. [24] proposed the improved empirical mode decomposition (IEMD), which is used to analyze normal and ALS. EMG signals contain multiple types of noise, such as electrical and electronic devices. The method of empirical mode decomposition (EMD) with median filter (MF) was used to eliminate the impulsive noise components of intrinsic mode function (IMF) produced by EMD. Additionally, the EMD technique is used to produce improved IMFs. Hence, a new technique for selecting the window size of MF is introduced in the IEMD algorithm. To this end, the features, namely, spectral moment of power spectral density, bandwidth frequency modulation, bandwidth amplitude modulation, and first instantaneous frequency derivative derived from the enhanced IMFs, are utilized to differentiate normal and ALS. Lastly, it is seen that the IEMD technique improves the capability of these features to discriminate against the EMD approach. Effective characterization of the EMG signals is critical for the identification of neuromuscular diseases. The classification algorithms based on machine learning patterns are generally employed to yield such representations. Numerous classifiers were examined to establish precise and computationally effective approaches for characterizing EMG signals. Yousefi and Hamilton-Wright [25] proposed a new system for filtering the ALS discrimination method on the market. Four characteristics are utilized for the study of normal and ALS signals. To eliminate those noises from the EMG signals, the EMD method is implemented. All the filtered IMF components are assembled to create a new signal, and the EMD technique is further utilized to detect EMG signals in order to produce a better IMF, the so-called improved EMD.

In recent years, the focus of a significant research effort has been a precise and computationally effective classification of EMG signals. A significant source of information for the treatment of neuromuscular diseases is the quantitative study of EMG signals. Subasi et al. [26] compared classifiers based on feedforward error backpropagation artificial neural networks (FEBANN) and wavelet neural networks (WNNs) with regard to their accuracy in the classification of EMG signal. An autoregressive (AR) model of the EMG signals is used as an input to the classification framework in these methods. The effective rate was 90.7% for the WNN technique and 88% for the FEBANN technique. The FEBANN counterpart was outperformed by the classifier centered at WNN. EMG signals contain valuable neuromuscular disease information such as ALS. ALS is a well-known disease of the brain that can gradually degenerate the motor neurons. Sengur et al. [27] developed a deep learning approach to distinguish ALS and normal EMG signals efficiently. For time-frequency representation of EMG signals, continuous wavelet transform, spectrogram, and smoothed pseudo-Wigner–Ville distribution were used. To identify certain characteristics, a convolutional neural network (CNN) is used. The feasibility of the suggested technique is tested on the EMG dataset, which is publicly available. This technique achieved 96.80% accuracy.

Subasi [28] used various extraction methods of features to describe the morphology of the MUP. In addition, soft computing techniques for the classification of EMG signals were presented. The suggested technique classifies the EMG signals automatically into normal, myopathic, and neuropathic signals. The multilayer perceptron neural network (MLPNN)-, adaptive neuro-fuzzy inference system (ANFIS)-, and dynamic fuzzy neural network (DFNN)-based classifiers were also used in the classification of EMG signals, and compared in terms of their classification performance. The presented study showed that the modeling of ANFIS achieved better classification performance than that of the DFNN and MLPNN. Naik et al. [29] proposed an approach for classifying neuromuscular diseases, which addresses the data recorded by employing an EMG sensor with a single channel. The ensemble EMD method breaks down the single-channel EMG signal into a series of noise-eliminated IMFs that are then disjointed by the FastICA algorithm. Using the linear discriminant analysis, a reduced set of five time-domain features extracted from the separate components are categorized, and the results are finely adjusted with a majority voting system. A clinical EMG database that yields a higher classification accuracy (98%) validates the efficiency of the proposed system.

The findings of traditional automated diagnostic methods using the MUAP functionality, however, are not yet convincing. The main objective of creating an MUAP classification scheme is to obtain high recognition accuracy for utilizing in clinical decision-making systems. Kamali et al. [22] suggested a robust classifier to boost the efficiency of MUAP classification in estimating a given MUAP class label (myopathic, neuropathic, and normal). The proposed scheme utilizes the MUAP's time and time-frequency characteristics with an ensemble of SVM classifier. Time-domain characteristics include the MUAP phase, change, low-to-low amplitude, field, and length. Time-frequency characteristics are the discrete MUAP coefficients of wavelet transformation. Evaluation results of the developed system achieved an average accuracy of 91%. Rasheed et al. [30] introduced an interactive framework to incorporate the supervised classification task of EMG signal decomposition utilizing a fuzzy k-nearest-neighbor (k-NN) classifier with MATLAB's interactive environment. The method uses an assertion-based recognition, which takes into consideration a combination of MUP shapes and passive and active modes of MU firing pattern information.

Parsaei and Stashuk [31] presented a method for solving an EMG signal in its motor unit potential train (MUPT) component. The scheme is anticipated primarily for clinical usage, where several physiological parameters of MUs of interest are included.

The scheme employs the K-means clustering algorithm and a certainty-based algorithm to implement the supervised classification. In order to achieve a robust recognition performance over a variety of EMG signals, supervised classification algorithms with clustering employed MU firing pattern information and MUP shape. The validity of extracted MUPTs is evaluated during classification utilizing several supervised classifiers. An average classification accuracy of 86.4% for simulated and 96.4% for real data is achieved; the developed system outperforms a previously proposed EMG decomposition technique. Multichannel surface EMG enables the evaluation of physiological and anatomical properties of a single MU. Gazzoni et al. [32] suggested an automatic scheme for detecting and classifying MUAP from multichannel surface EMG signals. The techniques are described for identifying and extracting action potentials using clustering. The matched continuous wavelet transform is utilized for the feature extraction, whereas a multichannel neural network, which is modified from the adaptive resonance theory, is employed for the classification. The proposed method was tested on simulated signals, at several force levels. Hence, the suggested system was able to classify the representation of the MU samples. The key drawback of the method is that whole firing patterns can only be attained in particular circumstances due to possible superimpositions of MU action.

Katsis et al. [33] suggested a novel scheme to extract and classify individual MUAPs from intramuscular EMG signals. The suggested scheme identifies the number of MUAP cluster templates automatically and classifies them into normal, myopathic, and neuropathic ones. The scheme includes three steps: (i) EMG signal preprocessing, (ii) identification and clustering of MUAPs, and (iii) classification of MUAPs. The scheme was tested by employing the EMG signals and an annotated MUAP's collection. The exact identification rate for MUAP clustering is 93%, 92%, and 95% for normal, neuropathic, and myopathic, respectively. Ninety-one percent of the MUAPs were identified correctly. The accuracy obtained for the classification of MUAP is around 86%. Apart from the efficient EMG decomposition, the proposed approach addresses the automatic MUAP classification directly from raw EMG signals to neuropathic, myopathic, or normal groups. Hazarika et al. [34] provided a real-time feature extraction with fusion model for the automated EMG identification utilizing canonical correlation analysis (CCA). The suggested approach can capture information from multiple feature matrices produced from the EMG signals. The presented technique utilizes an optimization method to extract a set of statistical characteristics between the paired views on the basis of which potential signal deviations have been revealed. The next step is to employ DWT on multiple views to achieve domain-independent views that are optimized with CCA. Results reveal that the suggested methodology outperforms several other approaches, 98.80%, 99.0%, and 98.0% in terms of precision, specificity, and accuracy, respectively.

Subasi et al. [35] developed the bagging ensemble classifier for automatic EMG signal recognition. Ensemble classifiers achieve better performance through the use of a weighted combination of different methods for classification. Several scientists have demonstrated the competences of ensemble classifiers in various domains. The process is composed of three steps. The DWT is used in the first step to extract features from any kind of EMG signals. Then, the statistical values of DWT sub-bands are calculated to represent the wavelet coefficients distribution. In the last step, the statistical values are utilized as an input to a Bagging ensemble classifier for the identification of neuromuscular diseases. In the experiments, the Bagging ensemble classifier with SVM achieved 99% accuracy for the identification of neuromuscular disorders. Interpretation of children EMG is challenging. Hafner et al. [36] evaluated and compared the diagnostic accuracy of the pediatric EMG signal with the results of muscle biopsy and clinical diagnosis. The muscle pathology has been divided into seven categories based on current studies of histopathology. The gold standard was used for the clinical diagnosis, containing neurogenic, myopathy, and normal groups. EMG as well as muscle biopsy was performed in the identification of myopathy.

4.3 EMG SIGNAL CLASSIFICATION FRAMEWORK

The proposed framework is composed of three main components. In the first stage, EMG signals are denoised by using MSPCA (multiscale principal component analysis) denoising technique. In the second stage, the coefficients of wavelet-based time-frequency methods are calculated for each type of EMG signal, and then statistical values of each sub-band are computed. In the last stage, the statistical values are utilized as an input to a classifier to diagnose different neuromuscular diseases (Figure 4.1).

4.3.1 WAVELET TRANSFORM

It is a widely used method for the study of time frequency, which facilitates the achievement of a multiresolution time-frequency study that the traditional short-term Fourier Transform cannot accomplish. The WT is a promising choice for processing signals with a changing spectrum as opposed to the short-time Fourier transform.

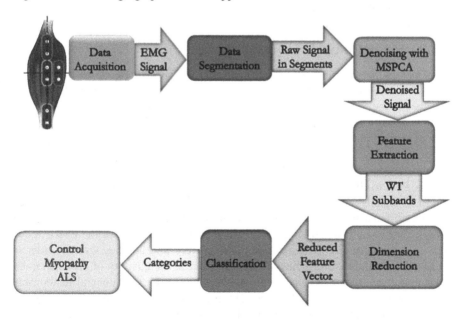

FIGURE.4.1 Proposed EMG signal classification framework.

It tells us not only which frequencies are present in a signal but also when these frequencies took place. Working with different scales allows achieving this. The WT has superior time resolution and inferior frequency resolution for high frequencies, and vice versa [37]. For the efficient evaluation, it generates a compact representation of a signal in time and frequency domains.

The basic functions are called wavelets, which are repartitioned in time and frequency. They can be signified mathematically by using Equation (3.1), where s and u are the dilation and translation factors, respectively [37]. The dilation and translation factor values are continuous, meaning there can be an infinite number of wavelets.

$$\Psi(t) = \frac{1}{\sqrt{S}} \psi\big((t-u)/s\big) \tag{4.1}$$

The process of analyzing a signal $x(t)$ with the help of wavelets is mathematically described as:

$$W_x(u,s) = \frac{1}{\sqrt{S}} \int_{-\infty}^{+\infty} x(t)\psi * \big((t-u)/s\big) dt \tag{4.2}$$

Any signal can be decomposed in various time-frequency bands by the application of wavelets. Since the wavelet is located in time, we can multiply our signal time-wise with the wavelet at various locations. It is the convolution procedure. Once the procedure is carried out with the mother wavelet, it can be scaled and the procedure can be repeated again at different scales.

4.3.2 Signal Denoising with MSPCA

Principal component analysis (PCA) combines the variables as a linear weighted sum of transforms. The hyperplane direction saves the maximum likely residual variance in the observed variables represented by the principal component loadings, while maintaining orthonormality. The eigenvalues identify the variance taken from the covariance matrix, and eigenvectors are the loadings of the associated eigenvector of input. The amount of nonzero eigenvalues is equivalent to the matrix rank if the measured variables are correlated or redundant. MSPCA includes the wavelet and the PCA to eliminate the cross-correlation among the variables. WT decomposes the EMG signals in order to combine the benefits of the PCA and wavelets. Wavelet decomposition achieves the deterministic parts of the EMG signals, whereas the stochastic part in each signal is nearly de-correlated with the WT of the EMG signal. The covariance matrix is transformed alternatively into another scale in order to utilize the multiscale properties of the signal. Since the quantity of principal components to be reserved at every scale doesn't change the fundamental relationship among the variables at any scale, it is not transformed by the wavelet decomposition [38,39]. In the implementation, Symlets four-wavelet function is utilized as a mother wavelet and five-level decomposition is applied.

4.3.3 FEATURE EXTRACTION

One of the critical steps in the diagnosis of neuromuscular disorders using EMG signals is the feature extraction. The EMG signals consist of several data points, and informative and distinguishing features. These informative and distinguishing features describe the characteristic of the EMG signal that might designate a particular muscle action and can be achieved by utilizing several feature extraction techniques. Informative and distinguishing features can describe neuromuscular disorders, nerve health, and muscle strength [1–3]. In this work, wavelet-based methods, namely, DWT, WPD, TQWT, DT-CWT, SWT, FAWT, and EWT, are used to extract the features from the EMG signals. In the implementation, different mother wavelets with different levels of decomposition and parameters are tested, and the one which achieves the best performance is used. Mostly, the decomposition levels are related to the signal type and sampling frequency. However, in order to be sure for achieving the best classification performance, different levels of decomposition in the feature extraction should be tried and the best one should be selected. Hence, different levels of decomposition are tried and tested; the optimum one is chosen for each method.

4.3.3.1 Discrete Wavelet Transform (DWT)

Wavelet-based techniques are considered as a suitable solution in order to investigate diverse nonstationary signals such as EMG. For example, DWT is utilized to decompose a discrete-time signal $x[k]$ into wavelet coefficients by shifting and scaling the mother wavelet. First, the right number of decomposition levels (j_m) must be chosen for the DWT decomposition. J is equivalent to 1 for the principal scale level and signal $x[k]$ is instantaneously passed to both the low-pass and the high-pass filters. After that comes the down-sampling phase. Presentation of j for each level's output is described as approximation and detail in two signal forms. The Daubechies four-wavelet function is employed as a mother wavelet, and the decomposition of seven levels is implemented.

4.3.3.2 Wavelet Packet Decomposition (WPD)

The WPD is identified as the extension of the DWT where the low-frequency components (approximations) are decomposed into sub-bands as well. Nevertheless, WPD utilizes both the low-frequency components (approximations) and the high-frequency components (details) [15–17]. DWT and WPD differ from each other as WPD divides both the high- and low-frequency components into their sub-levels. Thus, WPD achieves a better frequency resolution for the EMG signal decomposition. The WPD is beneficial to combine numerous decomposition levels to construct the original signal. In this research, the Daubechies four-wavelet function is utilized as a mother wavelet and four decomposition levels have been chosen.

4.3.3.3 Tunable Q-Factor Wavelet Transform (TQWT)

TQWT is an important way of analyzing oscillating signals [18,40]. As a feature of guided deployment, it can be conveniently modified. Q, r, and j are its major adjustable parameters. Q represents the Q-factor, r the amount of oversampling, and j the degree of decomposition. Q performs the count of wavelet oscillations. Unwanted overoscillation is governed by r. The TQWT is designed using digital filter banks, developed especially for the targeted application. Such functions are nonrational functions of transformation and are known for their practical implementation [18,19].

The oscillatory activity of the signal to be analyzed is used to pick an effective Q-factor. The wavelet transformations most commonly display a mild tendency to change the Q-factor, except for continuous wavelet transformation. Due to this restriction, their use is limited to the particular system. In this sense, the TQWT is conceived. The perfect restoration of oversampled filter banks is employed, with precise scaling factors, to realize the TQWT. The discrete-time TQWT handles finite length-sampled signals. The principle is similar to that of the short-time Fourier transform and could be effectively realized with the fast Fourier transform (FFT). Selesnick has shown that by using moderate oversampling ratios, the TQWT can be efficiently realized. Sampling rates of about 3- to 4-folds of the Nyquist sampling criteria are also used as a feature of the intended TQWT implementation [18,40].

The TQWT is a strong transform for the analysis of the oscillatory signals. Usually, the Q-factor of a TQWT is selected within the specific part of the signal in accordance with the oscillatory behavior of the signal. Nevertheless, a low Q-factor should be used by the TQWT for the duration of the processing of signals that do not have any oscillatory characteristic or have little oscillatory characteristic. Most wavelet transforms do not have such ability to regulate the wavelet's Q-factor apart from the continuous WT [18]. In the implementation of TQWT, the optimal parameters are chosen as $Q = 3, r = 4, J = 15$.

4.3.3.4 Dual-Tree Complex Wavelet Transform (DT-CWT)

The DT-CWT is an enhanced version of the DWT and possesses two features that have not advantage over the DWT. It is almost shift-invariant and directionally selective in two and higher dimensions. It is done with a redundancy factor of just 2 and is significantly smaller than the redundancy factors utilized in a stationary DWT [41]. It shows a fine movement-invariant property and improves the directional estimation in contrast with the DWT directional estimation. The DT-CWT provides a fantastic restoration by making use of two simultaneous-devastated channel bank trees with coefficients with adding value created at each tree. The DT-CWT calculates the complex signal transformation by means of two separate DWT decompositions, namely, tree 1 and tree 2. The filters used in tree 1 could be designed differently compared to filters designed in tree 2. In this way, the real coefficients can be produced by one DWT and the imaginary coefficients can be produced by the second DWT. It shows the DT-CWT with a second level of decomposition [42]. l_1 and h_1 are, respectively, the low- and high-pass half-band filters for tree 1. l_2 and h_2 are, respectively, low- and high-pass half-band filters for tree 2. $W_{110}(k)$ and $W_{111}(k)$ are the sub-band coefficients after the first level of decomposition for tree 1. $W_{120}(k)$, $W_{121}(k)$, $W_{122}(k)$, and $W_{123}(k)$ are the sub-band coefficients after the second level of decomposition for tree 1. $W_{210}(k)$ and $W_{211}(k)$ are the sub-band coefficients after the first level of decomposition for tree 2. $W_{220}(k)$, $W_{221}(k)$, $W_{222}(k)$, and $W_{223}(k)$ are the sub-band coefficients after the second level of decomposition for tree 2 [41].

DT-CWT includes two-real values. The normal DWT has a shift-variant feature due to the decimation method employed in the transform. Different set of wavelet coefficients that exist at the output can generate a minor variation in the input signal. Hence, the DT-CWT, a novel wavelet transform, was presented by Kingsbury [20] to eliminate this problem. The DT-CWT is an up-to-date improvement of the DWT devising substantial novel capabilities. DT-CWT is directionally discriminating in two dimensions and nearly shift-invariant. Apparently, the multidimensional DT-CWT is indivisible yet, since it is based on a filter bank, which is computationally sufficient and distinct [43]. Two real DWTs are used by the DT-CWT. The real part of the DT-CWT is given by DWT, whereas the imaginary part of the DT-CWT is given by another DWT. In order to have the whole transform that is approximately analytic, the two sets of filters are deliberated in a joint way [21]. In the implementation of dual-tree complex wavelet transform, seven levels of decomposition are used and the parameters are selected automatically.

4.3.3.5 Stationary Wavelet Transform (SWT)

The DWT is a decomposition of the time series $x(t)$ that is a consecutive band-pass filtering and down-sampling. $x(t)$ can be decomposed into the high-frequency parts and the low-frequency parts. The DWT can be implemented by successive filter banks, but it is not shift-invariant once implemented for biomedical signals such as EMG. If the biomedical signal is shifted, the subsequent coefficients might be significantly dissimilar. The stationary wavelet transform (SWT) does not have such problems. Principally, the SWT is the same as the DWT; however, the down-sampling phase is substituted by an up-sampling phase [44].

The SWT is an enhancement of the DWT and possesses one feature that has benefit over the DWT. It is almost shift-invariant. Any time series can be decomposed effectively by utilizing the DWT [45]. The process consists of the cascaded stages of band-pass filtering and decimation. It is a quick operation in computational terms and can be put into place by subsequent filter banks. Regrettably, as applied to discrete-time series, the DWT is not shift-invariant. If the incoming time series is shifted in time, then the DWT output coefficients could change significantly. This shortfall is treated up to a certain extent by utilizing the SWT. It is attained by replacing the down-sampling step of the DWT with the up-sampling process. l_1 and h_1 are, respectively, the low- and high-pass half-band filters used at the first level of decomposition. l_2 and h_2 are, respectively, the low- and high-pass half-band filters used at the second level of decomposition, and l_3 and h_3 are, respectively, the low- and high-pass half-band filters used at the third level of decomposition [45].

4.3.3.6 Flexible Analytic Wavelet Transform (FAWT)

The projection on coordinating wavelet basis function may isolate any signal as per its impact in the distinctive time and frequency domain. In this manner, quick calculation algorithms are required for the wavelet circulation relying upon the representation of signal projection as the separating or filtering activity. A line of scaling functions could characterize the signal's lowest frequency content. The wavelet coefficients of the scaling functions decrease since this dilation has parameters of dyadic interpretation and discretization. The FAWT makes use of the Hilbert transform and provides adaptability to reduce the dilation and the quality factor (QF). It also helps in the breakdown of the signals with appropriate dynamic parameters, down-sampling factor (t) and the up-sampling factor (s) for the low-pass channel, down-sampling factor (v) and the up-sampling factor (u) for the high-pass channel, and the parameter controlling QF (β). This transform will achieve the iterative dimensions by adding the iterative channel or filter banking assembly. At each dimension, the signal decomposition in this method produces two channels where one channel is related to low-pass and the other high-pass channels, with individual control of the inspection rate. The high-pass channels deal separately with both the positive and negative frequencies.

The QF is characterized as the relation of the center frequency and the bandwidth, which is defined as $QF = \dfrac{2-\beta}{\beta}$. Redundancy is the ratio of the input samples to the output samples and can be derived as $R = \dfrac{u}{v} * \dfrac{1}{1-(s/t)}$. The low-pass

filter frequency response $H(w)$ and the high-pass filter frequency response $G(w)$ are expressed mathematically as [46]

$$H(w) = \begin{cases} (st)^{\frac{1}{2}} & |w| < w_p \\[2ex] (st)^{\frac{1}{2}} \theta\left(\dfrac{w - w_p}{w_s - w_p}\right) & w_p \leq w \leq w_s \\[2ex] (st)^{\frac{1}{2}} \theta\left(\dfrac{\pi - (w - w_p)}{w_s - w_p}\right) & -w_s \leq w \leq -w_p \\[2ex] 0 & |w| \geq w_s \end{cases} \qquad (4.3)$$

$$G(w) = \begin{cases} (2uv)^{\frac{1}{2}} \theta\left(\dfrac{\pi - (w - w_0)}{w_1 - w_0}\right) & w_0 \leq w < w_1 \\[2ex] (2uv)^{\frac{1}{2}} & w_1 \leq w < w_2 \\[2ex] (2uv)^{\frac{1}{2}} \theta\left(\dfrac{w - w_2}{w_3 - w_2}\right) & w_2 \leq w \leq w_3 \\[2ex] 0 & w \in \left[(0, w_0) \bigcup (w_3, 2\pi)\right] \end{cases} \qquad (4.4)$$

Where, $w_s = \dfrac{\pi}{t}, w_0 = \dfrac{(1-\beta)\pi + \epsilon}{u}, w_1 = \dfrac{\pi}{tu}, w_2 = \dfrac{\pi - \epsilon}{u}$, and $w_3 = \dfrac{\pi + \epsilon}{u}, \epsilon \leq \dfrac{s - t + \beta t}{s + t}\pi$. w_p is the pass band frequency and w_s is the stop band frequency of the low-pass filter $H(w)$.

The $\theta(w)$ can be described as $\theta(w) = \dfrac{[1 + \cos(w)][2 - \cos(w)]^{\frac{1}{2}}}{2}$ for $w \in [0, \pi]$.

4.3.3.7 Empirical Wavelet Transform (EWT)

Empirical wavelet transform (EWT) was suggested by Jerome Gilles [47], who proposed a technique to construct a family of wavelets modified to the processed signal, which is equivalent to constructing a set of bandpass filters. In order to achieve the adaptability, the filters' supports are based on the information in the spectrum of the studied signal location. Actually, the IMF properties are equal to the spectrum of an IMF, which is centered around a specific frequency. The EWT is a time-frequency method to decompose signals based on information content of the signal using the adaptive wavelet. Before dividing the spectrum based on the detected maxima, EWT first detects the local maxima of the Fourier spectrum of the signal, and then it creates a consistent wavelet filter bank. This method works in the following three steps [48]:

- Define the frequency components of the employed signal by means of FFT.
- The different modes are produced by finding an appropriate separation of the Fourier spectrum.
- Employ scaling and wavelet functions related to every identified part. Separation of the Fourier spectrum is the most crucial step, which delivers the adaptability of this approach according to the examined signal.

4.3.4 DIMENSION REDUCTION

The statistical values, such as mean, standard deviation, kurtosis, and skewness of wavelet coefficients, are computed from the sub-bands of the DWT, WPD, TQWT, DT-CWT, SWT, FAWT, and EWT. The first- and second-order statistics are important in signal analysis. Moreover, for the representation of several signals, especially while working with nonlinear signals such as EMG, first- and second-order statistics are not sufficient. Hence, higher-order statistics (HOS) must be employed for a better representation of the EMG signals. Even though the first- and second-order statistics describe the mean and variance, the HOS (skewness and kurtosis) characterize the higher-order moments.

Since the wavelet-based feature extraction techniques generate the feature vector, which is too big in size to be used as an input to a classifier, the dimension reduction methods are employed to reduce the number of features from the wavelet coefficients. By utilizing mean, variance, skewness, and kurtosis of each level of sub-bands, the reduced feature sets are computed from the sub-bands of the wavelet decomposition. The ability of transforming the set of coefficients into a reduced feature set describes one of the critical stages in any recognition process, since this reduced feature set describes the characteristic of the EMG signals in a superior manner. The six statistical features are employed for the EMG signal classification, which are:

1. Average power of the coefficients,
2. Ratio of the absolute mean values of coefficients of adjacent sub-bands,
3. Mean values of coefficients,
4. Variance of the coefficients,
5. Skewness of the coefficients,
6. Kurtosis of the coefficients.

4.3.5 CLASSIFICATION METHODS

4.3.5.1 Artificial Neural Networks (ANNs)

A set of input and output units that are linked generates ANNs. A weight that is associated with each connected unit exists in that unit. The network obtains the information by adjusting the weights until it finds the right class label of the instances. One of the strengths of the ANNs is that they are highly tolerant of noisy instances. The other benefit is that they can classify the unseen instances. When you do not have enough knowledge about the relations between attributes and classes, they may be useful. Different types of neural network algorithms exist; the most famous algorithm is the backpropagation. Details are given in Ref. [49].

4.3.5.2 k-Nearest Neighbor (k-NN)

When we give large training sets to the k-NN technique, it needs more effort. Nearest-neighbor classifier learns with the help of analogy. Analogy means the comparison of given test instances with similar training instances. These instances should be the closest ones to the unseen instances. Closeness can be defined with a distance metric like Euclidean distance. The values of each attribute are generally normalized, and this provides prevention for attributes that have large ranges in the beginning. In order to classify k-NN, the instances, which are not seen, are chosen as the most common class among their k-NNs. Details are given in Ref. [49].

4.3.5.3 Support Vector Machine (SVM)

SVM is a classifier, which is employed for the classification process of both linear and nonlinear data. SVM employs a nonlinear transformation to map the original training data into a higher dimension. It looks for a linear optimal separating hyperplane like a decision boundary to discriminate the one class from another. If there were no straight line which separates the classes in the case of the linearly inseparable data, a feasible solution would not be found by the linear SVM. However, nonlinear SVM can be created by an extension of the linear SVM approach to classify linearly inseparable data. These kinds of SVMs are able to find nonlinear decision boundaries, and they can be realized in two main steps. A nonlinear mapping of the original input data into a higher dimensional space is done by transforming the data into the new higher space. Then, a linear separating hyperplane can be built in the second step [49].

4.3.5.4 C4.5 Decision Tree

C4.5 decision tree algorithm tests, including their decision tree for which the training instances have the same outcome, are removed because they are not very significant. Thus, unless they have at least two results that have a minimum number of instances, they are not included in the decision tree. We can monitor the minimum value given and boost it for noisy data instances. There is an MDL-based tuning to split numeric attributes. A heuristic is designed to avoid overfitting. The information gain might be negative after the subtraction. The tree will halt growing if the attributes do not have a positive information gain that is a sort of pre-pruning. This is expected at this step because a pruned tree may be unexpected while post-pruning is not active [50].

4.3.5.5 Classification and Regression Trees (CART)

One of the most successful and powerful classification techniques of decision tree induction is the CART. CART creates binary trees from the data that is defined by both continuous and discrete features. It is a step-by-step procedure where the nodes are split or not split into child nodes. The parent node is termed as the nonterminal node, and the child node is termed as the binary split. The node will become terminal if it is not split, and thereby, it is assigned with a class label. The Gini index measures the imperfection by which a randomly selected element will be mislabeled regarding the constructed tree. These missing instances are taken care of with a strategy of surrogate splits [51]. The default parameters are employed in the implementation.

4.3.5.6 Reduced Error Pruning Tree (REP) Tree

With the help of gain/variance reduction, a decision or regression tree is made in a REP Tree, and the constructed tree is pruned with a reduced error pruning. As in C4.5, missing values are dealt with by splitting instances into parts. It is possible to evaluate the number of pruning folds, the maximum tree size, the total number of examples on each leaf, and the minimum amount of training set variance for a split [49].

4.3.5.7 LogitBoost Alternating Decision (LAD) Tree

The LAD Tree learning algorithm is applied to find the error criterion and acquire the regression trees. It is a parallel binary classifier and henceforth can separate among positive and negative examples. In each iteration, a single feature is picked at the splitter node for the testing purpose. The effective response and weights are stored in each training instance. The working principle is to calculate the mean value of the instances by restraining the least-squares value in a particular subset. The best drop in this calculation is relevant by selecting the maximum gain carefully in the tests [52].

LAD Tree can deal with multiclass problems that employ the LogitBoost technique [53]. Moreover, LAD Tree determines the size of the produced tree, and it is able to adjust the number of boosting iterations to fit the data [49].

4.3.5.8 Random Tree Classifier

Random Tree is a supervised algorithm that employs collective learning to create several individual classifiers. It utilizes bagging samples to create a random arrangement of samples for generating a decision tree. In the standard tree, each node is split by utilizing the best division among all features. In the random Tree classifier, every node is divided by employing the best among the subset of predictors randomly picked at that node. The random Tree classifier receives the input feature vector, classifies it with each tree available in the forest, and produces the class name, which gets the majority of votes [54].

4.3.5.9 Random Forests (RF)

The Random Forest (RF) is an ensemble of decision tree classifiers that make them a category of learners called forest. We produce separate decision trees by employing a random collection of attributes at every node to define the splitting. The RF is formed by utilizing bagging together with the random attribute selection. The CART decision tree is used to form the forest [49].

4.4   RESULTS AND DISCUSSION

In this chapter, in order to classify EMG signals for the identification of neuromuscular diseases, different statistical features extracted from the sub-bands of wavelet decomposition were utilized. The extracted features from every signal frame were used to yield a total feature set to describe the EMG signal pattern. Because of the characteristic of the EMG signal, there is a big difference between individuals' signal characteristics. A robust classifier has to eliminate the individual variations. Therefore, a set of features is extracted from the sample of EMG signals by using different wavelet-based signal processing techniques, namely, DWT, WPD, TQWT, DT-CWT, SWT, FAWT, and EWT. The reduced feature vectors were computed by utilizing statistical values of the wavelet sub-bands. Then, diverse machine learning algorithms are trained in order to have more accurate classifier for MUAP classification.

4.4.1   SUBJECTS AND DATA ACQUISITION

In this study, clinical EMG data downloaded from EMGLAB website [55] is used. The clinical EMG signals were collected under normal settings for the MUAP analysis. The data collection was approved by the Stanford University Committee on the Use of Human Subjects in Research, and informed consent was given to each subject. The EMG signals are collected using a standard concentric needle electrode at the constant level of contraction. The EMG signals are composed of a control group and a group of patients with myopathy and ALS, are filtered between 2 Hz and 10 kHz. There were ten normal subjects (four females and six males) aged 21–37 in the control group. There were seven patients (two females and five males) aged between 19 and 63 years in the myopathy group. There were eight patients (four females and four males) aged between 35 and 67 years in the ALS group [56].

4.4.2   PERFORMANCE EVALUATION

Finding a classifier's prediction performance with insufficient data is difficult, interesting, and contentious. The performance of machine learning methods can be determined through the cost of misclassification error, if a negative example is misclassified as positive, or vice versa. Cross-validation is a simple approach utilized if a small number of data exists [50]. Thus, in this research, 10-fold cross-validation is used, which is the option for assessing the error rate of a classification technique in the most realistic limited data situations. In general, 10-fold cross-validation is used where the data samples are randomly divided into ten portions to characterize every class roughly the same sizes as in the whole dataset. Ultimately, the ten error evaluations are combined to get an overall error value [50].

The performance of a classifier can be evaluated by using the number of true positives (TP), true negatives (TN), false positives (FP), and false negatives (FN).

The overall performance measure is defined by the accuracy, which is represented as follows:

$$\text{Accuracy} = \frac{TP + TN}{TP + FP + TN + FN} \tag{4.5}$$

In order to assess the differentiation capacity of the classifiers, ROC curves are produced by using the cross-validation method [50]. The mean AUC is utilized to measure the classification performance. As AUC yields a single measure of the overall accuracy, which is not reliant on any specific threshold, it is generally taken as the performance index [57].

F-measure is also employed to describe the performance of a classifier, which is represented as:

$$\text{F-measure} = \frac{2*TP}{(2*TP) + FP + FN} \tag{4.6}$$

The kappa statistic (κ) utilizes the fact that the classifier will agree or disagree only by chance. Once there is no independent revenue for evaluating the probability of chance agreement among two or more learners, the kappa statistic is the most commonly utilized statistic for the assessment of categorical data. Cohen [58] described the kappa statistic as an agreement index and defined as follows:

$$K = \frac{P_0 - P_e}{1 - P_e} \tag{4.7}$$

where P_0 is the observed agreement and defined as

$$P_0 = \frac{TP + TN}{TP + TN + FP + FN} \tag{4.8}$$

and P_e measures the probability of random agreement [59]. Total random agreement probability is the probability, which they agreed on either Yes or No, i.e.:

$$P_e = P_{YES} + P_{NO} \tag{4.9}$$

where

$$P_{YES} = \frac{TP + FP}{TP + TN + FP + FN} * \frac{TP + FN}{TP + TN + FP + FN} \tag{4.10}$$

$$P_{NO} = \frac{FN + TN}{TP + TN + FP + FN} * \frac{FP + TN}{TP + TN + FP + FN} \tag{4.11}$$

4.4.3 EXPERIMENTAL RESULTS

The performances of the classifiers are assessed in terms of the total classification accuracy, AUC, F-measure, and kappa statistics (κ) that are described in the preceding section.

Each model's performance is assessed using 10-fold cross-validation that reduces the bias introduced by selecting the particular training and test instances. The performance of learners for the EMG data is summarized in Tables 4.1–4.4. All approaches are accomplished rationally well according to the total classification accuracy, AUC, F-measure, and kappa statistics. Since it can be realized from the tables, the maximum performance is achieved with SVM by using DT-CWT with a total classification accuracy of 99.6%, F-measure value of 0.996, and κ value of 0.994. The best performance in terms of AUC (1.0) is accomplished by utilizing the RF classifier with all feature extraction methods.

Table 4.1 shows the total classification accuracy for seven different wavelet-based feature extraction approaches, namely, DWT, WPD, TQWT, DT-CWT, SWT, FAWT, and EWT. The best result is accomplished with SVM classifier using DT-CWT with an accuracy of 99.6%, SWT with an accuracy of 99.5%, and DWT with an accuracy of 99.1%. All classifiers achieved a reasonable accuracy with all feature extraction methods, except FAWT and EWT.

TABLE 4.1
Total Classification Accuracies of Different Feature Extraction Methods

	DWT	WPD	DTCWT	TQWT	SWT	FAWT	EWT
SVM	0.991	0.989	0.996	0.989	0.995	0.884	0.941
k-NN	0.983	0.942	0.976	0.963	0.984	0.682	0.736
ANN	0.984	0.978	0.976	0.986	0.992	0.804	0.821
RF	0.986	0.977	0.99	0.987	0.986	0.757	0.874
CART	0.958	0.937	0.949	0.946	0.967	0.653	0.769
C4.5	0.963	0.942	0.954	0.947	0.968	0.646	0.766
REP Tree	0.958	0.924	0.946	0.942	0.964	0.644	0.773
Random Tree	0.936	0.875	0.923	0.889	0.919	0.582	0.718
LAD Tree	0.96	0.924	0.957	0.955	0.964	0.578	0.598

TABLE 4.2
AUC Results for Different Feature Extraction Methods

	DWT	WPD	DTCWT	TQWT	SWT	FAWT	EWT
SVM	0.994	0.993	0.997	0.994	0.997	0.93	0.962
k-NN	0.997	0.981	0.995	0.992	0.996	0.829	0.865
ANN	0.998	0.998	0.997	0.999	0.999	0.922	0.929
RF	1	0.999	1	1	0.999	0.905	0.967
CART	0.976	0.971	0.969	0.971	0.98	0.799	0.875
C4.5	0.975	0.959	0.967	0.964	0.975	0.747	0.831
REP Tree	0.985	0.984	0.983	0.977	0.989	0.796	0.88
Random Tree	0.952	0.906	0.942	0.917	0.94	0.686	0.789
LAD Tree	0.993	0.984	0.994	0.995	0.996	0.755	0.722

TABLE 4.3
F-Measure of Different Feature Extraction Methods

	DWT	WPD	DTCWT	TQWT	SWT	FAWT	EWT
SVM	0.991	0.993	0.996	0.989	0.995	0.884	0.941
k-NN	0.982	0.981	0.976	0.963	0.984	0.684	0.739
ANN	0.984	0.998	0.976	0.986	0.992	0.801	0.82
RF	0.986	0.999	0.99	0.987	0.986	0.756	0.873
CART	0.958	0.971	0.949	0.946	0.967	0.651	0.769
C4.5	0.962	0.959	0.954	0.947	0.968	0.647	0.766
REP Tree	0.958	0.984	0.946	0.942	0.964	0.641	0.772
Random Tree	0.936	0.906	0.923	0.889	0.919	0.582	0.718
LAD Tree	0.96	0.984	0.957	0.955	0.964	0.57	0.584

TABLE 4.4
Kappa Statistics of Different Feature Extraction Methods

	DWT	WPD	DTCWT	TQWT	SWT	FAWT	EWT
SVM	0.9858	0.9829	0.9938	0.9837	0.9921	0.8262	0.9112
k-NN	0.9737	0.9133	0.9646	0.9437	0.9767	0.5225	0.6042
ANN	0.9767	0.9663	0.9633	0.9792	0.9879	0.7054	0.7317
RF	0.9788	0.965	0.985	0.9808	0.9783	0.6358	0.8108
CART	0.9375	0.905	0.9233	0.9183	0.9508	0.4787	0.6542
C4.5	0.9437	0.9133	0.9304	0.9204	0.9521	0.4696	0.6483
REP Tree	0.9375	0.8867	0.9183	0.9133	0.9454	0.4663	0.6587
Random Tree	0.9042	0.8129	0.8837	0.8333	0.8792	0.3729	0.5775
LAD Tree	0.9404	0.8867	0.9358	0.9325	0.9463	0.3675	0.3967

Similarly, Table 4.2 presents the AUC with four different wavelet-based feature extraction techniques. It can be realized from Table 4.2 that the best AUC result is accomplished with RF (1.0), using DT-CWT, TQWT, and DWT, while the second-best AUC result (0.999) is accomplished with RF using WPD and SWT, and ANN using TQWT and SWT. k-NN, CART, C4.5, REP Tree, and LAD Tree achieved the reasonable AUC (>0.983). Random Tree achieved the lowest AUC value among the classifiers. FAWT and EWT achieved lower AUC value compared to the other feature extraction methods.

Another performance measure of a classifier is F-measure, which is given in Table 4.3. It can be realized from Table 4.3 that the F-measure values for almost all classifiers are similar to the total classification accuracy, so the best result is accomplished with RF method using WPD with an F-measure value of 0.999; then the second-best result is accomplished with ANN using WPD with an F-measure value of 0.998. k-NN, CART, C4.5, REP Tree, and LAD Tree achieved the reasonable F-measure values for DWT, DT-CWT, TQWT, and SWT. Random Tree achieved the lowest F-measure value among the classifiers. FAWT and EWT achieved lower F-measure value compared to the other feature extraction methods.

The similar situation occurred for kappa statistic as in the case of the total classification accuracy and F-measure, which is shown in Table 4.4. Thus, the best result is accomplished with SVM classifier using DT-CWT with a kappa value of 0.9938. Actually, the kappa values for all feature extraction methods are the highest with SVM. RF, ANN, k-NN, CART, C4.5, REP Tree, and LAD Tree achieved a reasonable kappa statistic. Random Tree achieved the lowest kappa value among the classifiers. FAWT and EWT achieved lower kappa value compared to the other feature extraction methods especially with weak classifiers.

This study confirms the superiority of SVM classifier compared to the other classifiers when dealing with statistical features extracted for each DT-CWT sub-band of EMG signals. Cohen's kappa coefficient was 0.9938, presenting an almost-perfect agreement with experts.

The performance revealed by the classifiers for the MUAP classification depends on the feature extraction and the chosen classification technique. The most effective parameters that are extracted from the MUAP that rely on the feature extraction and classification methods are used. The best-fitted parameters for MUAP classification should be used as the inputs of the classifier. Hence, the statistical features extracted from each sub-band of DWT, WPD, TQWT, DT-CWT, and SWT are selected, as they are suitable for the recognition of the nonlinear dynamics essential muscle activity in EMG.

Exact description of the EMG signal is crucial for the treatment and diagnosis. The proposed model classifies MUAPs by utilizing statistical features extracted from each sub-band of DT-CWT with an accuracy of 99.6%. This also results in an improvement of ROC area (AUC = 1), and F-measure (0.999) of RF is higher than that of C4.5-, ANN-, and SVM-based classifiers. The RF classifier as defined in this paper is comparable with SVM in the MUAP classification. Moreover, it is confirmed that the SVM with statistical features extracted from every sub-band of DT-CWT is robust enough to variations in the EMG signals for the diagnosis of the neuromuscular disorders.

4.4.4 DISCUSSION

In this study, we compared seven different feature extraction methods, namely, DWT, WPD, TQWT, DT-CWT, SWT, FAWT, and EWT, for EMG signal classification by using different machine learning techniques. It is seen that DT-CWT feature extraction method achieved the best performance with SVM in total classification accuracy. Also, all feature extraction methods yielded the best performance in AUC using RF. Although results were significantly better, the computational cost of DT-CWT was much higher than those of the other feature extraction methods. FAWT and EWT achieved lower values compared to the other feature extraction methods especially with weak classifiers. There is no big difference between k-NN, CART, C4.5, REP Tree, and LAD Tree classifiers with respect to the classification performance of the classifiers. Although the classification results are not significantly diverse, the ease of model construction is simpler for these classifiers. Random Tree always achieved

TABLE 4.5

Classification Accuracies of Different Types of EMG Signals Using DT-CWT

	Control	MYO	ALS	Average
SVM	0.994	0.995	0.998	0.996
k-NN	0.965	0.973	0.991	0.976
ANN	0.967	0.968	0.993	0.976
RF	0.98	0.991	0.999	0.99
C4.5	0.934	0.943	0.984	0.954
REP Tree	0.923	0.93	0.984	0.946
LAD Tree	0.941	0.95	0.981	0.957

lower performance. The classification accuracies of different groups of subjects are also given in Table 4.5. As it can be seen from the table, the accuracy of control subjects is the lowest and the accuracy of ALS subjects is the highest. Hence, the discrimination of neuromuscular disorders is higher compared to the control subjects.

Input variable selection is the crucial step in the MUAP classification. In this study, statistical values, including HOS, extracted for each sub-band are employed as an input to a classifier. These statistical values are used for the dimension reduction purposes to yield the smallest possible subset consistent with the full feature set. Therefore, the features selected for the model creation were those relevant to the signal statistics of different frequency bands. Neuroscientists in the clinics must be careful about the requirements of the model before using it. The consequence of this study implies that the DT-CWT with SVM achieved a better performance compared to the other models in diagnosing neuromuscular disorders. This model offers an easy, fast, and low-cost method to diagnose neuromuscular disorders precisely, but the concrete impact of local data will necessitate further research in the future. Furthermore, the complex algorithm is not easily understood by clinicians, which may hinder its persistent use. Potential external justification may extend the generalization ability of the proposed model, and it can be used as an efficient tool in clinical decision-making.

Based on the results achieved in this study and familiarity in EMG feature extraction and MUAP classification problems, the followings can be emphasized:

- There is an indication of supporting the privilege of high generalization capability of the wavelet-based feature subsets chosen by using the statistical values of wavelet sub-bands.
- DT-CWT feature subsets reveal a bit higher generalization and competence (in terms of classification performance) than the feature subsets extracted using DWT, WPD, TQWT, SWT, FAWT, and EWT.
- The high performance of the classifiers yields the understandings of the features used for describing the MUAPs. The results achieved in this study confirmed that the statistical features extracted from each sub-band are the features which characterize well the MUAP, and a good distinction between the classes can be achieved by using these statistical features.

- The classifier performances indicated that the SVM and RF have a significant achievement in the EMG signal classification as compared to the other classification techniques.
- The testing performance of the DT-CWT-based decision support system is found to be rational, and the proposed system can be utilized in the clinical studies for the diagnosis of neuromuscular disorders to support physicians.

This study conveys an independent assessment of EMG signals, and their computerized characteristics make it easy to be utilized in clinical practice as a decision support system. Moreover, the possibility of a real-time implementation of the expert diagnosis system accomplishes more precise decisions by increasing the variety and the number of parameters.

4.5 CONCLUSION

As the precise classification and recognition of numerous types of MUAPs is crucial for an accurate diagnosis of the patient, diagnosis of neuromuscular disorders from the EMG signal has been an imperative and widespread area for biomedical research. In this study, a novel framework that compares different feature extraction methods, namely, DWT, WPD, TQWT, DT-CWT, SWT, FAWT, and EWT, to characterize MUAP is evaluated. Different feature extraction methods are tested to eliminate the redundant features in the main feature set. The results confirmed that different pathological variations in EMG signals can be well represented by the statistical features (mean, variance, skewness, kurtosis etc.) taken from each sub-band of wavelet transform. Furthermore, the proposed framework is effective in discriminating the several types of MUAPs as compared to the similar methods in the literature. Furthermore, the DT-CWT feature extraction method for EMG signal classification has a superior generalization capability compared to other feature extraction techniques. This ability usually results in a better classification accuracy and a high computational complexity. Moreover, classification accuracies of different classifiers are tested with seven different wavelet-based feature extraction methods for the diagnosis of neuromuscular diseases. Additionally, the performance of different classifiers is compared by testing different feature extraction methods.

REFERENCES

1. Subasi A. Classification of EMG signals using combined features and soft computing techniques. *Appl Soft Comput* 2012;12:2188–2198.
2. Subasi A. Classification of EMG signals using PSO optimized SVM for diagnosis of neuromuscular disorders. *Comput Biol Med* 2013;43:576–586.
3. Begg R, Lai DT, Palaniswami M. *Computational intelligence in biomedical engineering*. Boca Raton, FL: CRC Press; 2008.
4. Gokgoz E, Subasi A. Effect of multiscale PCA de-noising on EMG signal classification for diagnosis of neuromuscular disorders. *J Med Syst* 2014;38:31.
5. Canal MR. Comparison of wavelet and short time Fourier transform methods in the analysis of EMG signals. *J Med Syst* 2010;34:91–94.

6. Koçer S. Classification of EMG signals using neuro-fuzzy system and diagnosis of neuromuscular diseases. *J Med Syst* 2010;34:321–329.

7. Subasi A, Yilmaz M, Ozcalik HR. Classification of EMG signals using wavelet neural network. *J Neurosci Methods* 2006;156:360–367.

8. Kamali T, Boostani R, Parsaei H. A multi-classifier approach to MUAP classification for diagnosis of neuromuscular disorders. *IEEE Trans Neural Syst Rehabil Eng* 2014;22:191–200.

9. Subasi A. A decision support system for diagnosis of neuromuscular disorders using DWT and evolutionary support vector machines. *Signal Image Video Process* 2015;9:399–408.

10. Naik GR, Arjunan S, Kumar D. Applications of ICA and fractal dimension in sEMG signal processing for subtle movement analysis: A review. *Australas Phys Eng Sci Med* 2011;34:179–193.

11. Phinyomark A, Phukpattaranont P, Limsakul C. Feature reduction and selection for EMG signal classification. *Expert Syst Appl* 2012;39:7420–7431.

12. Rasheed S, Stashuk D, Kamel M. A software package for interactive motor unit potential classification using fuzzy k-NN classifier. *Comput Methods Programs Biomed* 2008;89:56–71.

13. Subasi A. Medical decision support system for diagnosis of neuromuscular disorders using DWT and fuzzy support vector machines. *Comput Biol Med* 2012;42:806–815.

14. Dobrowolski AP, Wierzbowski M, Tomczykiewicz K. Multiresolution MUAPs decomposition and SVM-based analysis in the classification of neuromuscular disorders. *Comput Methods Programs Biomed* 2012;107:393–403.

15. Unser M, Aldroubi A. A review of wavelets in biomedical applications. *Proc IEEE* 1996;84:626–638.

16. Daubechies I. The wavelet transform, time-frequency localization and signal analysis. *IEEE Trans Inf Theory* 1990;36:961–1005.

17. Learned RE, Willsky AS. A wavelet packet approach to transient signal classification. *Appl Comput Harmon Anal* 1995;2:265–278.

18. Selesnick IW. Wavelet transform with tunable Q-factor. *IEEE Trans Signal Process* 2011;59:3560–3575.

19. Patidar S, Pachori RB. Classification of cardiac sound signals using constrained tunable-Q wavelet transform. *Expert Syst Appl* 2014;41:7161–7170.

20. Kingsbury NG. The dual-tree complex wavelet transform: A new technique for shift invariance and directional filters. Utah 1998, p. 86.

21. Selesnick IW, Baraniuk RG, Kingsbury NC. The dual-tree complex wavelet transform. *IEEE Signal Process Mag* 2005;22:123–151.

22. Kamali T, Boostani R, Parsaei H. A multi-classifier approach to MUAP classification for diagnosis of neuromuscular disorders. *IEEE Trans Neural Syst Rehabil Eng* 2014;22:191–200.

23. Kamali T, Boostani R, Parsaei H. A hybrid classifier for characterizing motor unit action potentials in diagnosing neuromuscular disorders. *J Biomed Phys Eng Internet* 2013 [cited 2018 April 29];3:145–154. Available from: https://www.ncbi.nlm.nih.gov/pmc/articles/PMC4204505/.

24. Mishra VK, Bajaj V, Kumar A, et al. An efficient method for analysis of EMG signals using improved empirical mode decomposition. *AEU Int J Electron Commun* 2017;72:200–209.

25. Yousefi J, Hamilton-Wright A. Characterizing EMG data using machine-learning tools. *Comput Biol Med Internet* 2014 [cited 2018 April 29];51:1–13. Available from: http://www.sciencedirect.com/science/article/pii/S0010482514001048.

26. Subasi A, Yilmaz M, Ozcalik HR. Classification of EMG signals using wavelet neural network. *J Neurosci Methods* 2006;156:360–367.

27. Sengur A, Akbulut Y, Guo Y, et al. Classification of amyotrophic lateral sclerosis disease based on convolutional neural network and reinforcement sample learning algorithm. *Health Inf Sci Syst* 2017;5:9.

28. Subasi A. Classification of EMG signals using combined features and soft computing techniques. *Appl Soft Comput* 2012;12:2188–2198.

29. Naik GR, Selvan SE, Nguyen HT. Single-channel EMG classification with ensemble-empirical-mode-decomposition-based ICA for diagnosing neuromuscular disorders. *IEEE Trans Neural Syst Rehabil Eng* 2016;24:734–743.

30. Rasheed S, Stashuk D, Kamel M. A software package for interactive motor unit potential classification using fuzzy k-NN classifier. *Comput Methods Programs Biomed* 2008;89:56–71.

31. Parsaei H, Stashuk DW. EMG signal decomposition using motor unit potential train validity. *IEEE Trans Neural Syst Rehabil Eng* 2013;21:265–274.

32. Gazzoni M, Farina D, Merletti R. A new method for the extraction and classification of single motor unit action potentials from surface EMG signals. *J Neurosci Methods* 2004;136:165–177.

33. Katsis CD, Goletsis Y, Likas A, et al. A novel method for automated EMG decomposition and MUAP classification. *Artif Intell Med* 2006;37:55–64.

34. Hazarika A, Dutta L, Boro M, et al. An automatic feature extraction and fusion model: Application to electromyogram (EMG) signal classification. *Int J Multimed Inf Retr* 2018;7:173–186.

35. Subasi A, Yaman E, Somaily Y, et al. Automated EMG signal classification for diagnosis of neuromuscular disorders using DWT and bagging. *Procedia Comput Sci* 2018;140:230–237.

36. Hafner P, Phadke R, Manzur A, et al. Electromyography and muscle biopsy in paediatric neuromuscular disorders: Evaluation of current practice and literature review. *Neuromuscul Disord* 2019;29:14–20.

37. Addison PS. *The illustrated wavelet transform handbook: Introductory theory and applications in science, engineering, medicine and finance.* Boca Raton, FL: CRC Press; 2017.

38. Bakshi BR. Multiscale PCA with application to multivariate statistical process monitoring. *AIChE J* 1998;44:1596–1610.

39. Alickovic E, Subasi A. Effect of multiscale PCA de-noising in ECG beat classification for diagnosis of cardiovascular diseases. *Circuits Syst Signal Process* 2015;34:513–533.

40. Subasi A, Qaisar SM. Surface EMG signal classification using TQWT, bagging and boosting for hand movement recognition. *J AMBIENT Intell Humaniz Comput* 2020. doi:10.1007/s12652-020-01980-6.

41. Wang S, Lu S, Dong Z, et al. Dual-tree complex wavelet transform and twin support vector machine for pathological brain detection. *Appl Sci* 2016;6:169.

42. Selesnick IW, Baraniuk RG, Kingsbury NC. The dual-tree complex wavelet transform. *IEEE Signal Process Mag* 2005;22:123–151.

43. Daubechies I. *Ten lectures on wavelets.* New Delhi: SIAM; 1992.

44. Nason GP, Silverman BW. The stationary wavelet transform and some statistical applications. In: *Wavelets and Statistics*, (Lecture Notes in Statistics; Vol. 103). New York: Springer; 1995, pp. 281–299.

45. Wang S, Du S, Atangana A, et al. Application of stationary wavelet entropy in pathological brain detection. *Multimed Tools Appl* 2018;77:3701–3714.

46. Gupta V, Priya T, Yadav AK, et al. Automated detection of focal EEG signals using features extracted from flexible analytic wavelet transform. *Pattern Recognit Lett* 2017;94:180–188.

47. Gilles J. Empirical wavelet transform. *IEEE Trans Signal Process* 2013;61:3999–4010.

48. Bhattacharyya A, Sharma M, Pachori RB, et al. A novel approach for automated detection of focal EEG signals using empirical wavelet transform. *Neural Comput Appl* 2018;29:47–57.
49. Han J, Pei J, Kamber M. *Data mining: Concepts and techniques.* Amsterdam: Elsevier; 2011.
50. Hall M, Witten I, Frank E. Data mining: Practical machine learning tools and techniques. *Kaufmann Burlingt* 2011.
51. Breiman L, Friedman J, Stone CJ, et al. *Classification and regression trees* [Internet]. New York: Taylor & Francis; 1984. Available from: https://books.google.com.sa/books?id=JwQx-WOmSyQC.
52. Holmes G, Pfahringer B, Kirkby R, et al. *Multi class alternating decision trees.* Berlin, Heidelberg: Springer; 2002, pp. 161–172.
53. Holmes G, Hall M, Prank E. *Generating rule sets from model trees.* Berlin, Heidelberg: Springer; 1999, pp. 1–12.
54. Kalmegh S. Analysis of WEKA data mining algorithm REPTree, Simple CART and RandomTree for classification of Indian news. *Int J Innov Sci Eng Technol* 2015;2:438–446.
55. EMGLAB [Internet]. [cited 2018 May 11]. Available from: http://www.emglab.net/emglab/index.php.
56. Nikolic M. Findings and firing pattern analysis in controls and patients with myopathy and amytrophic lateral sclerosis. PhD Thesis, Faculty of Health Science, University of Copenhagen, 2001.
57. Swets JA. ROC analysis applied to the evaluation of medical imaging techniques. *Invest Radiol* 1979;14:109–121.
58. Cohen J. A coefficient of agreement for nominal scales. *Educ Psychol Meas* 1960;20:37–46.
59. Yang Z, Zhou M. Kappa statistic for clustered physician–patients polytomous data. *Comput Stat Data Anal* 2015;87:1–17.

5 Prosthesis Control Using Undersampled Surface Electromyographic Signals

Hamid Reza Marateb, Farzad Ziaie Nezhad,
Marjan Nosouhi, Zahra Nasr Esfahani,
Farzaneh Fazilati, Fatemeh Yusefi,
Golnaz Amiri, Negar Maleki Far,
and Mohsen Rastegari
University of Isfahan

Mohammad Reza Mohebbian and
Khan A. Wahid
University of Saskatchewan

Mislav Jordanić, Joan Francesc Alonso López,
and Miguel Ángel Mañanas Villanueva
Universitat Politècnica de Catalunya – BarcelonaTech

Marjan Mansourian
Isfahan University of Medical Sciences

CONTENTS

5.1   INTRODUCTION

Functional loss as a result of amputations affects a patient both physically and men-
tally [1]. It not only has an impact on the entire life because of the loss of earned
income, but it also affects the family, society, and health systems [2]. Upper-limb
amputations are mostly related to work [3]. Such injuries include ~7% of the total
burden of disease [4]. They may be performed as distal (e.g., the fingertip) and as
proximal (e.g., shoulder) [5]. They could be the result of pathologies, such as traumatic injuries and malignancies [6]. Depression can also result in following the surgery after a sudden disability [7,8].

At a closer look, the prevalence of unilateral and bilateral upper-limb amputations from 1990 to 2017 for different countries is shown in Figures 5.1 and 5.2,
respectively. For example, the prevalence of unilateral upper-limb amputations is 141
per 100,000 Iranian population for both sexes, 25–29 years old, resulting in related
1,160,000 amputees in the entire Iranian population.

Prosthetic hands could be used to assist the amputees. Different prosthetic hands
include body-powered, pneumatic-powered, and electric-powered (e.g., myoelectric-
controlled) [9], some of which are complex and expensive. For example, those with
multiple gripping fingers and superior functions cost up to USD 50,000 [10,11]. Such
devices are not affordable by amputees because of the lack of funding and public
health coverage. Thus, there has been great interest in the design and fabrication of
low-cost active upper-limb prostheses [10,12].

This chapter discusses upper-limb myoelectric-controlled prostheses, their control strategy, and frameworks for low-cost surface electromyographic (sEMG) signal
recording in such prostheses. Various experiments are performed to identify the pros
and cons of such recordings in advanced prosthesis control as a proof of concept.

5.2   MYOELECTRIC CONTROLLED PROSTHESIS

The sEMG signal has been the most commonly used signal for active prosthesis
since 1948 [14,15]. Myoelectric control is usually performed by the activation of the
muscles of the residual limbs. In addition to surface electrodes and instrumentation

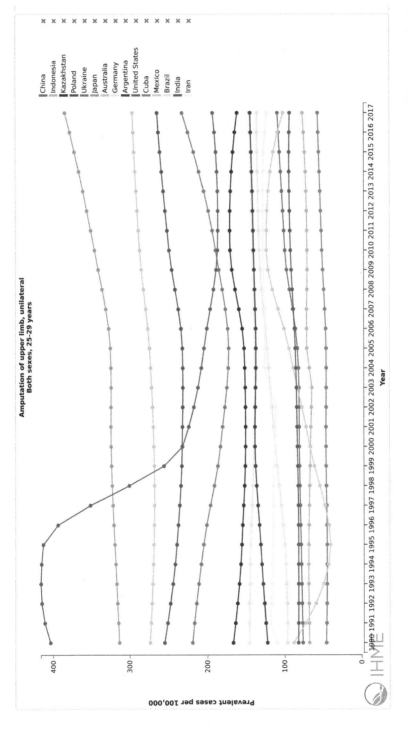

FIGURE 5.1 The prevalence of unilateral upper-limb amputations in both sexes, 25–29 years old per 100,000 population in different countries from 1990 to 2017, based on the Global Burden of Diseases. (Source: Institute for Health Metrics Evaluation. Used with permission. All rights reserved [13].)

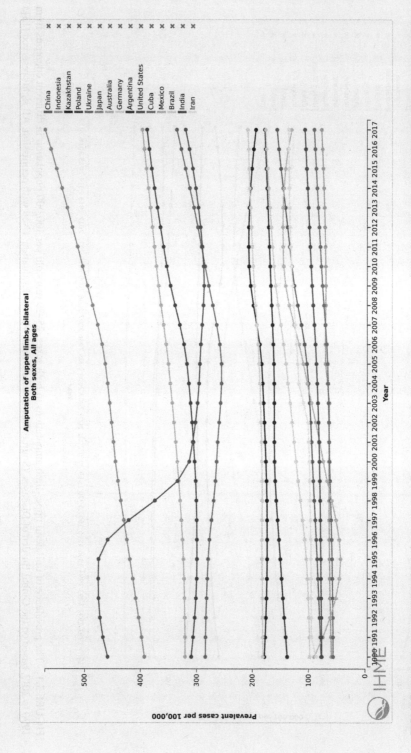

FIGURE 5.2 The prevalence of bilateral upper-limb amputations in both sexes, all ages per 100,000 population in different countries from 1990 to 2017, based on the Global Burden of Diseases. (Source: Institute for Health Metrics Evaluation. Used with permission. All rights reserved [13].)

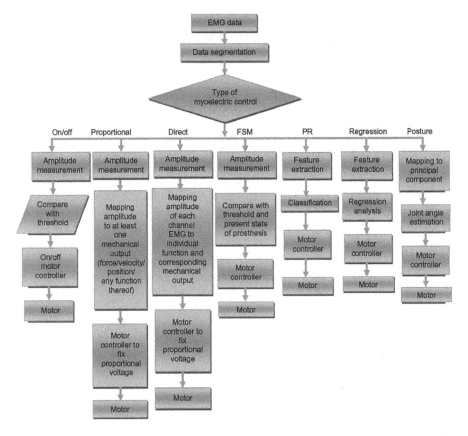

FIGURE 5.3 Types of myoelectric prosthesis control schemes. (Reproduced with permission from Dove Medical Press; Reference [9].) Abbreviations: EMG, electromyographic; PR, pattern recognition; and FSM, finite-state machine.

amplifiers, advanced myoelectric prosthesis requires powerful processing units [10,16]. Moreover, proper skin preparation is recommended to improve the quality of the recorded signal. Different schemes were used in the literature to control the myoelectric prosthesis (Figure 5.3).

In the following, some of these methods are discussed.

5.2.1 On/Off and Finite-State Myoelectric Control

Independent from the level of contraction, the prosthesis operates at a constant speed. While easy to implement, this method has many limitations, such as the number of degrees of freedom (DOF). It usually requires smoothing and thresholding sEMG signals to open/close the prosthesis. In this control scheme, different thresholds can be used, for example, low/high contractions to open/close the hand, and no contraction for stop mode (Figure 5.4a). This method can help separate sEMG from the background noise [9,17–19]. On the other hand, hand posture could be predefined as states, and the transitions between different states are performed using the input sEMG signal, which is suitable for a limited fixed number of movements [20,21]. It is known as the finite-state myoelectric control [9].

5.2.2 Proportional, Direct, and Posture Myoelectric Control

Since the prosthesis moves at a constant speed in the on/off control method, this method's first alternative is a proportional control scheme. In this method, the voltage sent to a hand controller is proportional to sEMG activity [9,18]. This control plan makes a move faster and improves the function of the prosthesis (Figure 5.4b). Although this method is suitable for large movements such as hand open/close, it is not sufficient for fine control, such as the flexion of individual fingers [18].

The proportional control could be further refined for fine movements, in which sEMG signals are mapped to the functions of the prosthetic devices. This procedure is entitled as the direct drive [22]. However, its performance reduces due to the crosstalk between sEMG signals [23]. A method to overcome this problem is the implementation of the electrodes into the muscles [24]. On the other hand, in the posture control, sEMG signals are mapped into the prosthesis's joint angles using transformations such as the principal component analysis [25]. It is also considered as one of the methods of simultaneous myoelectric control [9].

The proportional and direct myoelectric control suffers from a limitation that the number of movements is limited to the number of recorded sEMG channels, which is not the case for advanced controllers such as pattern recognition and regression methods [18,26,27].

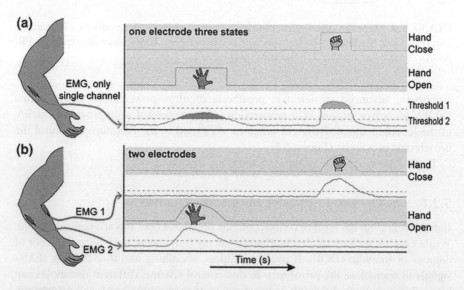

FIGURE 5.4 An illustration of the early control strategies using (a) one electrode in on/off strategy and (b) two electrodes in proportional control. (Reproduced with permission from Reference [18].)

5.2.3 Pattern Recognition-Based Myoelectric Control

In this control strategy, the sEMG signal is first filtered and segmented. Various features are then extracted from the segmented signal, and the output of a classifier is applied to the motor controller (Figure 5.5). Sometimes there is a feature reduction step to reduce the number of features used in the classifier. Various features and classifiers can be used in this control strategy [9,18,28]. Some of these are discussed in the following paragraphs.

In 1982, Saridis and Gootee [29] collected data from an amputation simulator. They considered 29 motion classes and one rest class (i.e., no movement). They extracted variance and zero crossings (ZCs) from the sEMG signal, and used the linear discriminant analysis (LDA) for classification. The misclassification error of their entire movements was <5%.

In 1993, Hudgins et al. [30] presented a new method based on pattern recognition for prosthetic control. Nine healthy subjects and six amputees participated in this study. They extracted the mean absolute value (MAV), MAV slope, number of ZCs, waveform length (WL), and number of slope sign changes from the sEMG signal. An artificial neural network (ANN) was used for classification. The average accuracy was $91.2\% \pm 5.6\%$ for healthy subjects and $85.5\% \pm 9.8\%$ for amputees. Almost three decades after the publication of this article, such features are still used in the literature, and the collection of these features is known as Hudgins' (time-domain) features [31,32].

Due to the stochastic and nonlinear nature of the sEMG signal, the autoregressive (AR) models have been used in myoelectric control [9]. In 2003, Soares et al. [33] used the combination of AR model and ANN classifier to control a virtual prosthesis. They achieved 100% and 95%–96% accuracy for the AR models of the order ten and four, respectively. In 2005, Huang et al. [34] investigated the application of the AR and Gaussian mixture models (GMMs) for myoelectric control. Twelve intact subjects participated in this study. The combination of AR, Hudgins' features, and the root mean square (RMS) of the sEMG signals was used as the feature set. The GMM classification error for the entire six-movement classes was 3.09%.

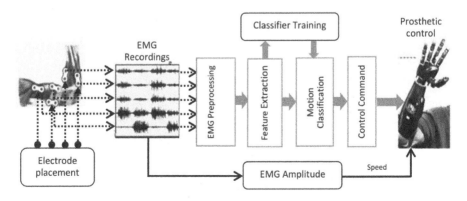

FIGURE 5.5 Schematic diagram of a pattern recognition-based myoelectric prosthetic control system. (Reproduced with permission from Reference [28].)

In 2008, Oskoei and Hu [35] used a support vector machine (SVM) classifier to discriminate between six classes, namely, five motion classes and a rest class. Data were recorded from 11 intact subjects. This study showed that among the analyzed features and classifiers, Hudgins' feature set and the SVM classifier had the best performance (95% ± 4%). Moreover, Hudgins' features were more stable in different sessions than AR models to change the segment length.

In 2013, Al-Timemy et al. [36] proposed an offline method for finger movement classification. Ten intact and six below-elbow amputees participated in this study. The time-domain AR features were extracted from EMG data, and orthogonal fuzzy neighborhood discriminant analysis was then used for feature reduction, and LDA classification was used to separate classes from each other. This study showed a significant difference between intact and amputees for 9 or 12 motion classes, and there is no significant difference for five motion classes. In this study, the accuracy obtained for amputees for most classes such as thumb abduction and flexion was higher than 90%. Also, the average accuracy obtained for intact subjects was 98%, which was higher than in previous studies.

In 2016, Goen and Tiwari [15] designed a study to evaluate different features and classifications for prosthetic hand control. Thirty people took part in the study. The extracted features included Hudgins', the discrete wavelet transform (DWT), the wavelet packet transform (WPT), the short-time Fourier transform (STFT), and the time-varying autoregressive model (TVAR). In this study, orthogonal LDA (OLDA) and sparse principal component analysis (SPCA) methods were used to reduce the features' dimensions. The classifiers used in this study were SVM ensemble, LDA, k-nearest neighbors (k-NN), and multilayer perceptron (MLP). The best accuracy was 98%, which was obtained by SVM. Also, the results showed that the lowest error rate was obtained using the WPT feature.

In 2020, Raheema et al. [37] designed an intelligent, low-cost prosthesis control system suitable for developing countries. The Thalmic Labs Myo Gesture Control Armband was used for sEMG signal recording, and five different hand gestures were considered in this study. RMS, standard deviation (SD), and the maximum and minimum values of the time samples of each 50-ms signal epoch were extracted. The MLP classifier was used, and the average accuracy was higher than 99%.

Although the classification accuracy of such pattern recognition-based myoelectric prostheses has been significantly improved over time, their usability is weak. The performance can more objectively be described by reporting indices such as the compilation rate in real-time applications. For example, the classification accuracy of 85% with the completion rate of 55% suggests that they are still not reliable [18,38]. Although the optimal myoelectric controller delay is within 100–125 ms [39], the pattern recognition-based methods usually analyze signals in 200- to 300-ms epochs [40,41], thus reducing the user's performance in practice [42]. Moreover, pattern recognition-based systems could not provide an effective simultaneous myoelectric control [40]. Moreover, adding extra classes for the simultaneous movements [43–48] dramatically increases the learning procedure and decreases the classification accuracy [18]. Thus, regression-based methods were proposed in the literature that could be used for proportional, simultaneous myoelectric prosthesis control (Figure 5.6).

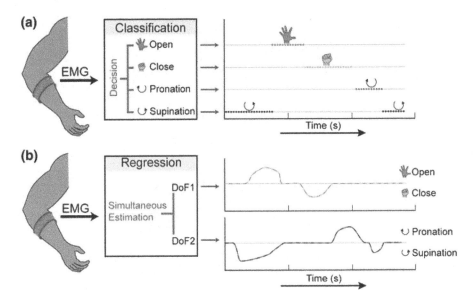

FIGURE 5.6 (a) The pattern recognition-based myoelectric prosthesis control used for sequence movements. (b) The regression-based method is used for simultaneous proportional control. (Reproduced with permission from Reference [18].)

5.2.4 REGRESSION-BASED MYOELECTRIC CONTROL

Linear regression [18] and nonlinear regression (e.g., non-negative matrix factorization (NMF) [49–54], autoencoder (AEN) neural networks [55], and ANN [56–59]) were used for simultaneous and proportional myoelectric prosthesis control in the literature.

In 2008, Jiang et al. [49] used the NMF method, for the first time in the literature, to estimate the force generated at two simultaneous DOFs. The authors solved this problem by assuming that the crosstalk between sEMG channels is minimal. The average R-square was $77.5\% \pm 10.9\%$. The results of this study showed that the force that is produced in two simultaneous DOFs could be estimated using the NMF method. In 2014, Jiang et al. [50] used the NMF method for online simultaneous and proportional myoelectric control. Seven intact and seven amputees took part in this study. Statistical tests showed no significant difference between the two groups in the indicators of completion rate, completion time, and accuracy. Moreover, the calibration phase was very short in this study.

In 2018, Lin et al. [51] used the NMF method to include sparseness restrictions for simultaneous and proportional myoelectric control of several hand motions in eight able-bodied subjects and two subjects with unilateral limb deficiencies. The results showed that this method has a shorter completion time than the conventional NMF or linear regressions. This method opened the path for the online myoelectric control with a minimal supervision.

In 2018, Vujaklija et al. [55] used a minimally supervised method based on the ANN and AEN (Figure 5.7) for two DOF (radial/ulnar deviation and flexion/

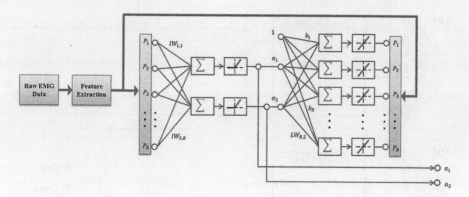

FIGURE 5.7 An autoencoder neural network to extract control signals in two opposite directions (a_1, a_2) for each DoF. The RMS values of 100-ms sEMG processing windows were simultaneously used as the input and output. The standard MLP structure (with one hidden layer and two neurons) was used, and it was trained using the Levenberg–Marquardt back-propagation algorithm. The lower-dimensional features (a_1, a_2) compared with the inputs were used to control each DOF. (Reproduced with permission from Reference [55].)

extension of the wrist) myoelectric control. Two eight-electrode rings were used around the forearm. The algorithms' online performance was assessed using six metrics, namely, completion rate and time, overshoots, throughput, speed, and path efficiency [55,60,61]. The performance of the proposed method was compared with the Fitts' law approach on seven intact subjects. Both methods had the completion rates close to 100%. However, the proposed method had 50% of the completion time and 200% of the other state-of-the-art controller's throughput.

5.2.5 Deep Learning-Based Myoelectric Control

Deep learning (DL) techniques have recently revolutionized numerous areas of machine learning. Thanks to the increasing release of many benchmark sEMG recording databases, such as NinaPro [62,63], CSL-HDEMG [64], BioPatRec [65], and CapgMyo [66], researchers have begun to investigate the ability of DL to process and decode sEMG data. The lack of classification robustness that can be overcome by better feature extraction or more robust classifiers is a typical constraint.

DL prosthesis control approaches can be categorized into various methods such as convolutional neural network (CNN), recurrent neural network (RNN) (e.g., long short-term memory (LSTM)), and AEN models. There are other new domains, such as deep reinforcement learning (DRL) and meta-learning. Recently, many researches started implementing prosthetic control based on such models [67,68].

CNN is the most popular method among all DL approaches. Atzori et al. [69] classified 50 hand gestures using CNN in 67 intact subjects and 11 transradial amputees. They used three NinaPro datasets and achieved 66.59% ± 6.4% on data-set 1, 60.27% ± 7.7% on dataset 2, and 38.09% ± 14.29% on amputees (dataset 3). Park and Lee [70] used CNN with four subsampling and two fully connected layers to decode six hand movements from NinaPro sEMG datasets and achieve

60% accuracy. Ghazaei et al. [71] used the machine vision powered by DL with object detection for grasp classification based on object type and achieved 84% accuracy. Côté-Allard et al. [72] used continuous wavelet transforms of sEMG instead of raw sEMG and could improve the CNN model for gesture detection from 68.98% to 98.31%.

The performance of DL relies on certain hyperparameters, such as the CNN architecture, the number of convolution-layer filters, using dropout, type of optimizer, and the number of training data. Therefore, hyperparameter optimization was performed by Triwiyanto et al. [73] for ten-class gesture classification using raw sEMG data and could achieve an average accuracy between 77% and 93% for all motion ranges. Zhai et al. [74] suggested a self-calibration classifier based on CNN that could be calibrated automatically over time. The accuracy of the CNN and SVM was 88.42%, and 87.86%, for the intact subjects, while they were 73.31% and 72.01% for amputees for ten-class movement classification. Ameri et al. [75] used transfer learning based on CNN for testing myoelectric patterns after the electrode displacement. They used ten channels and obtained $21.5\% \pm 2.3\%$ and $46.05 \pm 4.1\%$ error rate after 2.5-cm electrode shift for flexion–extension and pronation–supination.

Addressing the signal variability over time, Hu et al. [76] proposed an RNN model for gesture detection. It improved the accuracy of detection from 83.9% to 92.5% on a sub-database of BioPatRec (known as BioPatRec26MOV), from 73.4% to 74.8% on the second sub-database of NinaPro (known as NinaProDB2), from 92.1% to 94.9% on the CSL-HDEMG compared to the CNN framework, and from 97.7% to 99.7% on a sub-database of CapgMyo (known as CapgMyo-DBa). The NinaPro dataset was also utilized by Wang et al. [77] to predict 20 joint angles using LSTM and achieve a RMS error of $5.98° \pm 0.95°$.

An example of AEN-based models is the research performed by Li et al. [78], who used PCA for feature reduction, utilized two-layered stacked AEN and a softmax output classifier for DL architecture, and could achieve 95% accuracy on the recorded sEMG for gesture detection of 15 subjects. Also, Lv et al. [79] used the AEN architecture as a robust method against the electrode shift for myoelectric prosthetic control. With 192 channels, they used HDsEMG and obtained a $1.48\% \pm 1.43\%$ error rate without a shift scenario and $9.83\% \pm 8.15\%$ in a 1-cm shift scenario.

5.2.6 PERFORMANCE INDICES

A variety of performance indices were used in the literature for prosthesis applications. In this section, we provide the list of the indices along with their citation for the interested readers: completion rate, completion time, overshoots, throughput [80], speed [80,81], efficiency coefficient, path efficiency [55,80], task index of difficulty [55], lag [82], variance accounted for, information transmission rate, estimated gain parameters, delay parameter [83], tracking error, gain margin crossover frequency [83,84], percentage of time, task execution time [85], profiles of participants [86], motion completion times [87,88], relative error [89], path length, normalized jerk, resultant trajectories [90], symmetry index [91], R-square [58,92], RMS error [93], RMS difference [93], and cross-correlation [90,94].

5.2.7 CHALLENGES IN MYOELECTRIC PROSTHESIS CONTROL

Some of the challenges for the real-time myoelectric prosthesis control are listed in Table 5.1. Among the technical problems, the electrode shift [17,54,95–98] and the cost of the prostheses [10,12,16,99,100] are the two main issues. Injection molding and 3D-printing provided promising possibilities for personalized designs and easy manufacturing [99]. Two examples are You Bionic (*https://www.youbionic.com/*) and Open Bionics (*https://openbionics.com/*) in the low-cost prosthesis category. The comparison between different prosthesis hands is provided in online *Supplementary A1 and A2* (https://doi.org/10.6084/m9.figshare.13048067).

EMG-PR stands for EMG pattern recognition-based control. Note that the original table in Ref. [17] provided the challenges of the EMG-PR method for prosthesis control. However, such challenges are similar to other myoelectric prostheses.

5.3 sEMG SIGNAL RECORDING IN MYOELECTRIC PROSTHESIS CONTROL

The bandwidth of the sEMG signal is around 400–500 Hz [107], resulting in the recommendation of the sampling rate of 1 kHz or higher (Nyquist–Shannon sampling theorem) [108–110]. However, lower sampling rates were used in the literature for the movement detection [111–114]. Using low-cost commercial sEMG recording device Thalmic Labs Myo Gesture Control Armband with the sampling rate of 200 Hz, comparable results to the traditional sEMG recording devices for the movement detection were obtained [32,115,116]. However, there are still debates on this issue [32]. Since the sEMG signal is a stochastic signal and smoothing is usually performed before feature extraction (e.g., envelope detection, or RMS values), using undersampled signals could be justified. However, unlike image processing in which the theoretical concepts of super-sampling images using aliased/undersampled images [117–119], no theoretical concepts were provided for such a justification. Moreover, more experiments are required to identify whether such undersampled sEMG signals could be used for other applications rather than the movement detection. We provide the results of the experiments on undersampled sEMG signal processing for a variety of applications.

5.3.1 EMG-FORCE ESTIMATION

In this study, musculoskeletal torque models based on DL and sEMG signals recorded using Myo armband were examined. This study's results were presented in the Virtual Congress of the International Society of Electrophysiology and Kinesiology (ISEK) 2020, Japan.

It is possible to use nonlinear models to capture the relationship between sEMG signals and the joint's torque. DL or deep neural network (DNN) is a specific way for its formulation. The sEMG signals were recorded using a Thalmic Labs Myo-armband over the elbow, from five healthy subjects during one-minute submaximal isometric force-varying elbow flexion and extension. A KUKA LBR iiwa robot recorded the torque signal. The torque signal was low-pass-filtered, and the envelopes

TABLE 5.1

Some of the Challenges of Real-Time Myoelectric Prosthesis Control

Challenges	Description
Comfort	The socket that is the part of the upper-limb prosthesis may interfere with the elbow (a function of the residual joint). If the socket does not fit correctly, the patient may suffer from pain, sores, and blisters. Such prostheses will be experienced as heavy and cumbersome [101]. Even some prostheses with appropriately designed sockets face problems of heat, sweating, and chafing
Appearance	Most of the developed upper-limb prostheses do not look natural in appearance. Additionally, the user can find the prosthesis uncomfortable to wear. The user is still unable to control the multiple degrees of freedom simultaneously and consistently
Function	Nowadays, upper-limb prostheses perform almost all everyday activities. However, it remains challenging to obtain opening and closing positions of the hand from the residual limb. This is because residual muscles often used for hand prosthesis are biceps and triceps, which do not convey the information for closing and opening the hand [102]
Durability	Many of the upper-limb prostheses are heavy and have short battery life
Cost	Upper-limb prosthesis costs around $50,000, which is quite difficult to afford by amputees from all over the world
Technology	Developed prosthetic devices still lack intuitiveness and reliability between user motion volition and real motion of prosthesis. Similarly, much training is needed to operate those prosthetic hands
Processing delay	The embedded processor used exhibits some delay (around 3 s), which halts the acquisition of EMG for that delay period
EMG interferences	The transient changes in EMG often result from external interferences, changes in electrode impedance, muscle fatigue, and electrode shift, among others. During practical use, this transient change arising from variations (long and short term) in the acquisition environment caused a degradation of the clinical vitality of the device and limited its users' adoption [103]
Electrode displacement (shift)	Electrode displacement occurs each time when users use a prosthesis; electrodes slightly reconcile in a different position relative to the underlying musculature. When the user performs some tasks, due to the loading and positioning of the limb, a movement of electrodes occurs. Such an electrode shift can lead to a change in EMG characteristic (recording) of the limb, thus making it more difficult to decode the movements [104]
Amputee movement	EMG signal from the limb position is mostly recorded when the user is in a static position (sitting), but in a real-time scenario, prosthesis users have to use the device in different positions (walking, climbing stairs). However, the variation in limb position affects the classification performance of EMG-PR [105]
Muscle contraction forces	While performing everyday activities, the same limb assists different muscle contraction forces across different conditions. Thus, the variation in muscle contraction force occurs due to the same targeted limb, which results in myoelectric signal pattern classification inconsistency. Hence, it affects the EMG-PR control of prosthesis [106]
Limb position variation	Variation in limb position occurs while performing a different action in everyday life. For the upper-limb amputation, the effects are seen on residual muscle (located in a prosthetic socket) from which the EMG signal is collected. Additionally, various limb positions lead to the variation in gravitational force, which leads to the displacement of target muscles. These factors cause a variation in EMG signal patterns affecting the EMG-PR control of prosthesis performance

Source: Reproduced with permission from Reference [17].

TABLE 5.2

The Performance of the Analyzed Methods on the Validation Sets in the Entire Subjects in Mean ± SD

Method	MSE	R-square	ICC	ICC 95% CI	Significance
Time-delay NN with 20 hidden layers	0.0807 ± 0.1353	0.5656 ± 0.2486	0.7293±0.2691	0.7789 to 0.7842	P<0.001
DNN with LSTM layers with MA filter	0.0083 ± 0.0037	0.8913 ± 0.0527	0.9343±0.0346	0.9393 to 0.9409	P=0.074
DNN with LSTM layers	0.0144 ± 0.0055	0.7685 ± 0.1152	0.8811±0.0597	0.8968 to 0.8995	P=0.295
Linear LS	0.2807 ± 0.3645	0.2199 ± 0.1540	0.4871±0.2173	0.3401 to 0.3521	P<0.001
The second-order polynomial LS	0.3742 ± 0.1911	0.2365 ± 0.1523	0.5831±0.4092	0.4853 to 0.4957	P< 0.001

of the sEMG signals were estimated. The linear, second-order least-squares (LS), known as linear-in-the-parameters models, a time-delayed NN with 20 hidden layers, and DNN with LSTM layers with/without a moving average filter (MA) were used in this study. The (50%) holdout validation framework was used. We estimated that the musculoskeletal delay was estimated in the training set for the entire models. The indices intraclass correlation coefficient (ICC) with 95% confidence interval (CI), R-square as the goodness-of-fit, and mean-square error (MSE) were reported.

The DNN with LSTM layers with MA filters outperformed the other methods (Table 5.2). The best method (i.e., LSTM networks with smoothing) is shown in Figure 5.8. DNN is thus a promising method for sEMG force modeling.

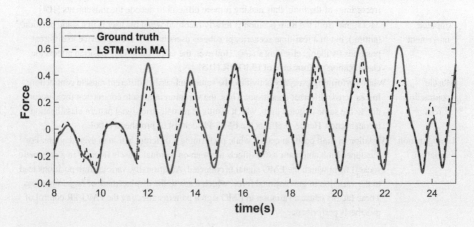

FIGURE 5.8   The estimated force using LSTM with MA and the ground truth.

The significance was calculated using the paired-sample t-test; when $P < 0.05$, the estimated and real force recordings were significantly different for ICC, MSE, CI, and LS.

5.3.2 ELBOW ANGLE ESTIMATION

The estimation of joint angles on the shoulder and elbow has been performed in the literature using the conventional sampling rates [120–124]. In this study, we examined how the undersampled sEMG signals could be used for elbow angle estimation. The sEMG signal of nine intact right-handed subjects (five women and four men; the average age was 39 years) was recorded using Myo armband device (with eight recording channels) with the elbow angle in a range from 0° to 150° in steps of 15°. Different classifiers were used to estimate the elbow angle from the RMS values of 100-ms nonoverlapping sEMG signals. The macroaveraged F-score index, which is better than the overall accuracy in terms of overestimation in multiclass classification [125]), was reported (Table 5.3). In these results, (50%) repeated-holdout validation ($N = 10$) framework was used. The results show that it is possible to estimate the elbow angle using undersampled sEMG signals.

5.3.3 FINGER GESTURE DETECTION

In this study, ten intact subjects participated. sEMG signals were recorded using Myo armband located on the forearm, during 6 repetitions of 12 finger movements. The RMS of the nonoverlapping 150-ms sEMG signal epochs was calculated. Different classifiers were used, and the macroaveraged F-score values were reported. In these results, (50%) repeated-holdout validation ($N = 10$) framework

---

TABLE 5.3

The Performance of Different Classifiers for Elbow Angle Estimation (Macro-Averaged *F*-Score %) in MEAN ± SD

Subject	LDA	SVM	k-NN ($k=7$)	Gradient Boosting
S_1	79.57 ± 0.56	82.46 ± 0.3	79.63 ± 0.61	80.58 ± 0.63
S_2	72.49 ± 0.54	75.53 ± 0.49	71.21 ± 0.31	73.06 ± 0.54
S_3	66.32 ± 0.3	70.03 ± 0.59	61.31 ± 0.64	60.26 ± 0.99
S_4	88.35 ± 0.55	90.65 ± 0.40	88.23 ± 0.69	88.86 ± 0.39
S_5	84.52 ± 0.44	87.94 ± 0.46	83.80 ± 0.49	67.48 ± 0.89
S_6	80.5 ± 0.49	85.54 ± 0.5	82.19 ± 0.57	82.76 ± 0.49
S_7	74.20 ± 0.49	82.19 ± 0.65	74.29 ± 0.71	79.93 ± 0.40
S_8	92.01 ± 0.32	94.58 ± 0.29	93.07 ± 0.3	92.86 ± 0.51
S_9	92.25 ± 0.23	94.92 ± 0.26	93.88 ± 0.32	93.85 ± 0.45
Mean ± SD	81.13 ± 9.01	84.87 ± 8.36	80.84 ± 10.63	79.96 ± 11.39

Nonlinear SVM with a radial basis function (RBF) kernel was used.

TABLE 5.4

The Performance of Different Classifiers for Finger Gesture Recognition (Macro-Averaged F-Score %) in Mean ± SD

Method	LDA	Linear SVM	Gaussian SVM	k-NN ($k=13$)
S_1	83.04 ± 0.72	89.05 ± 0.96	88.48 ± 0.64	86.33 ± 0.41
S_2	83.87 ± 0.88	87.84 ± 0.50	86.01 ± 0.90	85.30 ± 0.71
S_3	89.19 ± 0.79	94.81 ± 0.69	94.55 ± 0.81	93.44 ± 0.62
S_4	74.18 ± 1.35	83.41 ± 1.12	84.11 ± 0.57	79.92 ± 0.70
S_5	85.43 ± 0.95	92.35 ± 0.51	93.32 ± 0.44	90.39 ± 0.85
S_6	79.86 ± 0.88	83.80 ± 0.99	84.35 ± 0.92	82.14 ± 1.05
S_7	60.90 ± 0.96	69.52 ± 0.90	74.85 ± 0.78	71.74 ± 0.87
S_8	84.40 ± 0.71	89.85 ± 0.60	89.86 ± 0.73	87.89 ± 0.56
S_9	85.17 ± 0.72	89.55 ± 0.58	90.11 ± 0.86	88.62 ± 0.34
S_{10}	83.52 ± 1.17	88.97 ± 0.92	89.11 ± 0.73	85.23 ± 0.60
Mean ± SD	80.96 ± 8.07	86.92 ± 7.00	87.48 ± 5.61	85.10 ± 6.09

was used. The finger movements were similar to Exercise A in Ref. [63]. The results are shown in Table 5.4.

5.3.4 TIME AND FREQUENCY FEATURE EXTRACTION

In this study, ten one-minute sEMG signals were recorded using linear array electrodes on the bicep brachii using an EMG-USB2 amplifier (OT Bioelettronica, Italy) during isometric elbow flexion at 30% and 50% of maximum voluntary contraction (MVC), and at elbow angles of 105°, 135°, and 165. The experimental protocol was based on the SENIAM recommendations, and the IED was 8 mm to avoid spatial aliasing [126]. The sampling frequency was 2048 Hz, which is higher than such recommendations. The electrode was aligned with the muscle fibers, and the innervation zone (IZ) was detected, and the single differential channels not close to the muscle IZ in which the MUAP propagation was observed, were selected in this study.

The time feature RMS and frequency feature mean frequency (MNF), and median frequency (MDF) were extracted from 200 to 1 s epochs. The same procedure was performed on the undersampled sEMG signals, where the original signals were downsampled at 200 Hz. The extracted features (RMS, MNF, and MDF) were compared between the original and undersampled signals. For the RMS, the average correlation coefficient (r), adjusted R-square, and ICC were 0.83 [CI 95%: 0.82–0.84], 0.69, and 0.79 [CI 95%: 0.73–0.83], respectively. Such values were 0.40 [CI 95%: 0.39–0.42], 0.16, and 0.07 [CI 95%: −0.05,0.21] for MNF, and 0.33 [CI 95%: 0.31–0.34], 0.11, and 0.16 [CI 95%: −0.03, 0.32] for MDF. The results showed that the RMS values extracted from undersampled sEMG signals are reliable. However, the frequency features MNF and MDF are not reliable as expected; the difference is higher in frequency measures (about 90% of the spectrum was removed in downsampling), but is not as high in RMS.

5.3.5 PROSTHESIS CONTROL

The BRUNEL HAND 2.0 is a low-cost lightweight 3D-printing-based prosthesis hand with open-source design from Open Bionics (*https://openbionicslabs.com/shop/brunel-hand*). It has four actuators with nine DOFs. The prosthesis was prepared based on the open-source design files. Then, we built the control circuits with ATmega32 that also records the sEMG signals. The schematics were provided, and the real-time control of the BRUNEL HAND 2.0 was recorded in videos that are both available as the *Supplementary A3 and A4*.

5.4 CONCLUSION AND FUTURE SCOPE

In this chapter, the importance of prostheses and different types of myoelectric prostheses and their advantages or disadvantages were discussed. Moreover, a comprehensive list of performance indices was provided. The use of undersampled sEMG signals for prosthesis control was discussed, and the results of some experiments about using undersampled signals for sEMG force prediction, elbow angle estimation, and finger gesture classification were provided as the proof of concept. An example of myoelectric control was provided in practice. Moreover, the comparison between sEMG signals recorded based on the SENIAM recommendation and the undersampled signals was made for the time and frequency feature extraction. Not only more experiments are required to cover different isometric and dynamic contractions, but it is also necessary to provide a theoretical proof of the possibility of using undersampled sEMG signals.

Much work has been done in the literature to improve how amputees link with their prostheses. However, there are many technological and clinical challenges in this area, and with the advancements in machine learning, biomedical instrumentations, and lower-cost predictions, they will be improved, and thus, the prosthesis rejection rate is reduced.

ACKNOWLEDGMENTS

This work was supported by the Ministry of Economy and Competitiveness (MINECO), Spain, under contract DPI2017-83989-R and the Ministry of Science and Innovation (MICINN), Spain, under contract PRE2018-085387. CIBER-BBN is an initiative of the Instituto de Salud Carlos III, Spain. JF Alonso is a Serra Hunter Fellow. The research leading to these results has also received funding from the European Union's Horizon 2020 research and innovation program under the Marie Skłodowska-Curie Grant Agreement No 712949 (TECNIOspring PLUS) and from the Agency for Business Competitiveness of the Government of Catalonia.

REFERENCES

1. Sinha R, van den Heuvel WJA and Arokiasamy P (2011) Factors affecting quality of life in lower limb amputees. *Prosthet. Orthot. Int.* **35**, 90–6.
2. Rosberg H-E, Carlsson KS, Cederlund RI, Ramel E and Dahlin LB (2013) Costs and outcome for serious hand and arm injuries during the first year after trauma: A prospective study. *BMC Public Health* **13**, 501.

3. Liang H-W, Chen S-Y, Hsu J-H and Chang C-W (2004) Work-related upper limb amputations in Taiwan, 1999-2001. *Am. J. Ind. Med.* **46**, 649–55.

4. Ro J-S, Leigh J-H, Jeon I and Bang MS (2019) Trends in burden of work-related upper limb amputation in South Korea, 2004-2013: A nationwide retrospective cohort study. *BMJ Open* **9**, e032793.

5. Ovadia SA and Askari M (2015) Upper extremity amputations and prosthetics. *Semin. Plast. Surg.* **29**, 55–61.

6. Lusardi MM, Jorge M and Nielsen CC (2012) *Orthotics and Prosthetics in Rehabilitation*, 4th Edition, Edinburgh: Elsevier Health Sciences. ISBN: 9780323610193.

7. Bhuvaneswar CG, Epstein LA and Stern TA (2007) Reactions to amputation: Recognition and treatment. *Prim. Care Companion J. Clin. Psychiatry* **9**, 303–8.

8. Durmus D, Safaz I, Adıgüzel E, Uran A, Sarısoy G, Goktepe AS and Tan AK (2015) The relationship between prosthesis use, phantom pain and psychiatric symptoms in male traumatic limb amputees. *Compr. Psychiatry* **59**, 45–53.

9. Geethanjali P (2016) Myoelectric control of prosthetic hands: State-of-the-art review. *Med. Devices* **9**, 247–55.

10. Sreenivasan N, Ulloa Gutierrez DF, Bifulco P, Cesarelli M, Gunawardana U and Gargiulo GD (2018) Towards ultra low-cost myoactivated prostheses. *Biomed Res. Int.* **2018**, 9634184.

11. Inglis T and Maceachern L (2013) 3D printed prosthetic hand with intelligent EMG control. *Carleton University* Available Online: http://www.doe.carleton.ca/Course/4th_year_projects/Am4_Inglis_Timothy_2013.pdf-.pdf.

12. Tong Y, Kucukdeger E, Halper J, Cesewski E, Karakozoff E, Haring AP, McIlvain D, Singh M, Khandelwal N, Meholic A, Laheri S, Sharma A and Johnson BN (2019) Low-cost sensor-integrated 3D-printed personalized prosthetic hands for children with amniotic band syndrome: A case study in sensing pressure distribution on an anatomical human-machine interface (AHMI) using 3D-printed conformal electrode arrays. *PLoS One* **14**, e0214120.

13. Institute for Health Metrics and Evaluation (2016) GBD compare data visualization. Available Online: http://www.healthdata.org/data-visualization/gbd-compare.

14. Xiao ZG and Menon C (2014) Towards the development of a wearable feedback system for monitoring the activities of the upper-extremities. *J. Neuroeng. Rehabil.* **11**, 2.

15. Goen A (2016) Tiwari D: Classification of the myoelectric signals of movement of forearms for prosthesis control. *Journal of Medical and Bioengineering* **5**, 76–84.

16. Polisiero M, Bifulco P, Liccardo A, Cesarelli M, Romano M, Gargiulo GD, McEwan AL and D'Apuzzo M (2013) Design and assessment of a low-cost, electromyographically controlled, prosthetic hand. *Med. Devices* **6**, 97–104.

17. Parajuli N, Sreenivasan N, Bifulco P, Cesarelli M, Savino S, Niola V, Esposito D, Hamilton TJ, Naik GR, Gunawardana U and Gargiulo GD (2019) Real-time EMG based pattern recognition control for hand prostheses: A review on existing methods, challenges and future implementation. *Sensors* **19**, 1–30.

18. Roche AD, Rehbaum H, Farina D and Aszmann OC (2014) Prosthetic myoelectric control strategies: A clinical perspective. *Curr. Surg. Rep.* **2**, 44.

19. Iqbal NV, Subramaniam K and Asmi PS (2018) A review on upper-limb myoelectric prosthetic control. *IETE J. Res.* **64**, 740–52.

20. Kyberd PJ, Holland OE, Chappell PH, Smith S, Tregidgo R, Bagwell PJ and Snaith M (1995) MARCUS: A two degree of freedom hand prosthesis with hierarchical grip control. *IEEE Trans. Rehabil. Eng.* **3**, 70–6.

21. Cipriani C, Zaccone F, Micera S and Carrozza MC (2008) On the shared control of an EMG-controlled prosthetic hand: Analysis of user–prosthesis interaction. *IEEE Trans. Rob.* **24**, 170–84.

22. Parker P, Englehart K and Hudgins B (2006) Myoelectric signal processing for control of powered limb prostheses. *J. Electromyogr. Kinesiol.* **16**, 541–8.
23. De Luca CJ and Merletti R (1988) Surface myoelectric signal crosstalk among muscles of the leg. *Electroencephalogr. Clin. Neurophysiol.* **69**, 568–75.
24. Weir RF ff, Troyk PR, DeMichele GA, Kerns DA, Schorsch JF and Maas H (2009) Implantable myoelectric sensors (IMESs) for intramuscular electromyogram recording. *IEEE Trans. Biomed. Eng.* **56**, 159–71.
25. Segil JL and Weir RF ff (2014) Design and validation of a morphing myoelectric hand posture controller based on principal component analysis of human grasping. *IEEE Trans. Neural Syst. Rehabil. Eng.* **22**, 249–57.
26. Graupe D and Cline WK (1975) Functional separation of EMG signals via ARMA identification methods for prosthesis control purposes. *IEEE Trans. Syst. Man Cybern.* **SMC-5**, 252–9.
27. Hargrove LJ, Miller LA, Turner K and Kuiken TA (2017) Myoelectric pattern recognition outperforms direct control for transhumeral amputees with targeted muscle reinnervation: A randomized clinical trial. *Sci. Rep.* **7**, 13840.
28. Li G, Samuel OW, Lin C, Asogbon MG, Fang P and Idowu PO (2019) Realizing efficient EMG-based prosthetic control strategy. In: *Neural Interface: Frontiers and Applications*, ed. X Zheng. Singapore: Springer, pp. 149–66.
29. Saridis GN and Gootee TP (1982) EMG pattern analysis and classification for a prosthetic arm. *IEEE Trans. Biomed. Eng.* **29**, 403–12.
30. Hudgins B, Parker P and Scott RN (1993) A new strategy for multifunction myoelectric control. *IEEE Trans. Biomed. Eng.* **40**, 82–94.
31. Abbaspour S, Lindén M, Gholamhosseini H, Naber A and Ortiz-Catalan M (2020) Evaluation of surface EMG-based recognition algorithms for decoding hand movements. *Med. Biol. Eng. Comput.* **58**, 83–100.
32. Phinyomark A, N Khushaba R and Scheme E (2018) Feature extraction and selection for myoelectric control based on wearable EMG sensors. *Sensors* **18**, 1615.
33. Soares A, Andrade A, Lamounier E and Carrijo R (2003) The development of a virtual myoelectric prosthesis controlled by an EMG pattern recognition system based on neural networks. *J. Intell. Inf. Syst.* **21**, 127–41.
34. Huang Y, Englehart KB, Hudgins B and Chan ADC (2005) A Gaussian mixture model based classification scheme for myoelectric control of powered upper limb prostheses. *IEEE Trans. Biomed. Eng.* **52**, 1801–11.
35. Oskoei MA and Hu H (2008) Support vector machine-based classification scheme for myoelectric control applied to upper limb. *IEEE Trans. Biomed. Eng.* **55**, 1956–65.
36. Al-Timemy AH, Bugmann G, Escudero J and Outram N (2013) Classification of finger movements for the dexterous hand prosthesis control with surface electromyography. *IEEE J Biomed Health Inform* **17**, 608–18.
37. Raheema MN, Hussain JS and Al-Khazzar AM (2020) Design of an intelligent controller for myoelectric prostheses based on multilayer perceptron neural network. *IOP Conf. Ser. Mater. Sci. Eng.* **671**, 012064.
38. Li G, Schultz AE and Kuiken TA (2010) Quantifying pattern recognition-based myoelectric control of multifunctional transradial prostheses. *IEEE Trans. Neural Syst. Rehabil. Eng.* **18**, 185–92.
39. Farrell TR and Weir RF (2007) The optimal controller delay for myoelectric prostheses. *IEEE Trans. Neural Syst. Rehabil. Eng.* **15**, 111–8.
40. Englehart K and Hudgins B (2003) A robust, real-time control scheme for multifunction myoelectric control. *IEEE Trans. Biomed. Eng.* **50**, 848–54.
41. Chu J-U, Moon I and Mun M-S (2006) A real-time EMG pattern recognition system based on linear-nonlinear feature projection for a multifunction myoelectric hand. *IEEE Trans. Biomed. Eng.* **53**, 2232–9.

42. Ajiboye AB and Weir RF ff (2005) A heuristic fuzzy logic approach to EMG pattern recognition for multifunctional prosthesis control. *IEEE Trans. Neural Syst. Rehabil. Eng.* **13**, 280–91.

43. Herberts P, Almström C, Kadefors R and Lawrence PD (1973) Hand prosthesis control via myoelectric patterns. *Acta Orthop. Scand.* **44**, 389–409.

44. Jiang N, Vest-Nielsen JLG, Muceli S and Farina D (2012) EMG-based simultaneous and proportional estimation of wrist/hand kinematics in uni-lateral trans-radial amputees. *J. Neuroeng. Rehabil.* **9**, 42.

45. Ameri A, Scheme EJ, Englehart KB and Parker PA (2014) Bagged regression trees for simultaneous myoelectric force estimation. 2014 22nd Iranian Conference on Electrical Engineering (ICEE), Tehran, pp. 2000–3.

46. Ameri A, Kamavuako EN, Scheme EJ, Englehart KB and Parker PA (2014) Support vector regression for improved real-time, simultaneous myoelectric control. *IEEE Trans. Neural Syst. Rehabil. Eng.* **22**, 1198–209.

47. Ameri A, Akhaee MA, Scheme E and Englehart K (2018) Real-time, simultaneous myoelectric control using a convolutional neural network. *PLoS One* **13**, e0203835.

48. Young AJ, Smith LH, Rouse EJ and Hargrove LJ (2013) Classification of simultaneous movements using surface EMG pattern recognition. *IEEE Trans. Biomed. Eng.* **60**, 1250–8.

49. Jiang N, Englehart K and Parker P (2008) Estimating forces at multiple degrees of freedom from surface EMG using non-negative matrix factorization for myoelectric control. *2008 First International Symposium on Applied Sciences on Biomedical and Communication Technologies*, Aalborg, Denmark, pp. 1–5.

50. Jiang N, Rehbaum H, Vujaklija I, Graimann B and Farina D (2014) Intuitive, online, simultaneous, and proportional myoelectric control over two degrees-of-freedom in upper limb amputees. *IEEE Trans. Neural Syst. Rehabil. Eng.* **22**, 501–10.

51. Lin C, Wang B, Jiang N and Farina D (2018) Robust extraction of basis functions for simultaneous and proportional myoelectric control via sparse non-negative matrix factorization. *J. Neural Eng.* **15**, 026017.

52. Jiang N, Englehart KB and Parker PA (2009) Extracting simultaneous and proportional neural control information for multiple-DOF prostheses from the surface electromyographic signal. *IEEE Trans. Biomed. Eng.* **56**, 1070–80.

53. Rehbaum H, Jiang N, Paredes L, Amsuess S, Graimann B and Farina D (2012) Real time simultaneous and proportional control of multiple degrees of freedom from surface EMG: Preliminary results on subjects with limb deficiency. *Conf. Proc. IEEE Eng. Med. Biol. Soc.* **2012**, 1346–9.

54. Muceli S, Jiang N and Farina D (2014) Extracting signals robust to electrode number and shift for online simultaneous and proportional myoelectric control by factorization algorithms. *IEEE Trans. Neural Syst. Rehabil. Eng.* **22**, 623–33.

55. Vujaklija I, Shalchyan V, Kamavuako EN, Jiang N, Marateb HR and Farina D (2018) Online mapping of EMG signals into kinematics by autoencoding. *J. Neuroeng. Rehabil.* **15**, 21.

56. Nielsen JLG, Holmgaard S, Jiang N, Englehart KB, Farina D and Parker PA (2011) Simultaneous and proportional force estimation for multifunction myoelectric prostheses using mirrored bilateral training. *IEEE Trans. Biomed. Eng.* **58**, 681–8.

57. Jiang N, Muceli S, Graimann B and Farina D (2013) Effect of arm position on the prediction of kinematics from EMG in amputees. *Med. Biol. Eng. Comput.* **51**, 143–51.

58. Muceli S and Farina D (2012) Simultaneous and proportional estimation of hand kinematics from EMG during mirrored movements at multiple degrees-of-freedom. *IEEE Trans. Neural Syst. Rehabil. Eng.* **20**, 371–8.

59. Kamavuako EN, Englehart KB, Jensen W and Farina D (2012) Simultaneous and proportional force estimation in multiple degrees of freedom from intramuscular EMG. *IEEE Trans. Biomed. Eng.* **59**, 1804–7.

60. Fitts PM (1954) The information capacity of the human motor system in controlling the amplitude of movement. *J. Exp. Psychol.* **47**, 381–91.

61. Williams MR and Kirsch RF (2008) Evaluation of head orientation and neck muscle EMG signals as command inputs to a human-computer interface for individuals with high tetraplegia. *IEEE Trans. Neural Syst. Rehabil. Eng.* **16**, 485–96.

62. Atzori M, Gijsberts A, Heynen S, Hager AM, Deriaz O, van der Smagt P, Castellini C, Caputo B and Müller H (2012) Building the Ninapro database: A resource for the biorobotics community. 2012 4th IEEE RAS EMBS International Conference on Biomedical Robotics and Biomechatronics (BioRob), Rome, Italy, pp. 1258–65.

63. Atzori M, Gijsberts A, Castellini C, Caputo B, Hager A-G M, Elsig S, Giatsidis G, Bassetto F and Müller H (2014) Electromyography data for non-invasive naturally-controlled robotic hand prostheses. *Sci Data* **1**, 140053.

64. Amma C, Krings T, Böer J and Schultz T (2015) Advancing muscle-computer interfaces with high-density electromyography. *Proceedings of the 33rd Annual ACM Conference on Human Factors in Computing Systems CHI'15*. New York, Association for Computing Machinery, pp. 929–38.

65. Ortiz-Catalan M, Brånemark R and Håkansson B (2013) BioPatRec: A modular research platform for the control of artificial limbs based on pattern recognition algorithms. *Source Code Biol. Med.* **8**, 11.

66. Du Y, Wenguang J, Wentao W and Geng W (2020) CapgMyo: A high density surface electromyography database for gesture recognition. Available in http://zju-capg.org/myo/data/; Access date: 10/4/20.

67. Mathewson KW and Pilarski PM (2017) Actor-critic reinforcement learning with simultaneous human control and feedback. arXiv Available in https://arxiv.org/abs/1703.01274.

68. Proroković K, Wand M and Schmidhuber J (2020) Meta-learning for recalibration of EMG-based upper limb prostheses. Available in https://www.automl.org/wp-content/uploads/2020/07/AutoML_2020_paper_45.pdf.

69. Atzori M, Cognolato M and Müller H (2016) Deep learning with convolutional neural networks applied to electromyography data: A resource for the classification of movements for prosthetic hands. *Front. Neurorobot.* **10**, 9.

70. Park K and Lee S (2016) Movement intention decoding based on deep learning for multiuser myoelectric interfaces. 2016 4th International Winter Conference on Brain-Computer Interface (BCI), South Korea, pp. 1–2.

71. Ghazaei G, Alameer A, Degenaar P, Morgan G and Nazarpour K (2017) Deep learning-based artificial vision for grasp classification in myoelectric hands. *J. Neural Eng.* **14**, 036025.

72. Côté-Allard U, Fall CL, Drouin A, Campeau-Lecours A, Gosselin C, Glette K, Laviolette F and Gosselin B (2019) Deep learning for electromyographic hand gesture signal classification using transfer learning. *IEEE Trans. Neural Syst. Rehabil. Eng.* **27**, 760–71.

73. Triwiyanto T, Pawana IPA and Purnomo MH (2020) An improved performance of deep learning based on convolution neural network to classify the hand motion by evaluating hyper parameter. *IEEE Trans. Neural Syst. Rehabil. Eng.* **28**, 1678–88.

74. Zhai X, Jelfs B, Chan RHM and Tin C (2017) Self-recalibrating surface EMG pattern recognition for neuroprosthesis control based on convolutional neural network. *Front. Neurosci.* **11**, 379.

75. Ameri A, Akhaee MA, Scheme E and Englehart K (2020) A deep transfer learning approach to reducing the effect of electrode shift in EMG pattern recognition-based control. *IEEE Trans. Neural Syst. Rehabil. Eng.* **28**, 370–9.

76. Hu Y, Wong Y, Wei W, Du Y, Kankanhalli M and Geng W (2018) A novel attention-based hybrid CNN-RNN architecture for sEMG-based gesture recognition. *PLoS One* **13**, e0206049.

77. Wang C, Guo W, Zhang H, Guo L, Huang C and Lin C (2020) sEMG-based continuous estimation of grasp movements by long-short term memory network. *Biomed. Signal Process. Control* **59**, 101774.

78. Li C, Ren J, Huang H, Wang B, Zhu Y and Hu H (2018) PCA and deep learning based myoelectric grasping control of a prosthetic hand. *Biomed. Eng. Online* **17**, 107.

79. Lv B, Sheng X and Zhu X (2018) Improving myoelectric pattern recognition robustness to electrode shift by autoencoder. *Conf. Proc. IEEE Eng. Med. Biol. Soc.* **2018**, 5652–5.

80. Jiang N, Vujaklija I, Rehbaum H, Graimann B and Farina D (2014) Is accurate mapping of EMG signals on kinematics needed for precise online myoelectric control? *IEEE Trans. Neural Syst. Rehabil. Eng.* **22**, 549–58.

81. Cooper RA, Corfman TA, Fitzgerald SG, Boninger ML, Spaeth DM, Ammer W and Arva J (2002) Performance assessment of a pushrim-activated power-assisted wheelchair control system. *IEEE Trans. Control Syst. Technol.* **10**, 121–6.

82. Xiloyannis M, Gavriel C, Thomik AAC and Faisal AA (2017) Gaussian process autoregression for simultaneous proportional multi-modal prosthetic control with natural hand kinematics. *IEEE Trans. Neural Syst. Rehabil. Eng.* **25**, 1785–801.

83. Lobo-Prat J, Keemink AQL, Stienen AHA, Schouten AC, Veltink PH and Koopman BFJM (2014) Evaluation of EMG, force and joystick as control interfaces for active arm supports. *J. Neuroeng. Rehabil.* **11**, 68.

84. Corbett EA, Perreault EJ and Kuiken TA (2011) Comparison of electromyography and force as interfaces for prosthetic control. *J. Rehabil. Res. Dev.* **48**, 629–41.

85. Cavallaro EE, Rosen J, Perry JC and Burns S (2006) Real-time myoprocessors for a neural controlled powered exoskeleton arm. *IEEE Trans. Biomed. Eng.* **53**, 2387–96.

86. Maheu V, Frappier J, Archambault PS and Routhier F (2011) Evaluation of the JACO robotic arm: Clinico-economic study for powered wheelchair users with upper-extremity disabilities. *IEEE Int. Conf. Rehabil. Robot.* **2011**, 5975397.

87. Kuiken TA, Li G, Lock BA, Lipschutz RD, Miller LA, Stubblefield KA and Englehart KB (2009) Targeted muscle reinnervation for real-time myoelectric control of multi-function artificial arms. *JAMA* **301**, 619–28.

88. Gage WH, Zabjek KF, Hill SW and McIlroy WE (2007) Parallels in control of voluntary and perturbation-evoked reach-to-grasp movements: EMG and kinematics. *Exp. Brain Res.* **181**, 627–37.

89. Song R and Tong KY (2005) Using recurrent artificial neural network model to estimate voluntary elbow torque in dynamic situations. *Med. Biol. Eng. Comput.* **43**, 473–80.

90. Bloomer C, Wang S and Kontson K (2020) Kinematic analysis of motor learning in upper limb body-powered bypass prosthesis training. *PLoS One* 15, e0226563.

91. Huang H, Crouch DL, Liu M, Sawicki GS and Wang D (2016) A cyber expert system for auto-tuning powered prosthesis impedance control parameters. *Ann. Biomed. Eng.* **44**, 1613–24.

92. Xia P, Hu J and Peng Y (2018) EMG-based estimation of limb movement using deep learning with recurrent convolutional neural networks. *Artif. Organs* **42**, E67–77.

93. Luh JJ, Chang GC, Cheng CK, Lai JS and Kuo TS (1999) Isokinetic elbow joint torques estimation from surface EMG and joint kinematic data: Using an artificial neural network model. *J. Electromyogr. Kinesiol.* **9**, 173–83.

94. Chen C, Chai G, Guo W, Sheng X, Farina D and Zhu X (2019) Prediction of finger kinematics from discharge timings of motor units: Implications for intuitive control of myoelectric prostheses. *J. Neural Eng.* **16**, 026005.

95. Pan L, Zhang D, Jiang N, Sheng X and Zhu X (2015) Improving robustness against electrode shift of high density EMG for myoelectric control through common spatial patterns. *J. Neuroeng. Rehabil.* **12**, 110.

96. Kyranou I, Vijayakumar S and Erden MS (2018) Causes of performance degradation in non-invasive electromyographic pattern recognition in upper limb prostheses. *Front. Neurorobot.* **12**, 58.

97. Campbell E, Phinyomark A and Scheme E (2020) Current trends and confounding factors in myoelectric control: Limb position and contraction intensity. *Sensors* **20**, 1613.

98. Amsüss S, Goebel PM, Jiang N, Graimann B, Paredes L and Farina D (2014) Self-correcting pattern recognition system of surface EMG signals for upper limb prosthesis control. *IEEE Trans. Biomed. Eng.* **61**, 1167–76.

99. Kate J, Smit G and Breedveld P (2017) 3D-printed upper limb prostheses: A review. *Disability Rehabil. Assistive Technol.* **12**, 300–14.

100. Manero A, Smith P, Sparkman J, Dombrowski M, Courbin D, Kester A, Womack I and Chi A (2019) Implementation of 3D printing technology in the field of prosthetics: Past, present, and future. *Int. J. Environ. Res. Public Health* **16**, 1641.

101. González-Fernández M (2014) Development of upper limb prostheses: Current progress and areas for growth. *Arch. Phys. Med. Rehabil.* **95**, 1013–4.

102. Pasquina PF, Perry BN, Miller ME, Ling GSF and Tsao JW (2015) Recent advances in bioelectric prostheses. *Neurol. Clin. Pract.* **5**, 164–70.

103. Samuel OW, Asogbon MG, Geng Y, Al-Timemy AH, Pirbhulal S, Ji N, Chen S, Fang P and Li G (2019) Intelligent EMG pattern recognition control method for upper-limb multifunctional prostheses: Advances, current challenges, and future prospects. *IEEE Access* **7**, 10150–65.

104. Hargrove L, Englehart K and Hudgins B (2006) The effect of electrode displacements on pattern recognition based myoelectric control. 2006 International Conference of the IEEE Engineering in Medicine and Biology Society, New York, NY, USA, pp. 2203–6.

105. Samuel OW, Li X, Geng Y, Asogbon MG, Fang P, Huang Z and Li G (2017) Resolving the adverse impact of mobility on myoelectric pattern recognition in upper-limb multifunctional prostheses. *Comput. Biol. Med.* **90**, 76–87.

106. Scheme E and Englehart K (2011) Electromyogram pattern recognition for control of powered upper-limb prostheses: State of the art and challenges for clinical use. *J. Rehabil. Res. Dev.* **48**, 643–59.

107. Clancy EA, Morin EL and Merletti R (2002) Sampling, noise-reduction and amplitude estimation issues in surface electromyography. *J. Electromyogr. Kinesiol.* **12**, 1–16.

108. Igual C, Pardo LA, Hahne JM and Igual J (2019) Myoelectric control for upper limb prostheses. *Electronics* **8**, 1244.

109. Farina D, Merletti R and Enoka RM (2004) The extraction of neural strategies from the surface EMG. *J. Appl. Physiol.* **96**, 1486–95.

110. Backus SI, Tomlinson DP, Vanadurongwan B, Lenhoff MW, Cordasco FA, Chehab EL, Adler RS, Henn RF 3rd and Hillstrom HJ (2011) A spectral analysis of rotator cuff musculature electromyographic activity: Surface and indwelling. *HSS J.* **7**, 21–8.

111. Li G, Li Y, Yu L and Geng Y 2011) Conditioning and sampling issues of EMG signals in motion recognition of multifunctional myoelectric prostheses. *Ann. Biomed. Eng.* **39**, 1779–87.

112. Chen H, Zhang Y, Zhang Z, Fang Y, Liu H and Yao C (2017) Exploring the relation between EMG sampling frequency and hand motion recognition accuracy. *2017 IEEE International Conference on Systems, Man, and Cybernetics (SMC)*, Banff, AB, pp. 1139–44.

113. Li G, Li Y, Zhang Z, Geng Y and Zhou R (2010) Selection of sampling rate for EMG pattern recognition based prosthesis control. *2010 Annual International Conference of the IEEE Engineering in Medicine and Biology*, Buenos Aires, Argentina, pp. 5058–61.

114. Hassan HF, Abou-Loukh SJ and Ibraheem IK (2020) Teleoperated robotic arm movement using electromyography signal with wearable Myo armband. *J. King Saud Univ. Eng. Sci.* **32**, 378–87.

115. Zia ur Rehman M, Waris A, Gilani SO, Jochumsen M, Niazi IK, Jamil M, Farina D and Kamavuako EN (2018) Multiday EMG-based classification of hand motions with deep learning techniques. *Sensors* **18**, 2497.

116. Pizzolato S, Tagliapietra L, Cognolato M, Reggiani M, Müller H and Atzori M (2017) Comparison of six electromyography acquisition setups on hand movement classification tasks. *PLoS One* **12**, e0186132.

117. Vandewalle P (2006) Super-resolution from unregistered aliased images (EPFL). PhD Dissertation. Available in https://infoscience.epfl.ch/record/86006.

118. Vandewalle P, Sbaiz L, Vetterli M and Sustrunk S (2005) Super-resolution from highly undersampled images. *IEEE International Conference on Image Processing 2005*, Genova, Italy, vol. 1, pp. 1–889.

119. Vandewalle P, Sbaiz L, Vandewalle J and Vetterli M (2004) How to take advantage of aliasing in bandlimited signals. *2004 IEEE International Conference on Acoustics, Speech, and Signal Processing*, Quebec, Canada, vol. 3, pp. 3–948.

120. Mamikoglu U, Nikolakopoulos G, Pauelsen M, Varagnolo D, Röijezon U and Gustafsson T (2016) Elbow joint angle estimation by using integrated surface electromyography. *2016 24th Mediterranean Conference on Control and Automation (MED)*, Valletta, Malta, pp. 785–90.

121. Zhang Q, Liu R, Chen W and Xiong C (2017) Simultaneous and continuous estimation of shoulder and elbow kinematics from surface EMG signals. *Front. Neurosci.* **11**, 280.

122. Choi K (2013) Reconstructing for joint angles on the shoulder and elbow from non-invasive electroencephalographic signals through electromyography. *Front. Neurosci.* **7**, 190.

123. Chaya NA, Bhavana BR, Anoogna SB, Hiranmai M and Krupa BN (2019) Real-time replication of arm movements using surface EMG signals. *Procedia Comput. Sci.* **154**, 186–93.

124. Thongpanja S, Phinyomark A, Limsakul C and Phukpattaranont P (2015) Application of mean and median frequency methods for identification of human joint angles using EMG signal. In: Information Science and Application, ed. Kuinam J Kim. Berlin Heidelberg: Springer, pp. 689–96. https://www.springer.com/gp/book/9783662465776.

125. Sokolova M and Lapalme G (2009) A systematic analysis of performance measures for classification tasks. *Inf. Process. Manag.* **45**, 427–37.

126. Merletti R and Muceli S (2019) Tutorial surface EMG detection in space and time: Best practices. *J. Electromyogr. Kinesiol.* **49**, 102363.

6 Assessment and Diagnostic Methods for Coronavirus Disease 2019 (COVID-19)

M. B. Malarvili
Universiti Teknologi Malaysia (UTM)

Alexie Mushikiwabeza
Universiti Teknologi Malaysia (UTM)
University of Rwanda (UR)/Center of Excellence in
Biomedical Engineering and E-Health (CEBE)

CONTENTS

6.1   INTRODUCTION

Coronavirus disease 2019 (COVID-19) is an infectious disease caused by a novel coronavirus known as SARS-CoV-2 [1]. Coronaviruses are members of the Coronaviridae family of the order Nidovirales that mainly cause infection in the respiratory tract [2]. All viruses belonging to this order are enveloped, nonsegmented positive-sense RNA viruses [3]. The outbreak of the novel coronavirus (SARS-CoV-2) occurred in Wuhan, the largest city of the Hubei Province in China, in December 2019 [1,2,4]. Due to the movement of population, SARS-CoV-2 infection spread rapidly across China, and in many other countries around the globe, resulting in the ongoing pandemic. On January 30, 2020, the World Health Organization (WHO) announced that the outbreak fulfills the criteria for Public Health Emergency of International Concern [5]. The COVID-19 pandemic is taking a tremendous toll worldwide mainly in the families, societies, healthcare sectors, as well as economies [6]. On May 11, 2020, WHO report showed that the number of COVID-19-confirmed cases was more than 3.9 million people with 206,529 deaths. Malaysia has 6643 confirmed cases and 109 deaths [7].

The primary route of SARS-CoV-2 infection is via person-to-person transmission by direct contact or indirectly through respiratory droplets and fomites [4]. Therefore, affected countries took extensive measures in preventing and controlling the infection, including the detection of suspected cases at an early stage, isolation of infected persons from others during treatment, and quarantine [1]. Moreover, citizens were encouraged to stay home, work from home [8], wash hands regularly, and keep social distancing [9]. A health screening strategy is being used as a primary way of testing the presence of SARS-CoV-2 infection. In this, infrared thermometers are greatly used to screen the core body temperature, especially at the entrance of public buildings, including schools, hospital, shopping malls, airports, etc. [10]. Noncontact infrared thermometers gained popularity for screening fever since they are portable, easy to use, and cost-effective. However, their low sensitivity and accuracy may affect the effectiveness of the measure.

To date, there is no specific treatment for COVID-19 pneumonia. Early diagnosis of the SARS-COV-2 infection can help in providing an effective support treatment to the infected person and reduce further transmission of the virus. Recent published works show that the diagnosis of COVID-19 is mainly based on clinical symptoms in addition to the use of the real-time RT-PCR, and antibody tests, chest computed tomography (CT) imaging, and some laboratory findings. Although the real-time RT-PCR test is the main test used for diagnosing SARS-CoV-2 infection, the performance of this test is based on numerous factors such as the laboratory equipment, the skills of the technicians in performing the test and interpreting the results, in addition to the long period of time required to generate the results [8]. These factors can lead to a delay in the detection of the virus at the early stage of infection. Some studies reported that the combined use of the real-time RT-PCR test with either chest CT or serological test may increase the sensitivity in detecting SARS-CoV-2 infection [11,12]. In some hospitals, the diagnosis was only based on clinical and CT findings due to the shortage of RT-PCR kits [13]. However, some patients presented a normal CT in the first 2 days after the presentation of symptoms [13]. Therefore, the development of new tools can contribute to timely and accurate detection of this infectious disease.

The major challenges for developing such tool include identifying the best monitoring technology and the optimal parameters with a sufficient sensitivity and specificity to assess the respiratory function and its changes. Hence, a rigorous and extensive research was carried out from December 2019 to July 2020 through Google Scholar, the Web of Science, PubMed, and Scopus using different keywords (*corona-virus, severe acute respiratory syndrome monitoring device, respiratory CO_2 monitoring device, SARS-CoV-2 monitoring device, capnograph, COVID-19 capnogram*) to identify appropriate respiratory diseases assessment tools. From the literature, we believe that capnography and the parameters derived from a capnogram can meet this challenge. We also manually searched the references of the select articles for additional relevant articles. Based on our literature review, we propose the use of a capnograph device as a screening tool for SARS-CoV-2 infection as the virus causes a respiratory tract-related disease. The proposed system will be based on the analysis of CO_2 signal's features in COVID-19 patients.

This chapter is divided into seven sections. Section 6.1 deals with the introduction. Section 6.2 describes the various clinical findings of COVID-19 disease. Section 6.3 provides an overview of the current methods used to diagnose SARS-CoV-2 infection. Section 6.4 presents the existing tools used to screen SARS-CoV-2 infection. Section 6.5 describes the capnogram features. Section 6.6 presents the proposed system for early screening of SARS-CoV-2 infection by analyzing CO_2 patterns. Section 6.7 provides conclusion.

6.2   CLINICAL FINDINGS OF COVID-19

Despite the widespread of SARS-CoV-2 infection, there are nonspecific clinical manifestations of the disease. At the initial presentation, most COVID-19 patients manifest fever, dry cough, and myalgia or fatigue [1,8,14]. However, some patients are asymptomatic [15]. Fever and cough are the predominant symptoms at the early stage of the illness, whereas diarrhea, sore throat, and chest tightness are rare [4,16]. In the study carried out by Ref. [15] on 262 patients, 82.1% had fever, while 45.8% had cough. This is consistent with a study by Chen, Jun et al. [17], whereby 87.1% of the 249 patients presented fever, 36.5% had cough, and only 3.2% had diarrhea [17]. Dyspnea is uncommon in COVID-19 patients, though it may be considered while classifying the severity of the disease [15]. Older males and patients with comorbidities such as hypertension, diabetes, and coronary heart disease are more exposed to SARS-CoV-2 infection due to their low immune system function [18,19].

During the disease course, some patients develop acute respiratory distress syndrome (ARDS) and septic shock, which lead to a multiple-organ failure, including liver dysfunction, heart failure, and abnormalities in renal function associated with an increased blood urea nitrogen [1,18]. Besides this, the patient may experience a variation in normal values of blood elements such as increased neutrophils, elevated C-reactive protein, reduced lymphocyte counts (particularly T lymphocytes), and hemoglobin counts. The study of Chen, Nanshan et al. [18] showed that out of 99 patients, 9% presented low leucocyte counts, 38% had a high level of neutrophil counts, and 35% admitted patients manifested lymphocyte counts below the normal range [18]. The reduction in the number of T lymphocytes was also identified in patients with severe acute respiratory syndrome coronavirus infection (SARS-CoV), which emerged in November 2002. The absolute lymphocyte count, which is below the normal range, could be considered as a reference index in the clinical settings while assessing novel coronavirus [18].

COVID-19-infected patients are also more likely to encounter blood coagulation disorders, especially those with cardiac injury [20]. In the study of Chen et al. [18], various blood coagulation tests had been taken into account. Out of 99 COVID-19 patients, 16% had the activated partial thromboplastin time (APTT) below the normal range, 30% had a prolonged prothrombin time, whereas 36% had D-dimer above the normal range [18]. Further, 3 seconds in the prothrombin time and 5 seconds in the APTT were classified as a coagulopathy condition [19]. Thereafter, Wang et al. [21] reported 138 cases with COVID-19, in which 58% had extended prothrombin time, and the level of D-dimer was higher in ICU patients compared to non-ICU patients [21]. It was suggested that a level of D-dimer above 1 µg/mL could be used to assess the patients' poor prognosis at the onset of illness [21]. Venous thromboembolism is another complication faced by some of COVID-19-hospitalized cases as a result of a limited movement during illness, dehydration, or the presence of chronic underlying conditions such as hypertension, diabetes, or cardiac-related diseases [20]. Although SARS-CoV-2 was mainly identified as a respiratory tract infection [22], it affects numerous systems, including the gastrointestinal, cardiovascular, respiratory, and immune systems. Therefore, both clinical symptoms and laboratory findings should be considered for a proper detection of COVID-19 disease.

6.3   EXISTING DIAGNOSTIC TOOLS

6.3.1   REAL-TIME RT-PCR TEST MOLECULAR TEST

Currently, the clinical diagnosis of COVID-19 pneumonia is based on the use of real-time RT-PCR test, CT imaging, and analysis of some hematology parameters [8] such as leucocyte or lymphocyte count [23]. In Wuhan, where the first cases were identified, the diagnosis was based first on epidemiological factors by assessing whether the suspected patient had been in contact with wildlife, Wuhan exposure, or a history of close contact with Wuhan people or patients who had tested positive in the previous two weeks [1,4]. Thereafter, chest imaging and detection of infectious agents in the collected specimens, including respiratory, blood, and feces samples, have been carried out. By using the real-time RT-PCR method, SARS-CoV-2 infection was detected in the lower respiratory tract samples [24]. Various studies reported the presence of SARS-CoV-2 infection in different clinical specimens. Wang et al. [25] detected SARS-CoV-2 infection in various clinical samples, including bronchoalveolar lavage fluid (BLF), sputum, nasal swabs, pharyngeal swabs, fibrobronchoscope brush biopsy, blood, feces, and urine. In this study, 205 COVID-19 patients were involved, and all of them had been previously diagnosed according to symptoms and radiological features. In the majority of the patients, pharyngeal swabs were sampled between the first and third days after admission to the hospital. Both BLF and fibrobronchoscope brush biopsy were collected from severe cases or patients assisted by mechanical ventilation. (Adapted from: Wang et al. [25].)

BLF had the highest positive results (93%), whereas pharyngeal swabs had the lowest positive rates (32%) among the lower respiratory tract samples. Sputum and nasal swab samples had 72% and 63% positive results, while fibrobronchoscope brush biopsy had a positive rate of 46%. The presence of the infection in feces was at the rate of 29%. In the blood samples, SARS-CoV-2 infection was almost absent whereby the positive rate was only 1%. Among the 72 urine specimens sampled, all of them presented negative results [25].

BLF is collected for the diagnosis and detection of viral RNAs particularly in severe cases although the suction tool is required in sampling process and collection of this sample is painful to the patients. Sputum and nasal swabs are also mostly collected for the laboratory diagnosis of SARS-CoV-2 infection as their collection process is simple, fast, and safe [26]. Although sputum was identified as the most sensitive specimen in detecting the virus, a study conducted on 41 COVID-19 patients showed that only a small number of patients (28%) had sputum production as a symptom [24]. Therefore, nasal swabs may be the most commonly applicable specimen in the detection of SARS-CoV-2 infection [27]. The absence of the virus in urine was also reported in a comparative study on four different samples tested using the nucleic acid amplification method [22]. All 19 COVID-19 patients had negative results from blood and urine specimens. On the other hand, stool samples and oropharyngeal swab presented almost the same number of positive rates (42% and 47%, respectively) [22]. These results show that SARS-CoV-2 infection is rarely found in the blood. The sensitivity of nucleic acid tests in the diagnosis of COVID-19 may be enhanced when it is coupled with the serological test [12].

Real-time RT-PCR is a diagnostic test that relies on the nucleic acid amplification approach. The test is performed in vitro to detect the presence of viruses from sera and respiratory specimens, including nasopharyngeal swabs, lower respiratory tract aspirates, and sputum [28,29]. Real-time RT-PCR test remains the gold standard method in diagnosing COVID-19. However, it presents some limitations as follows: first, the test is performed in a certified and a well-equipped laboratory, by a well-trained professional capable of interpreting the results, and generating the results requires a long time (over 2–3 hours on average) [8]. Second, false-negative results may be produced due to either inappropriate collection, transportation, and handling of specimens; the presence of amplification inhibitors in the specimen; or insufficient organism numbers in the specimen [29]. Moreover, the real-time RT-PCR has a low detection rate at the initial presentation of the disease [1]. The identification of the presence of viral proteins using an antigen-based approach is an alternative mean for a rapid and qualitative detection of SARS-CoV-2 infection [30].

6.3.2 RAPID ANTIGEN DETECTION (RAD) TEST

An antigen test is a qualitative method of detecting certain proteins that are present on or within the virus. Similar to the RT-PCR test, the antigen test also uses respiratory samples, including nasal and nasopharyngeal swabs [30]. Throat saliva and sputum are not commonly used for the RAD tests [31]. Despite its low sensitivity, the antigen test has been reported to be cost-effective and faster compared to RT-PCR test. Different antigen test kits are being produced by the manufacturers of diagnostic tests from different countries. Sofia 2 SARS Antigen Fluorescent Immunoassay (FIA) is a lateral flow immunofluorescent sandwich assay developed by the Quidel Company, the United States. The Food and Drug Administration (FDA) issued an Emergency Use Authorization for this test, which detects an antigen from the nucleocapsid protein of the SARS-CoV-2 virus. This test was reported to be helpful in testing a great number of individuals per day as the results are generated within 15 minutes. However, the test needs to be performed in laboratories certified by the Clinical Laboratory Improvement Amendments (CLIA) or in patient care settings operating under a CLIA Certificate of Waiver [30]. The Adeptrix Corporation developed a bead-assisted mass spectrometry (BAMS) antigen test. For this, Avacta Life Sciences Limited supplied Affimer® reagents that are used to coat the beads that bind the particles of the virus present in patient specimens. Every bead is analyzed using the mass spectrometry to detect the presence of the virus. The BAMS antigen test is cost-effective, and no special laboratory equipment is required. Moreover, the test has an increased capacity that numerous samples can be taken and analyzed by a single laboratory technician every day [32,33].

COVID-19 Ag Respi-Strip is a type of RAD test developed by Coris BioConcept, Belgium. This test was authorized by the Belgian Federal Agency for Medicines and Health Products to be performed in public health institutes in Belgium [34]. This diagnostic method uses patient nasopharyngeal secretions, and the results are generated within 15 minutes. Despite the low sensitivity that also depends on the type of specimen and the level of viral loads [35], the COVID-19 Ag Respi-Strip test is used as the first-line method in diagnosing COVID-19 in Belgium. RAD tests are less expensive and easy to operate; however, their analytical performance is affected by a variety of factors such as the viral load, quality of the samples, and the way the samples were processed [35]. Rapid antigen tests are not suggested to be used as standalone diagnostic tools in clinical practice due to their low sensitivity, which leads to false-negative results [31].

6.3.3 ANTIBODIES TEST

Antibody test, also known as serology test, is a screening carried out using blood samples taken by a finger prick or from a vein in the arm [36]. This test determines whether antibodies were developed against the virus [30]. The immune system produces these antibodies, a type of proteins that are critical for fighting and clearing out the virus. When an infection is present in the body, the adaptive immunity is expected to increase. B lymphocytes produce specific antibodies and CD8+ T cytotoxic lymphocytes that participate in the elimination of infected cells [12]. COVID-19 patients develop antibodies against nucleoprotein and receptor binding domain (RBD) of SARS-CoV-2 infection. However, the period for antibody response varies depending on the type of antibody [37]. Zhao et al. [38] evaluated the dynamics of three different antibodies: total antibody (Ab), immunoglobulin M (IgM), and immunoglobulin G (IgG) in conformity with the disease progression in COVID-19 patients. (Adapted from: Zhao et al. [38].)

In the first week of illness presentation, the RNA test showed the greatest sensitivity; however, it decreased in the later phases. In the last two phases after onset (8—14) and (15—39), the total Ab test presented the highest sensitivity (90%, and 100%, respectively). Moreover, the IgM test had a higher sensitivity compared to the IgG from day 1 after onset to the last day (day 39) [38]. It was suggested that the Ab and IgG tests could help in identifying the level of humoral immunity in COVID-19 patients [38]. The combination of IgM and IgG antibodies provided an increased sensitivity compared to the individual antibodies (IgM or IgG) [36]. Out of 397 COVID-19-confirmed cases, 64.48% developed both IgM and IgG antibodies, whereas the number of patients that tested positive for only IgM antibodies was greater than those for IgG antibodies (18.13% and 6.04%, respectively) [36]. In the work of Guo et al. [39], IgA antibodies were considered besides IgM and IgG antibodies in 208 plasma samples collected from 82 COVID-19-confirmed cases and 58 probable cases [39]. In this study, probable cases were patients who had negative results for the quantitative polymerase chain reaction (qPCR) test but presented a typical clinical manifestation [39]. Almost all samples were positive for IgA antibodies (93.3%), while IgM and IgG antibodies were present in 90.4% and 77.9% of plasma samples, respectively. The median time for detecting IgA and IgM antibodies was 5 days, and that for IgG 14 days [39]. Various studies reported the potential of serological tests in the diagnosis of SARS-CoV-2 infection at different stages of illness. However, seroconversion is not the same to all individuals, and it depends on the time taken for the symptoms to onset and when the specimen was sampled [39]. In addition, false-negative results are produced as a result of the low concentration of antibodies [36]. Antibody tests paired with RNA-based test show an enhanced sensitivity for detecting the novel coronavirus [38].

6.3.4 CHEST COMPUTED TOMOGRAPHY

Many researchers highlighted the application of the chest CT in COVID-19 diagnosis and evaluation of the disease course based on different imaging features such as ground glass opacity (GGO), consolidation, crazy-paving pattern, the presence of a halo sign, and changes in the airways [1,40]. GGO is defined as a hazy opacity with the preservation of bronchial and pulmonary vessels' markings, whereas consolidation is a pathological process whereby the air that is normally present in the alveoli is replaced by fluids, blood, or cells, and is characterized by an increased pulmonary parenchymal density that causes obscuration in vessels and airway wall margins [40]. In the work [21], GGO was the most common feature identified in all patients. These results are compatible with those of Li et al. [41] who considered both clinical and CT findings in 83 COVID-19 cases. Of the 83 patients, 30.1% had severe/critical illness and 69.9% were nonsevere cases. GGO was common in all severe cases, while consolidation, bronchial wall thickening, and crazy-paving pattern were present in 88%, 64%, and 56% of severe cases, respectively [41].

Bronchial wall thickening as a mark of changes in the airways was among the CT findings in addition to consolidation, interlobular septal thickening, crazy-paving pattern, spider web sign, subpleural line, etc [42]. The number of lobes affected, the level of harm due to GGO and consolidation, the presence of nodules in the lungs, pleural effusion, and the distribution of opacities and patterns were also identified in COVID-19 patients [43]. All the 21 patients were free from pulmonary nodules and pleural effusion, 6 (28%) patients manifested both GGO and consolidation, while 4 (19%) patients presented crazy-paving pattern. In 3 (14%) patients, the initial chest CT findings were normal, although the laboratory testing of respiratory samples showed that all patients were positive for SARS-CoV-2 infection [43]. The presence of negative imaging results in COVID-19-confirmed cases showed that chest CT has a limited sensitivity and reliability in detecting the infection especially at the onset of illness [43]. A combined use of chest CT and real-time RT-PCR test was reported to be an appropriate means of achieving accurate results in early diagnosis of COVID-19 [11,44].

6.4 CURRENT SCREENING TOOLS FOR COVID-19

6.4.1 THERMOMETERS

At present, infrared thermometers are highly used to screen fever as a primary means of detecting the presence of SARS-CoV-2 infection. Monitoring core body temperature has become a requirement before entering to the public buildings such as a shopping complex, clinics, schools, airports, etc [10]. This method was used to detect SARS infection [45], which has clinical symptoms similar to those of SARS-CoV-2 infection, including fever, cough, and fatigue [46]. Since most of COVID-19 patients present fever at the onset of illness, screening body temperature is of crucial importance in a rapid detection of suspect cases for the management of the health condition of the citizens [47]. Noncontact infrared thermometers gained popularity in detecting fever as they are portable, easy to use, and do not cause discomfort, and the screening process does not depend on direct contact of the device and the fore-head of subject being monitored [48]. Although the device is not expensive and no constant recalibration is required, its sensitivity and accuracy are low compared to oral thermometers. The low accuracy can result from the distance between the operator and the subject, which is greater than the required proximity (3–15 cm) [48]. On the other hand, near-infrared thermometers have a high accuracy in measuring body temperature, but require a direct contact with the subject and the probe is frequently replaced as a mean of avoiding the spread of the disease among the individuals being screened [49].

6.4.2 THERMAL IMAGING SYSTEMS

Thermal imaging cameras are alternative noncontact tools to screen fever. As the subjects pass in the field of view of the camera, their thermal images are captured and analyzed [49]. Appropriate use of thermal imaging systems provides an accurate measurement of the surface skin temperature of an individual. However, this accuracy is affected by various factors such as improper setup of the system, the environment where the system is installed, the skills of the operator, and the preparation of the person who is being screened [50]. As SARS-CoV-2 infection continues to spread, a variety of temperature screening tools were developed as a means of timely and easily identifying the suspect cases.

Rokid Company, China, has developed T1 thermal glasses that have the capability of simultaneously screening the temperature of up to 200 people in a short period of time (within 2 minutes). These smart glasses are equipped with an infrared sensor, a Qualcomm CPU, and a camera of 12 MP. The glasses have the particularity of recording both live photos and videos. Furthermore, these devices can even detect the temperature of a person located in 3 m away. These thermal imaging detection tools are being used in China by national authorities, national park staff, and schools [51]. Forward-looking-infrared (FLIR) systems introduced two different configurations of smart cameras, namely, A400/A700 Thermal Smart Sensor solutions and Thermal Image Streaming fixed camera solutions. The Configurable Thermal Smart Sensor has measurement tools of high quality, and it can detect the elevated skin temperature, particularly at the target area such as the forehead and the corner of the eye. Further screening is recommended for individuals identified with skin temperature that is above the average value. FLIR thermal cameras are being used at the airports to detect body temperature of passengers and flights crews [52]. Compared to non-contact infrared thermometers, thermal imaging systems demonstrated an increased screening capacity. However, they have a lower precision in screening fever, and they require a regular calibration, besides their initial cost, which is high [49]. We cannot, nonetheless, undermine the usefulness of these systems in the initial detection of body temperature. However, they should not be considered as definitive diagnostic tools to confirm the presence of SARS-CoV-2 infection since some individuals may have COVID-19 disease without having fever as a symptom [50].

6.5   CAPNOGRAM FEATURES

The respired CO_2 provides a significant information that can assist physicians in identifying spot ventilation derangements, extubation outcomes, bronchospasm, and the effectiveness of therapy in the clinical environment [53]. Furthermore, features extracted from CO_2 signals, such as end-tidal CO_2($EtCO_2$), respiratory rate (RR), time spent at $EtCO_2$, exhalation duration, Hjorth parameters (activity, mobility, and complexity), slope of phase II, end-exhalation slope, the slope ratio (SR), and area ratio, can be used to monitor and diagnose cardiopulmonary diseases, such as chronic obstructive pulmonary disease (COPD), asthma, congestive heart failure (CHF), pulmonary edema, and pneumonia [54,55]. To date, to the best of our knowledge, the capnograph is the only device on the market that serves this purpose. A capnograph is a noninvasive device that uses infrared technology to measure CO_2 in the expired gases and generates an incessant plot of exhaled CO_2 over time, known as a capnogram. A capnogram illustrates the variation in the partial pressure of expired CO_2 during the respiratory process [56]. Time-based capnogram is a graphical display of the instantaneous CO_2 concentration (mmHg) versus time (seconds). It can show the progress of respiratory condition of a patient. Unlike a peak flow meter or spirometer where the patient must follow a set of instructions, capnogram is taken while the patient is breathing as comfortably as possible. A normal capnogram has four phases

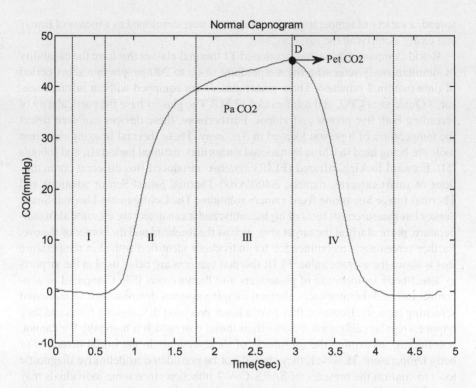

FIGURE 6.1   A normal capnogram with 4 consecutive phases (I, II, III and IV).

and an end-tidal point (Figure 6.1). Each phase reflects the usual process of CO_2 elimination [57].

The flat phase I represents early exhalation, which is relatively CO_2-free. As exhalation continues, there is a very rapid increase in expired CO_2 and this creates the near-vertical rising phase II. Near the termination of the normal exhalation is phase III. At the end of this plateau phase is D, the point in which the measured alveolar CO_2 levels best approximate the partial pressure of CO_2 in the arteries ($PaCO_2$). At this point, the level of sampled CO_2 is known as $PetCO_2$. As inspiration occurs, a near-vertical rapidly falling phase IV is observed. When ventilation and perfusion function normally, $PetCO_2$ should read 2—5 mmHg higher than the $PaCO_2$. In a healthy subject, the CO_2 waveform has a square shape. Figure 6.2 shows the morphological changes of this waveform due to various abnormal respiratory conditions such as asthma and COPD.

The change in capnogram shape was also observed in patients with ARDS (Figure 6.3) [58]. ARDS is a form of noncardiogenic pulmonary edema caused by acute injury to the lungs, leading to the flooding of small air sacs of the lungs that are involved in gas exchange [59]. ARDS leads to respiratory failure resulting from improper oxygenation and excretion of CO_2. Patients with ARDS have also the risk of experiencing arterial hypoxemia due to a ventilation-to-perfusion mismatch [60]. There is also an increase in minute ventilation and elevation of pulmonary dead

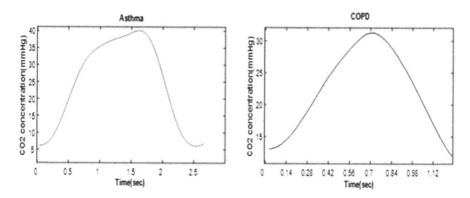

FIGURE 6.2 Morphology of a CO_2 waveform in asthmatic and COPD patients.

space resulting from an impaired elimination of CO_2 from the body, which causes hypercapnia.

COVID-19 patients with ARDS, especially those admitted to the ICU with low tidal volume ventilation, may experience hypercapnia [61]. These COVID-19 patients were clinically diagnosed using high positive end-expiratory pressure (PEEP) method, which is also used for ARDS cases. Nonetheless, the patients infected with SARS-CoV-2, despite meeting the Berlin definition of ARDS, manifest an atypical form of the syndrome [61]. In patients with ARDS, PEEP is frequently applied as a means of restoring proper oxygenation and preserving alveolar recruitment as well. However, at the bedside, this approach has a limited reliability in terms of identifying the risk-to-benefit ratio of various PEEP levels in a patient [60]. Volumetric capnography was proposed as a measurement tool to determine the pulmonary dead space, which is one of the predictors of mortality in ARDS. In contrast to time-based capnography, volumetric capnography provides information regarding both the CO_2 signal and the expired volume (Figure 6.3).

Features mentioned in Ref. [62], including $EtCO_2$, RR, time spent at $EtCO_2$, exhalation duration, Hjorth parameters (activity, mobility, and complexity), end-exhalation slope, the SR, and area ratio, have not been incorporated into capnography in a clinical setting as the findings are performed in an offline mode. Besides this, commercially existing capnograph devices are bulky and expensive, and provide a poor estimation of the ventilation and perfusion (V/Q) status of the lung. Therefore, a real-time human respiration CO_2 measurement device that can be used for a cardiorespiratory assessment has been developed in Ref. [62] based on the sidestream technology. This device is noninvasive and user-operable, and it displays the quantified features such as $EtCO_2$, RR, ICO_2, activity, and CO_2 signal on a small and single-screen thin-film transistor (TFT) module, hence minimizing the size of the device. More information on the hardware configuration of the device can be found in Ref. [62]. The proposed system has an average accuracy of 94.52%, a sensitivity of 97.67%, and a specificity of 90% for the discrimination of asthma and non-asthma using the CO_2 features, namely, AR1 (upward expiratory phase), AR2 (downward inspiratory phase), AR1 +AR2 (sum of the area of upward expiratory and downward

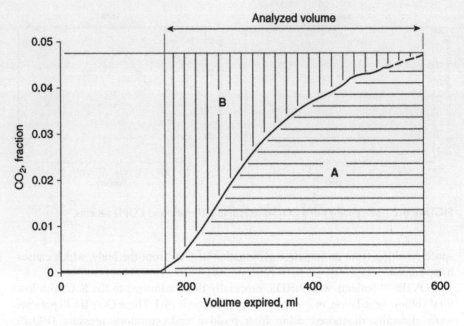

FIGURE 6.3 Illustration of a volumetric capnogram of a patient with ARDS. The graph represents the variation in expired volume (mL) in function of CO_2 fraction. Section A indicates the volume of exhaled CO_2 during the tidal breath, whereas the combined regions A and B show the theoretical volume of CO_2 that would be exhaled from an ideal perfectly homogenous lung over the analyzed volume interval. (Adapted from B. Jonson, "Volumetric capnography for noninvasive monitoring of acute respiratory distress syndrome," *American Journal of Respiratory and Critical Care Medicine,* vol. 198, no. 3, pp. 396–398, 2018 [58].)

inspiratory phases), and dCO_2/dt as detailed in Ref. [63]. Therefore, in this chapter, we propose the use of this device in order to differentiate the asthma, COPD, and SARS-CoV-2 conditions.

6.6 PROPOSED TOOL FOR EARLY SCREENING OF COVID-19 USING RESPIRED CO_2 FEATURES

Development of capnography is based on the following two types of technology: sidestream technology and mainstream technology. The proposed CO_2-based measurement device in Ref. [62] was developed using the sidestream technology due to its advantages such as being free of complexity in connection and cleaning, and the fact that the CO_2 samples can be collected even when the patient is in unusual position [64]. The device has the four main components as follows: the exhaled breath sample collection part, the processing part, a real-time clock (RTC), and a TFT to display the results (Figure 6.4). The device works first by collecting CO_2 samples through a nasal cannula connected to the nose of the patient. The quantity of exhaled CO_2 is measured by an infrared CO_2 sensor that is connected to the microcontroller. Thereafter, various features, including $EtCO_2$, RR, ICO_2, and

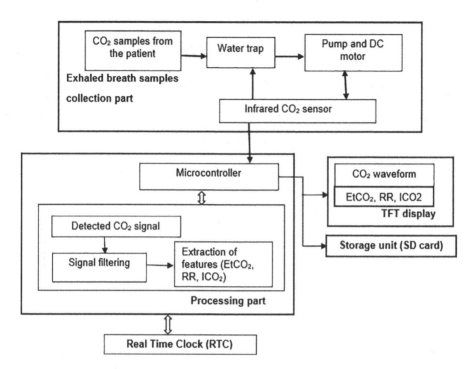

FIGURE 6.4 Block diagram of the proposed device to screen SARS-CoV-2 infection; CO_2 samples collection part, signal processing part, real-time clock (RTC), and TFT display and storage unit.

Hjorth parameters, are determined from the sensed CO_2 signal. Partial Et CO_2 is the peak value of exhaled CO_2 during each breathing cycle, the RR represents the number of breaths taken per minute, and ICO_2 indicates the CO_2 concentration inhaled in the respiratory process, and it provides information regarding the rebreathing of CO_2 caused by defects in the setup of the breathing circuit or an unusual condition of the patient. These features and the CO_2 waveform are visible on the TFT display. Besides this, a secured digital (SD) card is incorporated in the device to store the recorded data for future use.

6.6.1 EXHALED BREATH SAMPLES COLLECTION PART

The process of measuring CO_2 samples starts with the collection of the CO_2 samples from the patient using a nasal cannula. The movement of the CO_2 samples from the nasal cannula to the CO_2 sensor is facilitated by a pump and a DC motor (Figure 6.5). A water trap located between the pump and the nasal cannula has the capability of capturing any kind of respiratory secretions present in the CO_2 samples. Hence, the morphology of CO_2 signal is preserved. The communication between the infrared CO_2 sensor and the microcontroller is enabled by a 3-pin Universal Asynchronous Receiver/Transmitter (UART) incorporated in the sensor.

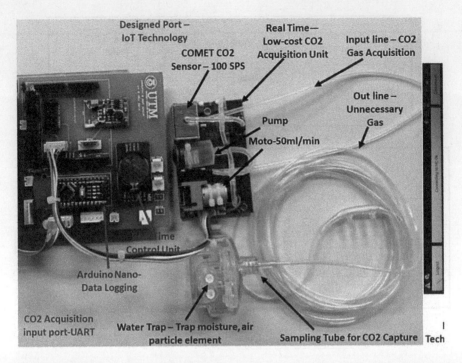

FIGURE 6.5 Illustration of different components of exhaled breath samples collection part of the proposed device.

6.6.2 PROCESSING PART

The computation task is performed by the Arduino connected to the sensor. In particular, a microcontroller board of the type Atmega 2560 is opted [65]. The Arduino mega 2560 is equipped with various components to support the microcontroller. It has 16 pins specific for analog input and 54 digital input/output pins driven by 5 V; each pin can serve as either input or output. In addition, the mega board has a 16-MHz crystal oscillator, a power jack, an In-Circuit Serial Programming (ICSP), header, and a reset button. The board can be powered by a computer via a USB connection or with AC to DC on the external supply ranging between 6 and 20 V. Serial communication is possible through four hardware UART serial ports, and the board can be connected to external devices (another Arduino or computer) [66]. Arduino mega 2560 requires an Arduino software to operate. The Arduino integrated development environment (IDE) is the software of choice as it simply runs based on C-programming language. The Arduino software is incorporated with libraries, which facilitates the programming process. After transferring the CO_2 data from the sensor to the microcontroller, the process of signal filtering and feature extraction is performed based on the code inserted in the program. The time-domain features are extracted from the segmented part of the CO_2 signals in order to differentiate respiratory conditions. More information on the features extracted and algorithms for feature extraction and classification has been reported in Ref. [63].

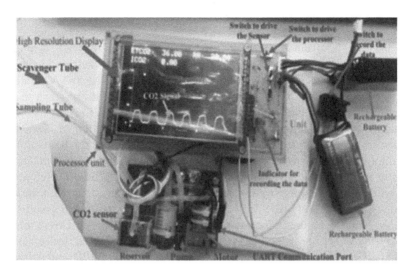

FIGURE 6.6 Display of a CO_2 waveform and its associated features ($EtCO_2$, RR, and ICO_2) on a TFT module.

6.6.3 DISPLAY UNIT

The extracted features and CO_2 waveform are displayed on a TFT-based liquid-crystal display, also known as the active-matrix display (Figure 6.6). The brightness of the screen is regulated by a transistor incorporated in each pixel of the display [67]. The TFT display is a 3.5″ diagonal with a resolution of 480 × 320 pixels and a four-wire-resistive touchscreen. The display can be connected to the microcontroller in a serial mode, known as serial peripheral interface (SPI), or in an 8-bit mode, which requires eight digital lines. Although a small number of pins is required, transferring data in the SPI mode takes longer compared to the 8-bit mode. Therefore, the SPI mode is suitable to the applications where data transmission speed is not prioritized. Furthermore, in the SPI mode, the display can be interfaced to different microcontrollers without difficulties [68]. There is also the option of using a microSD card on the same SPI bus. Based on the mentioned features, the TFT liquid-crystal module was used as a display in the proposed device. Besides this, SD card was incorporated in the design for storing the data.

6.6.4 REAL-TIME CLOCK (RTC)

A RTC module (DS3231) is an inter-integrated circuit, which serves to maintain the time on which the information is recorded, including seconds, minutes, hours, day, date, month, and year. When the month has ended, the date is automatically regulated according to the number of days in the month. It has a 3-V lithium battery that helps to accurately keep track of time when the main power of the device is shut down. In addition, the RTC can simply be connected to the most microcontrollers in a serial mode [69]. The RTC is a part of the developed CO_2 measurement device as it helps to store the recorded information of the patients on the SD card.

6.7 CONCLUSION

Early and effective diagnosis of the SARS-CoV-2 infection may greatly help in providing an effective support treatment to COVID-19 patients and in reducing further spread of the virus, which can even put an end to the SARS-CoV-2 infection in the human society. In this endeavor in mind, this chapter reviews and discuses current methods available to diagnose COVID-19 disease. This review has several limitations. First, the information regarding SARS-CoV-2 is limited. Second, the information provided here is based on current evidence, but may be modified as more information becomes available. Different literatures reported the limitations of currently available methods in diagnosing SARS-CoV-2 infection. Therefore, deep research is still needed to develop alternative tools with an enhanced accuracy in detecting the presence of SARS-CoV-2 infection at an early stage. In this preliminary study, we propose the use of a capnograph device to screen SARS-CoV-2 infection by analyzing CO_2 patterns of a COVID-19 patients. The proposed device has been tested to differentiate respiratory conditions such as asthma, COPD, and pulmonary edema. It was shown to have the ability of classifying asthmatic conditions based on capnogram features. In a future study, the feasibility of the same features will be verified on COVID-19 patients. On the other hand, adding an amalgamation of capnogram features, such as the slope of phase II, α-angle, SR, area ratio, and frequency components from the expiratory segment of the CO_2 signal, while developing the device, will provide a breakthrough in understanding COVID-19. Hence, the viability of the claimed features while developing a capnograph device should be verified in the future work.

DISCLOSURE

The authors report no conflicts of interest in this work.

ACKNOWLEDGMENTS

This research is conducted as part of the flagship grant of Prototype Development Research Grant Scheme (PRGS/2/2020/TK04/UTM/02/1), vot no R.J130000.7851.4L919 supported by Ministry of Higher Education (MOHE), Malaysia. We are also grateful to University Teknologi Malaysia for providing the facilities and laboratory equipment for the completion of the research. Our gratitude also goes to the Center of Excellence in Biomedical Engineering and E-Health, Rwanda (CEBE), for providing financial support to authors' PhD studies.

REFERENCES

1. Z. Y. Zu et al., "Coronavirus disease 2019 (COVID-19): A perspective from China," *Radiology*, vol. 296, p. 200490, 2020.
2. L. J. Saif, Q. Wang, A. N. Vlasova, K. Jung, and S. Xiao, "Coronaviruses". In: J. J. Zimmerman, L. A. Karriker, A. Ramirez, K. J. Schwartz, G. W. Stevenson, and J. Zhang (eds), Diseases of Swine, pp. 488–523. Hoboken, NJ: John Wiley & Sons, 2019.

3. A. R. Fehr, S. Perlman. "Coronaviruses: An Overview of Their Replication and Pathogenesis". In: H. Maier, E. Bickerton, and P. Britton (eds), Coronaviruses. Methods in Molecular Biology, vol. 1282, pp. 1–23. New York, NY: Humana Press, 2015.

4. W.-J. Guan et al., "Clinical characteristics of 2019 novel coronavirus infection in China," MedRxiv, 2020.

5. World Health Organization. Statement on the second meeting of the International Health Regulations (2005) Emergency Committee regarding the outbreak of novel coronavirus (2019-nCoV), 2020, April 28. Available from: https://www.who.int/news-room/detail/30-01-2020-statement-on-the-second-meeting-of-the-international-health-regulations-(2005)-emergency-committee-regarding-the-outbreak-of-novel-coronavirus-(2019-ncov).

6. World Health Organization. Global leaders unite to ensure everyone everywhere can access new vaccines, tests and treatments for COVID-19, 2020, April 28. Available from: https://www.who.int/news-room/detail/24-04-2020-global-leaders-unite-to-ensure-everyone-everywhere-can-access-new-vaccines-tests-and-treatments-for-covid-19.

7. World Health Organization. Coronavirus disease (COVID-2019) situation reports, 2020, May 11. Available from: https://www.who.int/emergencies/diseases/novel-coronavirus-2019/situation-reports.

8. L. Pan et al., "Clinical characteristics of COVID-19 patients with digestive symptoms in Hubei, China: A descriptive, cross-sectional, multicenter study," *The American Journal of Gastroenterology,* vol. 115, p. 790, 2020.

9. World Health Organization. Coronavirus disease (COVID-19) advice for the public, 2020, April 29. Available from: https://www.who.int/emergencies/diseases/novel-coronavirus-2019/advice-for-public.

10. C. -C. Liu, R.-E. Chang, and W.-C. Chang, "Limitations of forehead infrared body temperature detection for fever screening for severe acute respiratory syndrome," *Infection Control and Hospital Epidemiology*, vol. 25, no. 12, pp. 1109–1111, 2004.

11. W. Zhao, Z. Zhong, X. Xie, Q. Yu, and J. Liu, "Relation between chest CT findings and clinical conditions of coronavirus disease (COVID-19) pneumonia: A multicenter study," *American Journal of Roentgenology*, vol. 214, no. 5, pp. 1072–1077, 2020.

12. M. Infantino et al., "Serological assays for SARS-CoV-2 infectious disease: Benefits, limitations and perspectives," *Israel Medical Association Journal,* vol. 22, no. 4, pp. 203–210, 2020.

13. A. Bernheim et al., "Chest CT findings in coronavirus disease-19 (COVID-19): Relationship to duration of infection," *Radiology*, vol. 20, p. 200463, 2020.

14. Y. -H. Jin et al., "A rapid advice guideline for the diagnosis and treatment of 2019 novel coronavirus (2019-nCoV) infected pneumonia (standard version)," *Military Medical Research,* vol. 7, no. 1, p. 4, 2020.

15. S. Tian et al., "Characteristics of COVID-19 infection in Beijing," *Journal of Infection*, vol. 80, pp. 401–406, 2020.

16. S. Wan et al., "Clinical features and treatment of COVID-19 patients in northeast Chongqing," *Journal of Medical Virology*, vol. 92, no. 7, pp. 797–806, 2020.

17. J. Chen et al., "Clinical progression of patients with COVID-19 in Shanghai, China," *Journal of Infection,* vol. 80, pp. e1–e6, 2020.

18. N. Chen et al., "Epidemiological and clinical characteristics of 99 cases of 2019 novel coronavirus pneumonia in Wuhan, China: A descriptive study," *The Lancet*, vol. 395, no. 10223, pp. 507–513, 2020.

19. F. Zhou et al., "Clinical course and risk factors for mortality of adult inpatients with COVID-19 in Wuhan, China: A retrospective cohort study," *The Lancet,* vol. 395, pp. 1054–1062, 2020.

20. E. Terpos et al., "Hematological findings and complications of COVID-19," *American Journal of Hematology,* vol. 95, pp. 834–847, 2020.

21. D. Wang et al., "Clinical characteristics of 138 hospitalized patients with 2019 novel coronavirus–infected pneumonia in Wuhan, China," *JAMA,* vol. 323, no. 11, pp. 1061–1069, 2020.

22. C. Xie et al., "Comparison of different samples for 2019 novel coronavirus detection by nucleic acid amplification tests," *International Journal of Infectious Diseases,* vol. 93, pp. 264–267, 2020.

23. Z. Cheng et al., "Clinical features and chest CT manifestations of coronavirus disease 2019 (COVID-19) in a single-center study in Shanghai, China," *American Journal of Roentgenology,* vol. 215, pp. 1–6, 2020.

24. C. Huang et al., "Clinical features of patients infected with 2019 novel coronavirus in Wuhan, China," *The Lancet,* vol. 395, no. 10223, pp. 497–506, 2020.

25. W. Wang et al., "Detection of SARS-CoV-2 in different types of clinical specimens," *JAMA,* vol. 323, no. 18, pp. 1843–1844, 2020.

26. A. Tahamtan and A. Ardebili, *"Real-Time RT-PCR in COVID-19 Detection: Issues Affecting the Results".* Abingdon: Taylor & Francis, 2020.

27. Y. Yang et al., "Laboratory diagnosis and monitoring the viral shedding of 2019-nCoV infections," MedRxiv, 2020.

28. V. M. Corman et al., "Detection of 2019 novel coronavirus (2019-nCoV) by real-time RT-PCR," *Eurosurveillance,* vol. 25, no. 3, p. 2000045, 2020.

29. Centers for Disease Control and Prevention. CDC 2019-Novel Coronavirus (2019-nCoV) Real-Time RT-PCR Diagnostic Panel, 2020.

30. FDA. Coronavirus (COVID-19) update: FDA authorizes first antigen test to help in the rapid detection of the virus that causes COVID-19 in patients, 2020, June 22. Available from: https://www.fda.gov/news-events/press-announcements/coronavirus-covid-19-update-fda-authorizes-first-antigen-test-help-rapid-detection-virus-causes.

31. G. C. Mak et al., "Evaluation of rapid antigen test for detection of SARS-CoV-2 virus," *Journal of Clinical Virology,* vol. 129, p. 104500, 2020.

32. News Medical. Developing a COVID-19 antigen test, 2020, May 15. Available from: https://www.news-medical.net/news/20200515/Developing-a-COVID-19-Antigen-Test.aspx.

33. S. Mahapatra and P. Chandra, "Clinically practiced and commercially viable nano-bio engineered analytical methods for COVID-19 diagnosis," *Biosensors and Bioelectronics,* vol. 165, p. 112361, 2020.

34. P. Mertens et al., "Development and potential usefulness of the COVID-19 Ag Respi-Strip diagnostic assay in a pandemic con text," *Frontiers in Medicine,* vol. 7, p. 225, 2020.

35. A. Scohy, A. Anantharajah, M. Bodéus, B. Kabamba-Mukadi, A. Verroken, and H. Rodriguez-Villalobos, "Low performance of rapid antigen detection test as frontline testing for COVID-19 diagnosis," *Journal of Clinical Virology,* vol. 129, p. 104455, 2020.

36. Z. Li et al., "Development and clinical application of a rapid IgM-IgG combined antibody test for SARS-CoV-2 infection diagnosis," *Journal of Medical Virology,* vol. 92, pp. 1518–1524, 2020.

37. K. K.-W. To et al., "Temporal profiles of viral load in posterior oropharyngeal saliva samples and serum antibody responses during infection by SARS-CoV-2: An observational cohort study," *The Lancet Infectious Diseases,* vol. 20, pp. 565–574, 2020.

38. J. Zhao, et al. "Antibody responses to SARS-CoV-2 in patients with novel coronavirus disease 2019." *Clinical Infectious Diseases,* vol. 71, no.16, pp. 2027–2034, 2020.

39. L. Guo et al., "Profiling early humoral response to diagnose novel coronavirus disease (COVID-19)," *Clinical Infectious Diseases,* vol. 71, no. 15, pp. 778–785, 2020.

40. Z. Ye, Y. Zhang, Y. Wang, Z. Huang, and B. Song, "Chest CT manifestations of new coronavirus disease 2019 (COVID-19): A pictorial review," *European Radiology,* vol. 30, pp. 4381–4389, 2020.

41. K. Li et al., "The clinical and chest CT features associated with severe and critical COVID-19 pneumonia," *Investigative Radiology,* vol. 55, pp. 327–331, 2020.

42 J. Wu et al., "Chest CT findings in patients with coronavirus disease 2019 and its relationship with clinical features," *Investigative Radiology,* vol. 55, no. 5, pp. 257–261, 2020.

43. M. Chung et al., "CT imaging features of 2019 novel coronavirus (2019-nCoV)," *Radiology,* vol. 295, no. 1, pp. 202–207, 2020.

44. J. -L. He et al., "Diagnostic performance between CT and initial real-time RT-PCR for clinically suspected 2019 coronavirus disease (COVID-19) patients outside Wuhan, China," *Respiratory Medicine,* vol. 21, p. 105980, 2020.

45. A. Wilder-Smith, C. J. Chiew, and V. J. Lee, "Can we contain the COVID-19 outbreak with the same measures as for SARS?" *The Lancet Infectious Diseases,* vol. 20, pp. e102–e107, 2020.

46. N. Petrosillo, G. Viceconte, O. Ergonul, G. Ippolito, and E. Petersen, "COVID-19, SARS and MERS: Are they closely related?," *Clinical Microbiology and Infection,* vol. 26, pp. 729–734, 2020.

47 M. Tay, Y. Low, X. Zhao, A. Cook, and V. Lee, "Comparison of infrared thermal detection systems for mass fever screening in a tropical healthcare setting," *Public Health,* vol. 129, no. 11, pp. 1471–1478, 2015.

48. J. Aw, "The non-contact handheld cutaneous infra-red thermometer for fever screening during the COVID-19 global emergency," *The Journal of Hospital Infection,* vol. 104, p. 451, 2020.

49. L. Gold, E. Balal, T. Horak, R. L. Cheu, T. Mehmetoglu, and O. Gurbuz, "Health screening strategies for international air travelers during an epidemic or pandemic," *Journal of Air Transport Management,* vol. 75, pp. 27–38, 2019.

50. Food and Drug Administration. Thermal imaging systems (infrared thermographic systems/thermal imaging cameras), 2020, June 26. Available from: https://www.fda.gov/medical-devices/general-hospital-devices-and-supplies/thermal-imaging-systems-infrared-thermographic-systems-thermal-imaging-cameras.

51. techcrunch. Chinese startup Rokid pitches COVID-19 detection glasses in US, 2020, April 16. Available from: https://techcrunch.com/2020/04/16/chinese-startup-rokid-pitches-covid-19-detection-glasses-in-u-s/?guccounter=1&guce_referrer=aHR0cH M6Ly9jc2UuZ29vZ2xlLmNvbS9jc2U_cT10aGVybWFsK2ltYWdpbmcrc3lzdGVt cyZzYT1TZWFyY2gmaWU9VVRGLTgmY3g9cGFydG5lci1UUHB1YiUyRDY2Mzg-yNDc3Nzk0MzM2OTTAlM0EzODczMzg0OTTkx&guce_referrer_sig=AQAAAJQqc 0ni3gq0PLU4OJISavwkMKZKQa3I0iwAzOnOHhyQ0bz-2RFTHnDufbqoJxr969_ V4eQ6Tk5tGhhS86qNuJnc9-x3EHjigfzBSSYmFr4J0DOr0rJO_A_it3QgessBCFi-zYvptC4FL5k2jNu7n3ZV9jbsxdl4rvp0AwUFE9gij.

52. Lab Manager. Thermal sensor screens for skin temperatures to monitor COVID-19, 2020, April 16. Available from: https://www.labmanager.com/product-news/thermal-sensor-screens-for-skin-temperatures-22355.

53. M. B. Jaffe and J. Orr, "Continuous monitoring of respiratory flow and Co_2," *IEEE Engineering in Medicine and Biology Magazine,* vol. 29, no. 2, pp. 44–52, 2010.

54. T. A. Howe, K. Jaalam, R. Ahmad, C. K. Sheng, and N. H. N. Ab Rahman, "The use of end-tidal capnography to monitor non-intubated patients presenting with acute exacerbation of asthma in the emergency department," *The Journal of Emergency Medicine,* vol. 41, no. 6, pp. 581–589, 2011.

55. R. J. Mieloszyk et al., "Automated quantitative analysis of capnogram shape for COPD–normal and COPD–CHF classification," *IEEE Transactions on Biomedical Engineering,* vol. 61, no. 12, pp. 2882–2890, 2014.

56. A. Abid, R. J. Mieloszyk, G. C. Verghese, B. S. Krauss, and T. Heldt, "Model-based estimation of respiratory parameters from capnography, with application to diagnosing obstructive lung disease," *IEEE Transactions on Biomedical Engineering,* vol. 64, no. 12, pp. 2957–2967, 2017.

57. C. Rhoades and F. Thomas, "Capnography: Beyond the numbers," *Air Medical Journal,* vol. 21, no. 2, pp. 43–48, 2002.

58. B. Jonson, "Volumetric capnography for noninvasive monitoring of acute respiratory distress syndrome," *American Journal of Respiratory and Critical Care Medicine,* vol. 198, no. 3, pp. 396–398, 2018.

59. A. J. Boyle, R. Mac Sweeney, and D. F. McAuley, "Pharmacological treatments in ARDS: A state-of-the-art update," *BMC Medicine,* vol. 11, no. 1, p. 166, 2013.

60 M. A. Matthay et al., "Acute respiratory distress syndrome," *Nature Reviews Disease Primers,* vol. 5, no. 1, pp. 1–22, 2019.

61 L. Gattinoni, S. Coppola, M. Cressoni, M. Busana, S. Rossi, and D. Chiumello, "Covid-19 does not lead to a "typical" acute respiratory distress syndrome," *American Journal of Respiratory and Critical Care Medicine,* vol. 201, no. 10, pp. 1299–1300, 2020.

62. O. P. Singh, T. A. Howe, and M. Malarvili, "Real-time human respiration carbon dioxide measurement device for cardiorespiratory assessment," *Journal of Breath Research,* vol. 12, no. 2, p. 026003, 2018.

63. O. P. Singh, R. Palaniappan, and M. Malarvili, "Automatic quantitative analysis of human respired carbon dioxide waveform for asthma and non-asthma classification using support vector machine," *IEEE Access,* vol. 6, pp. 55245–55256, 2018.

64. O. P. Singh and M. Malarvili, "Review of infrared carbon-dioxide sensors and cap-nogram features for developing asthma-monitoring device," *Journal of Clinical & Diagnostic Research,* vol. 12, no. 10, pp. 1–6, 2018.

65. M. Fezari, R. Rasras, and I. M. El Emary, "Ambulatory health monitoring system using wireless sensors node," *Procedia Computer Science,* vol. 65, pp. 86–94, 2015.

66. Robotshop. Arduino mega 2560 datasheet, 2020, June 14. Available from: https://www.robotshop.com/media/files/pdf/arduinomega2560datasheet.pdf.

67. TechTerms. TFT definition, 2020, June 14. Available from: https://techterms.com/definition/tft.

68. Adafruit Industries. Adafruit 2.4″ color TFT touchscreen breakout, 2015, June 15. Available from: https://learn.adafruit.com/adafruit-2-4-color-tft-touchscreen-breakout.

69. Maxim Integrated. DS3231 extremely accurate I²C-integrated RTC/TCXO/crystal, 2020, June 15. Available from: https://www.maximintegrated.com/en/products/analog/real-time-clocks/DS3231.html#.

7 Predictive Analysis of Breast Cancer Using Infrared Images with Machine Learning Algorithms

Aayesha Hakim and R. N. Awale
Veermata Jijabai Technological Institute

CONTENTS

7.1 INTRODUCTION

Breast cancer has been baffling our society for years because of the inability to find a complete cure. According to the World Health Organization (WHO), one of the major reasons of death among women globally is breast cancer. It impacts 2.1 million women per year worldwide [1]. The latter the disease is diagnosed, the lesser is the probability of survival of the patient. Approximately 60% deaths are due to a delay in diagnosis. For better survival of patients and lesser use of the treatments, many imaging systems are continually being developed to diagnose this disease as early as possible. Although the gold standard for imaging breasts is mammography, its performance is poor in younger women with dense breasts. The tumour needs to be of a certain size to be detected in mammogram [2]. There is also a possibility of rupture of tumour due to breast compression, releasing cancer cells into the bloodstream. Mammography fails to detect micro-tumours [2] and can detect tumours of the size of a cherry. Hence, it is not able to detect cancer at an early stage. Patients who repeatedly undergo mammography, for the evaluation of suspected lesions, are exposed to harmful X-ray radiations [3]. Being a structural imaging modality, ultrasound (USG) [4] is used to find the size, shape, texture and density of a breast lump. Its diagnostic performance is poor in fatty breasts due to the poor penetration of sound waves, and images have a poor spatial resolution. Magnetic resonance imaging (MRI) [4] machine uses a large magnet and radio waves to create images of detailed cross sections of the breast. However, the rate of false positives (FPs) is very high. MRI is often used in conjunction with mammography. Figure 7.1 shows the breast screening done using various modalities.

E.Y.K. Ng et al. [5] suggested that human skin temperature pattern is symmetric bilaterally. Neo-angiogenesis [6] is the formation of new blood vessels that develop to feed cancerous tumours. It is more intense in the early stages, and proliferating tissues generate more infrared radiation. This leads to high vascularity and production of heat, which gets transferred to the skin surface and is detected by thermal imaging. Cancer increases the breast temperature by 1°C–2°C, which leads to asymmetrical patterns in the thermogram [7]. The thermal camera [8] is analogous to a thermometer for breast health assessment that gives the temperature distribution over the skin surface in the form of an image. The primary advantage of thermography is its efficiency in detecting nonpalpable breast cancer early for women between 30 and

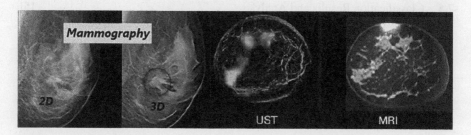

FIGURE 7.1 Methods for screening breasts, viz. mammography, USG, MRI.

50 years with dense breasts. It is a private, painless, contactless process and does not expose the patient to any radiation hazard. This opens the potential for thermography to be used as a safe risk marker for a routine examination of breasts.

7.1.1 RELATED WORKS

Colour analysis helps in the clinical interpretation of breast thermograms [9]. Healthy breasts indicate low heat level and appear purple on a thermograph. Warmer areas emit more heat than cooler ones. Spots appearing red, orange, or yellow in a breast thermograph indicate the presence of abnormality as shown in Figure 7.2. Food and Drug Administration (FDA) in 1982 [10] approved the usage of thermal imaging with mammography for the screening of breasts. The reappraisal of the use of thermal imaging to indicate inflammatory disorders in medicine was presented in Ref. [11].

Head et al. [13] used the infrared index as a metric to quantify breast abnormalities for infrared images of 220 patients who were screened with both first- and second-generation IR technology. They analysed the possibility of a link between family history and hormone therapy with the results obtained, and the study showed no correlation. Studies in Ref. [14,15] used Canny edge detector and Hough transform [16] to identify breast contours. Considering breast shape to be elliptical, edge detection gave false results in case of sagging breasts, i.e. flat lower part. Hough transform is a slow operation and takes more than 96% of processing time. Studies conducted in Refs. [17–19] used an SVM classifier by feeding it the extracted statistical and texture features from thermograms to classify them into malignant and benign. The accuracy obtained was 88.1% in Ref. [17]. The obtained sensitivity and specificity in Ref. [19] were 90% and 94.3% as compared to 85.71% and 90.48% from Ref. [18]. A similar work [20] reported the use of a free LibSVM classifier. Kuruganti and Qi [21] used the extracted features from thermograms to measure the asymmetry between left and right breasts using k-means clustering and k-nearest neighbour (k-NN) methods. However, the dataset used was too small to confirm their results.

A case study [22] was conducted on two patients – one who had undergone mammography and the other who had cancer (proven by biopsy), respectively. Segmented

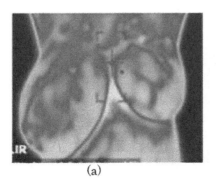 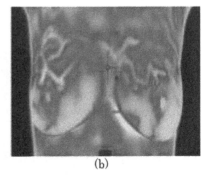

(a) (b)

FIGURE 7.2 Thermograms of (a) benign case and (b) malignant case [12].

tumour regions extracted from mammogram and thermogram were compared for the diagnosis of breast cancer. SVM with radial basis function (RBF) kernel was used in Ref. [23,24] to classify breast thermograms as healthy, benign and malignant, and hotspots were categorized quadrant-wise in breasts. The accuracy reported was 90%. In Ref. [25], SVM and ANN were used to segregate the thermal images into three classes – normal, benign and malignant. Their results were promising as many studies reported in the literature classified thermograms into two classes. Work conducted in Ref. [26] used principal component analysis (PCA) to reduce the data dimensions and fed the transformed feature set to an SVM. This classifier reported a sensitivity of 83.3%, which indicated that rotational thermography can be potentially used for screening breast cancer. Studies in Refs. [27,28] used Gabor filter to extract the texture features from the left and right breasts. SVM was used to classify images, and they obtained an accuracy of 84.5% and 92.06%, respectively. A. Sh et al. [29] used different training–testing data partitions, with an SVM classifier and their results proved that the 80%–20% data partition gave the best accuracy of 99.51%.

Decision-making for breast cancer diagnosis scientifically by using patterns within large medical databases has the ability to provide new knowledge as they include information about patients and their medical circumstances. This study explores the use of thermal imaging to predict breast pathologies coupled with machine learning algorithms. The rest of this chapter is organized as follows: Sections 7.2 and 7.3 describe the methods and materials used to conduct the investigation and various machine learning algorithms used, respectively. Section 7.4 describes the performance evaluation metrics. Classification results and discussions are presented in Section 7.5. Section 7.6 concludes this chapter with some discussion on scope for future work.

7.2 METHODS AND MATERIALS

There is a scarcity of radiologists in comparison with the cancer cases in India. Reports are read differently by different radiologists at a given time. Good use of computers to sort thermograms into benign and malignant can flag those most in need of the radiologist's attention. The objective of this study is to determine with a high degree of certainty if a tumour is malignant or benign using thermal images and machine learning algorithms. The flowchart of proposed methodology is presented in Figure 7.3.

7.2.1 REGION OF INTEREST (ROI) EXTRACTION

The RGB images acquired from Database of Mastology Research (DMR) are converted into greyscale images as they require less space to store and are simple to process. It is important to segment only the breast region, i.e. region of interest (ROI) from each image in order to eliminate errors from unnecessary warm areas other than breast tissues like background, arm-pits, head, neck portion and area underneath the breast [30]. Only the breast region contains information regarding any abnormality leading to breast cancer. Canny edge detector is used with a threshold to detect the significant edges and neglect the weak edges of the breast. A unique breast

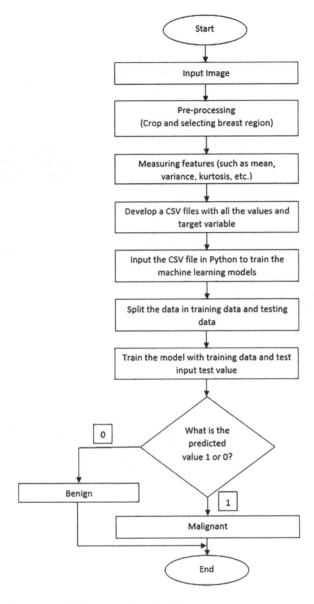

FIGURE 7.3 Flowchart of the proposed methodology.

mask is generated to extract the breast region [31] for each thermogram from the edge image as shown in Figure 7.4. We fixed the size of all the segmented breast images as 256 × 256 for the ease of computation as the original input format is very large. The individual breast masks are then multiplied with their corresponding greyscale images obtained after the removal of the irrelevant regions. Thus, we obtain only the breast region from each breast thermogram.

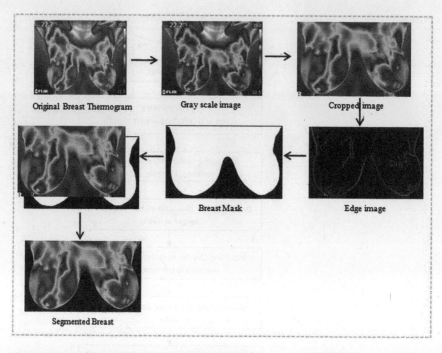

FIGURE 7.4   Segmentation of breast region (ROI).

7.2.2 DATABASE

We used a set of 287 images from the DMR [12], which were captured by FLIR SC-620 IR camera with a resolution of 640×480 pixels. This database contains anonymous, static, frontal breast images of varying size and shape of women aged 29 to 85 years. The images used were collected as a part of experiments conducted at the Hospital of the Federal University of Pernambuco, Brazil [32]. The demographic data of subjects are mentioned in Table 7.1.

7.2.3 FEATURE EXTRACTION

Thermal imaging is a functional imaging system, and hence, statistical features [33–35] are extracted from the segmented breast region. High-dimensional images

TABLE 7.1
Demographic Data of Subjects in DMR [12]

Age Range	No. of Malignant Cases (Pathology)	No. of Benign Cases (Healthy)
29–50	26	134
51–70	18	79
71–85	03	27
Total	47	240

are huge data to process. To minimize the data and computational cost, we have extracted the features from segmented thermograms and used them for classification. Before building the machine learning models, we scale the features in Python using StandardScaler() to ensure that no independent variable is dominating other variables in the model. This is a part of data pre-processing before applying machine learning techniques to it. The standardized data implicitly weighs all the variables in the model **equally**. A.csv (Comma Separated Values) file is prepared with all the feature values and the target variable, i.e. malignant (1) or benign (0).

7.2.3.1 First-Order Statistics

Histograms of the images are computed, and the mean (m) and variance (μ_v) will be calculated from them.

- **Mean:** Mean is defined as the average colour in the image and is calculated as,

$$\text{Mean} = m_1 = \sum_{X=0}^{N_g-1} XP(X)$$

- **Variance:** The variance measures the deviation of grey levels from the mean value. Standard deviation is the square root of variance.

$$\text{Variance} = \mu_v = \sum_{X=0}^{N_g-1} (X-m)^2 P(X)$$

7.2.3.2 Second-Order Statistics

- **Skewness:** Skewness is a measure of asymmetry of a pixel distribution around the mean value.

$$\text{Skewness} = \mu_s = \frac{1}{\sigma^3} \sum_{X=0}^{N_g-1} (X-m)^3 P(X)$$

- **Kurtosis:** Kurtosis is the fourth moment and characterizes the peakedness of the distribution in comparison with a normal distribution. Kurtosis is calculated as,

$$\text{Kurtosis} = \mu_k = \frac{1}{\sigma^4} \sum_{X=0}^{N_g-1} (X-m)^4 P(X)$$

7.2.3.3 Texture Features

Texture features explain the way the intensity varies within a given image and follows a pattern [36]. The texture information is obtained in a pixel domain using grey-level co-occurrence matrix (GLCM) [37]. We averaged every value obtained from the four

GLC matrices corresponding to four directions ($\theta = 0°$, $45°$, $90°$, and $135°$) keeping $d = 1$ pixel.

- **Contrast:** Contrast is a measure of grey-level variations between a pair of pixels.

$$\text{Contrast} = f_1 = \sum_{i=0}^{N_g-1}\sum_{j=0}^{N_g-1}(i-j)^2 P(i \cdot j)$$

- **Correlation:** Correlation presents the linear dependency of grey-level values in GLCM.

$$\text{Correlation} = f_2 = \sum_{i=0}^{N_g-1}\sum_{j=0}^{N_g-1}\frac{(i \times j) \times P(i \cdot j) - \{\mu_x \times \mu_y\}}{\sigma_x \times \sigma_y}$$

- **Energy:** Energy measures the local uniformities of grey levels. Images with similar pixels have large energy values.

$$\text{Energy} = f_3 = \sum_{i=0}^{N_g-1}\sum_{j=0}^{N_g-1}P(i \cdot j)^2$$

- **Homogeneity:** Homogeneity gives the distribution of elements with respect to the diagonal of GLCM.

$$\text{Homogeneity} = f_4 = \sum_{i=0}^{N_g-1}\sum_{j=0}^{N_g-1}\frac{P(i \cdot j)}{1+(i-j)^2}$$

- **Entropy:** Entropy is a measurement of randomness present in an image. It represents the degree of disorder present in an image.

$$\text{Entropy} = f_5 = \sum_{i=0}^{N_g-1}\sum_{j=0}^{N_g-1}P(i \cdot j)\log(P(i \cdot j))$$

7.3 CLASSIFICATION USING MACHINE LEARNING MODELS

Classification and predictions are a form of the powerful strategies, which are used to categorize datasets, especially in a medical field, so that the analysis can be utilized in prognosis to make faster decisions. The target of this study is to match and perceive a correct model to predict the prevalence of breast carcinoma that supports various patients' medical records. The extracted features from thermograms are fed into the classification algorithms to analyse the breast thermal images. Each image has nine biostatistical features. So, we have a set of 9×287 features for the classification purpose. We apply PCA, reduce the number of features of the dataset and then fit machine learning algorithms like support vector machine (SVM), k-NN, Naïve Bayes and logistic regression on the reduced dataset to classify thermograms as malignant or benign. Their results are compared for breast pathology classification. The training data learns the association between features and the outcome, while the test data assesses the classifier's generalization ability. The algorithms iteratively make predictions on training data and map input to predefined discrete binary classes, namely benign and malignant, after being trained on a dataset of breast thermal images. For the proposed work, we used open source tool Python® and Jupyter notebook [38] for machine learning and data visualization. We divided the dataset into training (80%) and test sets (20%) for all algorithms.

7.3.1 PRINCIPAL COMPONENT ANALYSIS (PCA)

PCA is one of the robust techniques for dimension reduction, feature selection and data visualization. Training the model with a large database and many features is not easy. Some of the features in the datasets are more selective and decisive than other features that are redundant. Two highly correlated variables bias the output and should not be both used in model. They make samples of both classes look the same and should be removed. PCA is a dimension reduction tool, not a classifier. It does not discard any variables. PCA reduces the overwhelming number of dimensions by constructing principal components (PCs) based on the maximum variance along the axis [39]. PCs are calculated only from the knowledge of features and not classes. Hence, PCA is an unsupervised method. PCA reduces the dimensionality, provides an efficient visualization of our nine-dimensional feature set and speeds up the machine learning algorithms. We use it as a pre-processing step for supervised learning tasks and fit a classifier on the PCA-transformed data.

Boxplot in Figure 7.5 shows the distribution by classification of 'kurtosis' as it has the highest correlation with the diagnosis dummy variable. There is a clear difference in distributions of 'benign' and 'malignant' classes. For comparison, the same boxplot is constructed with 'entropy', which is minimally correlated with diagnosis.

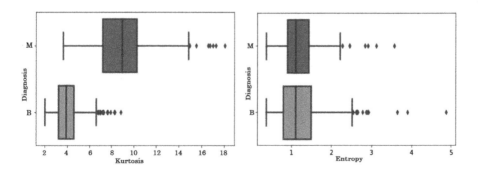

FIGURE 7.5   Boxplot distribution of 'kurtosis' and 'entropy' with respect to diagnosis.

The PCs are orthogonal to each other and are uncorrelated. Hence, observing the malignant and benign classes distinctly is possible. Covariance matrix identifies the correlations and patterns between different variables in the dataset. The eigenvectors comprise coefficients corresponding to each variable's relative weight. This gives the proportion of dataset's explained variance that lies along the axis of each PC. All the nine components capture 100% variance in data. Eigen vectors and eigen values are ordered in descending order where eigen vector with the highest eigen value is the most significant (i.e. the first PC). As seen from Figure 7.6 and Table 7.2, the first two PCs contribute to 86.46% of the total variance and can be chosen for classification. At least 80% of original dataset's information should be retained [39]. Eigen vectors with the lowest eigen value can be dropped as they represent very less information.

A scree plot helps to identify the variation contributed by each PC. Here, the first two PCs are sufficient to describe the essence of the data. After eigen value 1, the scree plot turns steep and bends quickly to flatten out as seen in Figure 7.7. We pick

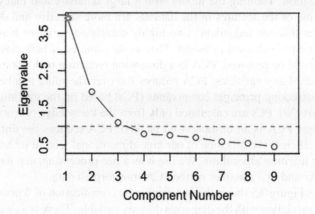

FIGURE 7.6   Scree plot for principal components and eigen values.

TABLE 7.2
Eigen Values and Explained Variance for Various Principal Components

Principal Component #	Eigen Value	Proportion of Variance
1	4.0	0.6295
2	2.0	0.2351
3	1.1	0.0721
4	0.8	0.0238
5	0.8	0.0231
6	0.7	0.010
7	0.6	0.0045
8	0.5	0.0010
9	0.4	0.0009

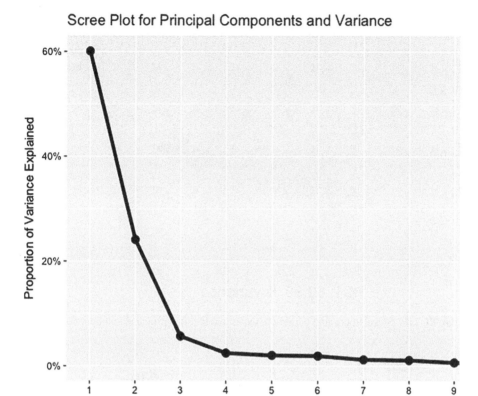

FIGURE 7.7   Scree plot for principal components and explained variance.

the number of PCs as 2 and use them to construct dimensions for new feature space as they give away most variation in data.

PCA extracts the axes on which data shows the highest variability. Using the first two components, we get the score plot to assess the class clusters, data distribution and outlier samples. With a score plot, we can therefore visualize nine dimensions using a 2D-plot as shown in Figure 7.8.

A loading plot helps to identify how strongly each variable influences a PC. It maps the coefficients of each feature for the first component vs the second component. Range of loadings can be from −1 to 1. When two vectors are close, with a small angle between them, the two variables they represent are positively correlated. If vectors form 90°, they are not likely to be correlated. When vectors diverge away from each other with 180° angle, they are negatively correlated. The loading plot for our data shown in Figure 7.9 shows that variables – variance, kurtosis, contrast, correlation and homogeneity, have large positive loadings (values close to 1) on the first PC. Skewness and energy have large negative loadings (values close to −1) on the second component. Mean and skewness are negatively correlated. The pairs of 'variance and entropy' and 'contrast and correlation' are positively correlated with a very small angle between them. Homogeneity is uncorrelated with mean and skewness.

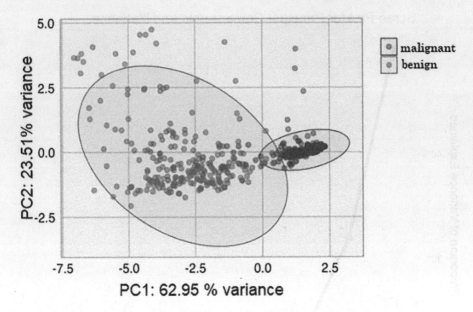

FIGURE 7.8   Score plot to find outliers based on the first two principal components.

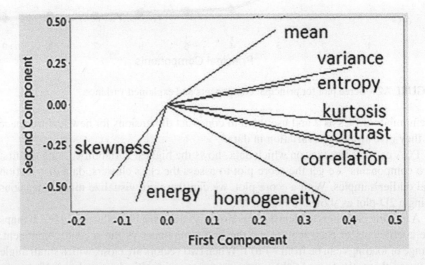

FIGURE 7.9   Loading plot for first two principal components.

7.3.2 SUPPORT VECTOR MACHINE (SVM) WITH GRID SEARCH

SVM is a supervised learning method that implements the classification by constructing a hyperplane as a decision boundary [40,41]. Based on this boundary, the class of new samples can be predicted. The hyperplane [42] maximizes the distance (margin) between the classes that lowers the generalization error of the classifier. SVM works well for unbalanced dataset [43] as shown in Figure 7.10.

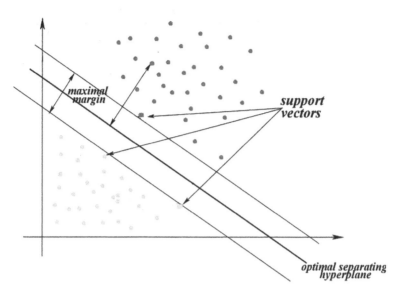

FIGURE 7.10   Illustration of support vector machine.

We build a two-class SVM classifier over a feature vector $\varphi(X, Y')$ obtained from the features extracted from thermograms and the class of the samples.

$$y = \arg \max_{y'} \bar{W}^T \varphi\left(\bar{X}, Y'\right)$$

The slack variable ξi states the misclassifications. Without it, the model is overfit. The objective function is written as,

$$\arg \min_{W,\xi,b} \left\{ \frac{1}{2}\|W\|^2 + C \sum_{i=1}^{n} \xi_i \right\}$$

We used a linear kernel (mapping function) to make a decision boundary. C is a regularization parameter that imposes a penalty to the model for misclassifications. Higher the penalty, it is less possible to misclassify a sample. But a very large C may also lead to overfitting and may bring a risk of losing the generalization properties of the classifier. We performed a 10-fold cross-validation grid search and computed the model accuracy for different values of C. We select the optimal C as 1.16 as it maximizes the training model's 10-fold cross-validation accuracy (87.70%) and gives the kappa value (k) as 0.754 as compared to the kappa value of 0.68 for $C = 0.4$ (84% accuracy). The plot of accuracy vs cost for training model is shown in Figure 7.11.

$$k = \left(a_{\text{obs}} - a_{\text{chance}}\right)/\left(1 - a_{\text{chance}}\right)$$

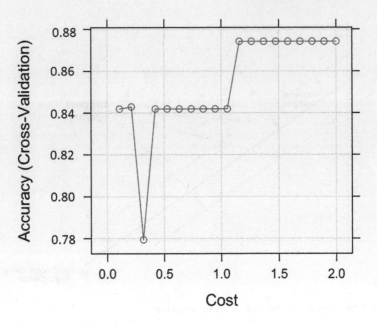

FIGURE 7.11   Plot of SVM training model's accuracy vs cost.

7.3.3 LOGISTIC REGRESSION

Logistic regression is a supervised learning technique where labelled data is provided for the classifier to make decisions rationally for the new data. There exists an issue of learning only majority class in unbalanced data. Probabilities [44] give a better understanding of a sample's membership to a particular class. A linear decision boundary is rigid and fails to work if the distribution varies. Thus, decision boundary is converted to probabilities, and a threshold of 0.5 is set in a sigmoid curve as shown in Figure 7.12. The probability greater than 0.5 is considered as 'malignant (1)', and that lower than 0.5 is considered as 'benign (0)'. We performed 50 iterations of training with a tolerance of 0.0001 after which convergence is reached and training stops. For prediction, the probabilities are transformed to binary values. Here, $p(X)$ is a sigmoid function of X that is bounded between 0 and 1.

$$p(X) = \frac{e^{(\beta_0 + \beta_1 X)}}{1 + e^{(\beta_0 + \beta_1 X)}}$$

$$\text{or } \log\left(\frac{p(X)}{1 - p(X)}\right) = \beta_0 + \beta_1 X$$

The algorithm finds decision parameters β_0 and β_1 to get the best possible classification (hyperplane) for the inputs from both the classes and maximize the likelihood objective function:

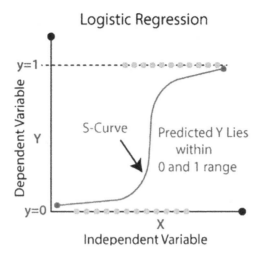

FIGURE 7.12   Sigmoid curve decision boundary for logistic regression.

$$L(\beta_0, \beta_1) = \prod_{i=1}^{n} (p(x_i))^{y_i} (1 - p(x_i))^{(1-y_i)}$$

When x_i belongs to class 0, $y_i = 0$
   When x_i belongs to class 1, $y_i = 1$

7.3.4   K-NEAREST NEIGHBOURS (K-NN)

k-NN is a classification technique that assigns an unknown sample to the class to which the majority of its 'k'-nearest neighbours belong to [45]. k-NN can deal with complex and arbitrary decision boundaries. 'Closeness' is defined in terms of Euclidean distance. It is a lazy learning algorithm as all the computation is deferred until classification. k is always a positive integer. Euclidean distance metric is computed between test datapoint and all labelled datapoints. The similarities are closer and differences are distant. For every attribute of two datapoints, we calculate the distance between them. The top 'k'-labelled datapoints are selected from the ascending order of distance, and their class labels are looked at. The test data is assigned to the class label that the majority of 'k'-labelled datapoints belong to, as seen in Figure 7.13.

$$d(x_i, x_i) = \sqrt{(x_{i1} - x_{i1})^2 + (x_{i2} - x_{i2})^2 + \cdots + (x_{ip} - x_{ip})^2}$$

$$A\ Ri = \left\{ X \in R_p : d(x, x_i) \leq d(x, xm), \forall i \neq m \right\}$$

k is empirically chosen by varying values between 1 and 15 by fitting the model for various tuning parameters 'k' on the trained dataset. The accuracy plot shown in Figure 7.14 is studied. Predictions for test values were checked, and the misclassification

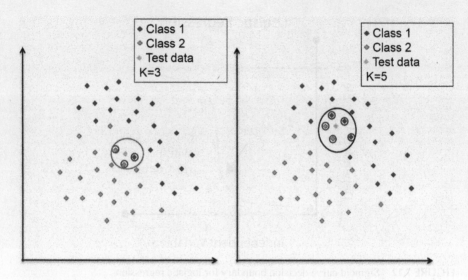

FIGURE 7.13   Illustration of testing the k-NN classifier with $k = 3$ and $k = 5$.

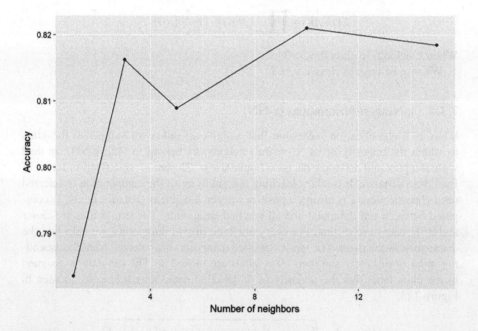

FIGURE 7.14   Plot of accuracy vs the number of neighbours.

error for each 'k' value between 1 and 15 was calculated. The number of misclassified samples was least for 'k' = 10, and the best performance with an accuracy of 82.2% was achieved. Using PCA as a precursor to k-NN, we apply k-NN to the matrix corresponding to extracted PCs. We therefore remove irrelevant features that are not important for classification and keep only the ones having discriminating capacity as they contribute in measuring distance and spoil k-NN results. k-NN on its own is slow at classification time. PCA-k-NN saves the computation complexity and time.

7.3.5 NAIVE BAYES CLASSIFIER

Naïve Bayes classifier [46] is a probabilistic classifier based on Bayes' theorem. It assumes the **conditional**ly independence of features. This classifier works well on a small training set. In Bayes theorem, posterior probability of samples in 'c' class is calculated as shown below:

$$p(c|x_1,\ldots,x_n) = \frac{p(x_1,\ldots,x_n|c)p(c)}{p(x_1,\ldots,x_n)}$$

$$p(c|x_1,\ldots,x_n) = \frac{p(c)\prod_{i=1}^{n}p(x_i|c)}{p(x_1,\ldots,x_n)}$$

For binary class dataset, a sample is classified in the class which has higher probability [44]. To decide between two class labels L_1 and L_2, the ratio of the posterior probabilities for each label is computed.

$$\frac{P(L_1|\text{features})}{P(L_2|\text{features})} = \frac{P(\text{features}|L_1)}{P(\text{features}|L_2)}\frac{P(L_1)}{P(L_2)}$$

During training, the prior probability of each class is computed by counting the occurrence of its samples in the training dataset, i.e. $P(L_1)$ and $P(L_2)$. The Gaussian Naïve Bayes classifier fits the model by finding the mean and standard deviation of the samples within each label, assuming each sample follows Gaussian distribution. The ellipses in Figure 7.15 represent the Gaussian generative model for each label, with larger probability towards the centre of the ellipses. With this generative model in place for each class, we computed the likelihood probability P (features|L_1) and P (features|L_2) for each datapoint. The posterior ratio is finally computed, thus determining which class label is the most probable for a given point.

7.4 PERFORMANCE EVALUATION PARAMETERS

Comparison of the performance of a classifier with other classifiers to categorize thermograms is done using evaluation metrics [47]. Accuracy, sensitivity, specificity, PPV and NPV are calculated using the data in the confusion matrix, which uses the values of true-positive (TP), true-negative (TN), FP, and false-negative (FN). TP: malignancy is correctly classified; TN: benign cases correctly identified as benign; FP: benign cases incorrectly classified as malignant; FN: malignancy incorrectly identified as benign.

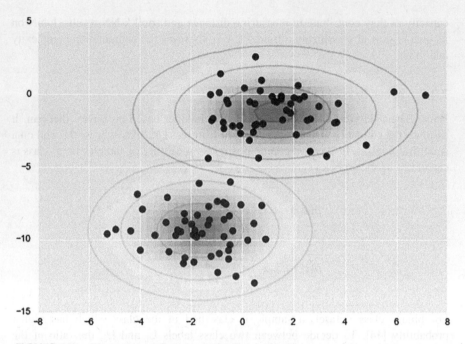

FIGURE 7.15   Illustration of Gaussian Naive Bayes classifier.

$$\text{Precision} = \frac{\text{True positive}}{\text{True positive} + \text{false positive}}$$

$$\text{Recall} = \frac{\text{True positive}}{\text{True positive} + \text{false positive}}$$

$$F_1 = 2 \times \frac{\text{Precision} * \text{recall}}{\text{Precision} + \text{recall}}$$

The confusion matrix is a 2×2 matrix that helps to evaluate the performance of supervised classification algorithms. The diagonal grid values in the matrix show the number of cases that are correctly classified, and the off-diagonal values show the falsely classified cases. Accuracy gives the percentage of correct classification. However, it is not enough alone to reveal how well the model predicted 'benign' and 'malignant' cases independently. Sensitivity is the ability of a classifier to detect malignancy, while specificity is the ability to detect benign cases. PPV reflects the malignant possibility of positive result, while NPV reflects the benign possibility of negative result as shown in Figure 7.16. F_1-score is the harmonic mean of precision and recall. It ranges between 0 (worst value) and 1 (best value).

The area under the receiver-operating characteristics (ROC) curve is an evaluation metric to compare the efficacy of the models. For a good model, the curve rises sharply covering a large area and then reaches the top-right corner at high sensitivity

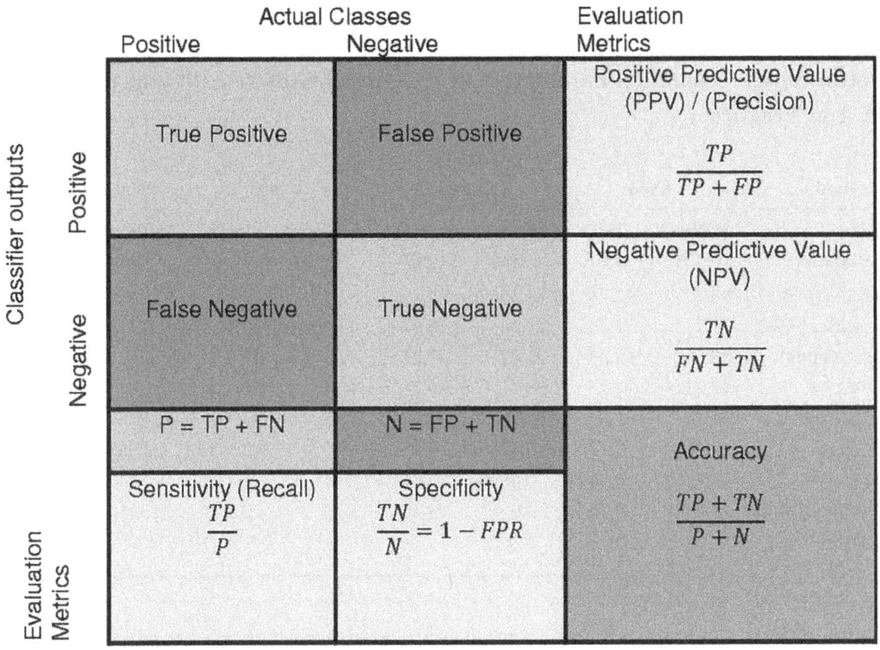

		Actual Classes		Evaluation Metrics
		Positive	Negative	
Classifier outputs	Positive	True Positive	False Positive	Positive Predictive Value (PPV) / (Precision) $$\frac{TP}{TP + FP}$$
	Negative	False Negative	True Negative	Negative Predictive Value (NPV) $$\frac{TN}{FN + TN}$$
Evaluation Metrics		P = TP + FN	N = FP + TN	Accuracy
		Sensitivity (Recall) $\frac{TP}{P}$	Specificity $\frac{TN}{N} = 1 - FPR$	$\frac{TP + TN}{P + N}$

FIGURE 7.16   Confusion matrix and evaluation metrics of a classifier.

and low FPR. The results are considered more precise, when the area under ROC curve (AUC) is large. A high AUC value represents low false-positive rate (FPR) and low false-negative rate (FNR). A study [48] suggests that an AUC of 0.5 reflects almost no discrimination, 0.7–0.8 is acceptable, 0.8–0.9 is excellent, and more than 0.9 is outstanding for medical diagnosis [48].

7.5   CLASSIFICATION RESULTS AND ANALYSIS

In our study as seen in Table 7.3, the highest classification accuracy of 92.74% is achieved by using two PCs with the SVM classifier followed by an accuracy of 92.5% with PCA-logistic regression. Without PCA as a precursor, SVM alone shows the best accuracy of 88.22%. SVM and logistic regression classifiers show remarkably high accuracy in separating malignant cases from benign when the axes are rotated with PCA. Due to imbalance in data, the ratio between positive and negative support vectors becomes more imbalanced; therefore, samples at the boundary of hyperplane are more likely to be classified as negative and favour predictions of the majority class (benign) on the test samples. Naive Bayes classifier is good at predicting negative class tuples, but it is worst among all classifiers at predicting positive class tuples. Naive Bayes is fastest in terms of computation speed. After performing feature reduction using PCA, features become uncorrelated. This satisfies the basic independence assumption of Naïve Bayes, and as a result, NB performs much better and has robust results with PCA as a precursor step. Loss in accuracy of Naïve Bayes

TABLE 7.3

Performance Evaluation Parameters of Various Classifiers with and without PCA as Precursor

Method	SVM		Logistic Regression		k-NN		Naive Bayes	
Parameters	Without PCA	With PCA	Without PCA	With PCA	Without PCA	With PCA	Without PCA	With PCA
Accuracy (%)	88.22	92.74	84.35	92.5	82.2	84.11	81.69	88
Sensitivity (%)	77.77	77.77	44.44	66.67	55.55	66.67	22.22	44.44
Specificity (%)	89.58	95.83	91.67	97.91	87.5	87.5	87.5	95.83
PPV (%)	58.33	77.77	50	85.71	45.45	50	33.33	66.66
NPV (%)	95.55	95.83	89.79	94	91.3	93.33	86.27	90.19
F1 score	0.6667	0.7777	0.4706	0.75	0.5	0.5714	0.2667	0.5333
AUC	0.8368	0.8699	0.6806	0.8229	0.7153	0.7708	0.5486	0.7013
Processing time (seconds)	1.5	0.75	0.9	0.48	1.35	0.37	0.36	0.17

as compared to other classifiers is a result of the assumption of class conditional independence. Classifiers that are based on comparing the pairwise distances of samples like k-NN are hardly affected when the axes are rotated using PCA because the pairwise Euclidean distances remain exactly the same.

Comparison of our results with past results of the literature is shown in Table 7.4. These results point out that accuracy and specificity of this work are appreciably higher as compared to other works, with a smaller number of misclassifications. Tiago B. Borchartt et al.'s [20] results have high sensitivity but very low specificity value is observed due to an unbalanced and small sample set used.

The accuracy and F_1 score of all machine learning algorithms are plotted with and without PCA as a precursor as shown in Figures 7.17 and 7.18. PCA-SVM model has less information available compared to the model using the original data. The information is condensed in fewer variables (i.e. only two PCs). The average accuracy of the two best models is about the same (PCA-SVM model has a slightly better average accuracy than PCA-LR), i.e. 92.74% vs 92.50%. As seen from Figure 7.19, standard deviation is in favour of the PCA-SVM model, i.e. 3.2% vs 2.3%.

Higher diagonal values of the confusion matrix imply many correct predictions of both the classes. Both PCA-SVM and PCA-LR misclassify only four samples. However, in case of cancer diagnosis, FP is more acceptable than FN. Thus, PCA-SVM performs better in terms of this aspect than PCA-LR. The confusion matrices for various machine learning algorithms with and without PCA as a precursor are shown in Table 7.5.

ROC analysis shown in Figure 7.20 was employed to evaluate the performance of all algorithms with PCA as a precursor. PCA-SVM has the best AUROC value of 0.8699, followed by PCA-LR with an AUROC of 0.8229.

TABLE 7.4

Comparison between the Results Obtained in Our Study and in Past Literature Involving Statistical Features

Author(s)/Method	Sensitivity (%)	Specificity (%)	Accuracy (%)	Area under ROC Curve (AUC)	No of Images Used
Our results (PCA-SVM)	77.77	95.83	92.74	0.8699	240 Benign, 47 malignant
Tang et al. [49]/Localized temperature increase (LTI)	93.6	55.7	-	-	70 Benign, 47 malignant
Schaefer et al. [35]/Fuzzy classifier & statistical features	79.86	79.49	79.53	-	29 Malignant, 117 benign
Tiago B. Borchartt et al. [20]/ Statistical temperature features & SVM classifier	95.83	25	85.71	0.604	24 Unhealthy, 4 healthyf
Acharya et al. [17]/Texture features & SVM classifier	85.71	90.48	88.1	-	50
Gaber et al. [28]/Gabor coefficients statistical & SVM RBF	-	-	92.06	-	29 Healthy, 34 malignant
Vijaya Madhavi and Christy Bobby [50]/BEMD & URLBP	92	73	86	0.82	43 Normal, 24 abnormal
Sathish et al. [24]/Statistical texture features/SVM RBF	87.5	92.5	90	-	40 Normal, 40 abnormal
A. A Khan & A.S. Arora [27]/ Gabor filter & SVM	90.52	82.47	84.5	-	35 Normal, 35 abnormal

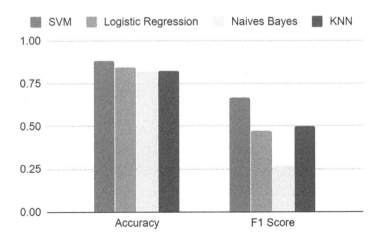

FIGURE 7.17 Accuracy and F_1 score plots for various machine learning algorithms.

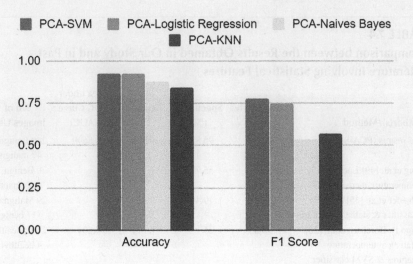

FIGURE 7.18   Accuracy and F_1 score plots for machine learning algorithms with PCA as precursor.

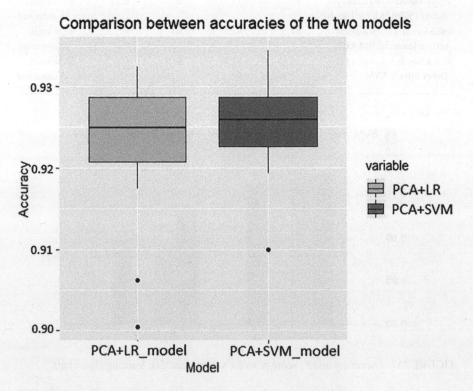

FIGURE 7.19   Box plot comparison of accuracies of two best models.

TABLE 7.5

Confusion Matrices of Various Classifiers without and with PCA as a Precursor

	Predicted Class				**Predicted Class**		
Algorithm Used	**Actual Class**	**Malignant**	**Benign**	**Algorithm Used**	**Actual Class**	**Malignant**	**Benign**
Naive	Malignant	2	7	PCA-Naive	Malignant	4	5
Bayes	Benign	4	44	Bayes	Benign	2	46
Logistic	Malignant	3	6	PCA-logistic	Malignant	6	3
regression	Benign	3	45	regression	Benign	1	47
k-NN	Malignant	5	4	PCA-k-NN	Malignant	6	3
	Benign	6	42		Benign	6	42
SVM	Malignant	7	2	PCA-SVM	Malignant	7	2
	Benign	5	43		Benign	2	46

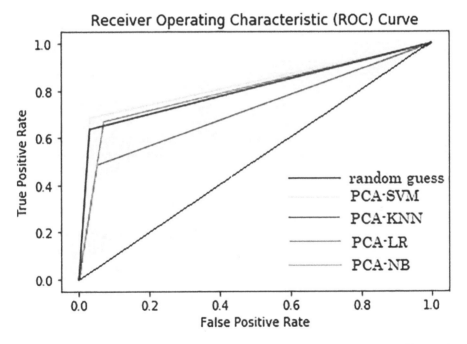

FIGURE 7.20 Receiver-operating characteristic (ROC) graph for various classifiers with PCA as a precursor.

7.6 CONCLUSION AND FUTURE WORK

Breast cancer is an important health problem globally. Thermography is a low-cost potential solution for the early prognosis of breast cancer. Despite the presence of mammography as a gold imaging standard in the diagnosis of breast cancer, there is a need for promoting additional research in thermography to increase the sensitivity of prognosis in young women with dense breasts. The manual assessment of disease is time-consuming, requiring minute inspection, and varies with the perception and the level of expertise of the radiologists. This study presents an approach to classify breast thermograms based on biostatistical features using various machine learning algorithms and increase the reliability of this technique for diagnostic purpose. In general, we observed from the literature that only accuracy is used as a metric for cancer prediction in most studies. But accuracy alone does not give a perfect prediction. It is further identified that AUC value is proven significant for the correct prediction of breast cancer. In this work, SVM classifier with and without PCA as a precursor significantly outperforms other decision-making models, viz., Naïve Bayes, k-NN and logistic regression on the DMR database. PCA-SVM-based model interprets thermograms as benign or malignant with an accuracy of 92.74% for training with 80% data and testing with 20% data, which is the most superior as compared to other classifiers, followed by accuracy of 92.5% with PCA-logistic regression. Both PCA-SVM and PCA-LR misclassify only four samples. However, in case of cancer diagnosis, FP is more acceptable than FR. Thus, PCA-SVM performs better in terms of this aspect than PCA-LR. PCA-SVM has the best AUROC value of 0.8699. By reducing the dimension of data and time complexity, PCA helps to improve the results for all classifiers in interpreting thermograms as benign or malignant. Results obtained point out at the viability of using simple statistical measures extracted from breast thermograms as a first approach to aid the diagnosis of breast disease.

Improving the accuracy of detection of malignant cases and thus the sensitivity using thermography is an ongoing process. Incorrect outcomes observed in our analysis could be because of small sample of patients with malignant breast. More texture features can be extracted as they help in identifying the abnormality regions in a thermogram. With improved accuracy on large database, thermography can be used in adjunct to USG or mammography in a clinical practice. In future work, nonlinear or kernel-based SVM can also be tested on the database. An appropriate combination of feature extraction technique, segmentation method and classification algorithm on a larger database can minimize the FPR and FNR of thermal imaging technique. Future works including (i) consideration of fractal and Hurst features for analysing thermal asymmetry between the breasts and (ii) use of deep learning approaches like convolutional neural networks (CNN) to classify the thermal breast images are already in progress. However, a database that is inclusive of more malignant cases will be helpful for training of deep learning classifiers.

REFERENCES

1. Breast Cancer India: Pink Indian Statistics. Available at: http://www.breastcancerindia.net/statistics/stat_global.html.
2. Keyserlingk, J., Ahlgren, P., Yu, E. and Belliveau, N. 1998, Infrared imaging of the breast: Initial reappraisal using high-resolution digital technology in 100 successive cases of stage I and II breast cancer. *The Breast Journal*, **4**, 245–251.
3. Khandpur, R. S. 1994, *Handbook of Biomedical Instrumentation*, 2nd edn. New York: Tata McGraw-Hill Education, pp. 670–684.

4. Prasad, P. and Houserkova, D. 2007, The role of various modalities in breast imaging. *Biomedical Papers of the Medical Faculty of the University Palacky, Olomouc, Czechoslovakia,* **151**(2), 209–218.

5. Ng, E. Y. K. and Sudharsan, N. M. 2001, Numerical computation as a tool to aid thermographic interpretation. *Journal of Medical Engineering & Technology,* **25**(2), 53–60.

6. Ng, E., Ung, L., Ng, F., and Sim, L. S. G. 2001, Statistical analysis of healthy and malignant breast thermography. *Journal of Medical Engineering & Technology,* **25**, 253–263.

7. Fok, S. C., Ng, E. Y. K., and Tai, K. 2002, Early detection and visualization of breast tumor with thermogram and neural network. *Journal of Mechanics in Medicine and Biology,* **2**, 185–195.

8. Bronzino, J. D. 2006, *Medical Devices and Systems,* 3rd edn. Boca Raton, FL: CRC Press, pp. 25-1-25-17.

9. Shahari, S. and Wakankar, A. 2015, Color analysis of thermograms for breast cancer detection. *Proceedings of 2015 International Conference on Industrial Instrumentation and Control (ICIC),* Pune, India.

10. U.S. Food and Drug Administration. Breast cancer screening: Thermography is not an alternative to mammography: FDA safety communication. Available at: https://www.fda.gov/NewsEvents/Newsroom/ PressAnnouncements/ucm257633.htm. Date posted: 6/2/2011. Accessed March 3, 2020.

11. Jones, B. F. 1998, A reappraisal of the use of infrared thermal image analysis in medicine. *IEEE Transactions on Medical Imaging,* **17**(6), 1019–1027.

12. [dataset] Visual Lab. A methodology for breast disease computer-aided diagnosis using dynamic thermography. Available online: http://visual.ic.uff.br/dmi (accessed on 15 April, 2020).

13. Head, J. F., Wang, F., Lipari, C. A., and Elliott, R. L. 2000, The important role of infrared imaging in breast cancer. *IEEE Engineering in Medicine and Biology Magazine,* **19**(3), 52–57.

14. Qi, H., Snyder, W., Head, J., and Elliott, R. 2000, Detecting breast cancer from infrared images by asymmetry analysis. *Proceedings of 22nd Annual International Conference of the IEEE, Engineering in Medicine and Biology Society,* vol. 2, Chicago, IL, USA, pp. 1227–1228.

15. Scales, N., Herry, C., and Frize, M. 2004, Automated image segmentation for breast analysis using infrared images. *Proceedings of Annual International Conference of the IEEE Engineering in Medicine and Biology Society,* Chicago, IL, USA, vol. 3, pp. 1737–1740.

16. Borchartt, T. B., Conci, A., Rita, C. F. L., Resmini, R., and Sanchez, A. 2013, Breast thermography from an image processing viewpoint: A survey. *Signal Processing,* **93**, 2785–2803.

17. Acharya, U. R., Ng, E. Y. K., Tan, J., and Sree, S. V. 2012, Thermography based breast cancer detection using texture features and support vector machine. *Journal of Medical Systems,* **36**, 1503–1510.

18. Acharya, U. R, Ng, E., Tan, J. H., and Sree, S.V. 2010, Thermography based breast cancer detection using texture features and support vector machine. *Journal of Medical Systems,* **36**, 1503–1510.

19. Madhu, H., Kakileti, S. T., Venkataramani, K., and Jabbireddy, S. 2016, Extraction of medically interpretable features for classification of malignancy in breast thermography. *Proceedings of 38th Annual International Conference of the IEEE Engineering in Medicine and Biology Society (EMBC),* Chicago, IL, USA, pp. 1062–1065.

20. Borchartt, T., Resmini, R., Conci, A., Martins, A., Silva, A., Diniz, E., Paiva, A., and Lima, R. 2011, Thermal feature analysis to aid on breast disease diagnosis. *Proceeddings of the 21st Brazilian Congress of Mechanical Engineering,* Natal, RN, Brazil.

21. Kuruganti, P. T. and Qi, H. 2002, Asymmetry analysis in breast cancer detection using thermal infrared images. *Proceedings of Engineering in Medicine and Biology, 24th Annual Conference and the Annual Fall Meeting of the Biomedical Engineering Society EMBS/BMES Conference*, Chicago, IL, USA, vol. 2, pp. 1155–1156.

22. Angeline Kirubha, S. P., Anburajan, M., Venkataraman, B., and Menaka. M. 2018, A case study on asymmetrical texture features comparison of breast thermogram and mammogram in normal and breast cancer subject. *Biocatalysis and Agricultural Biotechnology*, **15**, 390–401.

23. Gogoi, U., Majumdar, G., Mrinal, B., and Ghosh, A. 2019, Evaluating the efficiency of infrared breast thermography for early breast cancer risk prediction in asymptomatic population. *Infrared Physics & Technology*, **99**, 201–211.

24. Sathish, D., Kamath, S., Prasad, K., Kadavigere, R., and Martis, R. 2017. Asymmetry analysis of breast thermograms using automated segmentation and texture features. *Signal, Image and Video Processing*, **11**(4), 745–752.

25. Silva, L. F., Saade, D. C. M., Sequeiros, G. O., Silva, A. C., Paiva, A. C., Bravo, R. S. and Conci, A. 2014, A new database for breast research with infrared image. *Journal of Medical Imaging and Health Informatics*, **4**(1), 92–100.

26. Francis, S. V., Sasikala, M., Bharathi, G. B., and Jaipurkar, S. D. 2014, Breast cancer detection in rotational thermography images using texture features. *Infrared Physics & Technology*, **67**, 490–496.

27. Khan, A. A. and Arora, A. 2018, Breast cancer detection through gabor filter based texture features using thermograms images. *Proceedings of 2018 First International Conference on Secure Cyber Computing and Communication (ICSCCC)*, Jalandhar, India, pp. 412–417.

28. Gaber, T., Ismail, G., Anter, A., Soliman, M., Ali, M., Semary, N., Hassanien, A. E., and Snasel, V. 2013, Thermogram breast cancer prediction approach based on neutrosophic sets and Fuzzy C-means algorithm [Online]. Available: https://doi.org/10.5281/zenodo.34913.

29. Sh, A., Shahraki, H., Rowhanimanesh, A.R., and Eslami, S. 2016, Feature selection using a genetic algorithm for breast cancer diagnosis: An experiment on three different datasets. *Iranian Journal of Basic Medical Sciences*, **19**, 476–482.

30. Qi, H. and Head, J. F. 2001, Asymmetry analysis using automatic segmentation and classification for breast cancer detection in thermograms. *Proceedings of 23rd Annual International Conference of the IEEE Engineering in Medicine and Biology Society*, Chicago, IL, USA, vol. 3, pp. 2866–2869.

31. Gonçalves, C. B., Leles, A. C. Q., Oliveira, L. E., Guimaraes, G., Cunha, J. R., and Fernandes, H. 2019, Machine learning and infrared thermography for breast cancer detection. *Proceedings* 2019, **27**, 45.

32. Hakim, A. and Awale, R. N. 2020, Thermal imaging: An emerging modality for breast cancer detection: A comprehensive review. *Journal of Medical Systems*, **44**(136), 1–18.

33. Hakim, A. and Awale, R. N. 2020, Detection of breast pathology using thermography as a screening tool. *Proceedings of 15th Quantitative InfraRed Thermography Conference*, Portugal.

34. Image central moments computed using formulae given in: https://itl.nist.gov/div898/handbook/eda/section3/eda35b.htm (accessed on 24 April, 2020).

35. Schaefer, G., Závišek, M., and Nakashima, T. 2009, Thermography based breast cancer analysis using statistical features and Fuzzy classification. *Pattern Recognition*, **42**(6), 1133–1137.

36. Haralick, R. M., Shanmugam, K., and Dinstein, I. H. 1973. Textural features for image classification. *IEEE Transactions on Systems, Man and Cybernetics*, **3**(6), 610–621.

37. Bajaj, V., Pawar, M., Meena, V. K., et al. 2019, Computer-aided diagnosis of breast cancer using bi-dimensional empirical mode decomposition. *Neural Comput & Applic* **31**, 3307–3315.

38. scikit-learn.org, sklearn.svm.LinearSVC. http://scikit-learn.org/stable/modules/generated/sklearn.svm.LinearSVC.html (accessed on 24 May, 2020).

39. Lashkari, A., Pak, F., and Firouzmand, M. 2016, Full intelligent cancer classification of thermal breast images to assist physician in clinical diagnostic applications. *Journal of Medical Signals and Sensors*, **6**(1), 12–24.

40. Nunes, A. P., Silva, A. C., and Paiva, A. C. 2010, Detection of masses in mammographic images using geometry, Simpson's Diversity Index and SVM. *International Journal of Signal and Imaging Systems Engineering*, **3**(1), 40–51.

41. Deniz, E., Şengür, A., Kadiroğlu, Z., et al. 2018, Transfer learning based histopathologic image classification for breast cancer detection. *Health Information Science and Systems* **6**, 18.

42. Ireaneus Anna Rejani, Y. and Thamarai Selvi, S. 2009, Early detection of breast cancer using SVM classifier technique. *International Journal of Computational Science and Engineering*, **1**, 127–130.

43. Wakankar, A. T. and Suresh, G. R. 2016, Automatic diagnosis of breast cancer using thermographic color analysis and SVM classifier. *Proceedings of Advances in Intelligent Systems and Computing Intelligent Systems Technologies and Applications*, Jaipur, India, pp. 21–32.

44. Pearl, J. 1998, Probabilistic Reasoning in Intelligent Systems. Burlington, MA: Morgan Kaufmann Publisher.

45. Mejia, T., Perez, M. G., Andaluz, V., and Conci, A. 2015, Automatic segmentation and analysis of thermograms using texture descriptors for breast cancer detection. *Proceedings of Computer Aided System Engineering (APCASE) Asia-Pacific Conference*, Quito, Ecuador, pp. 24–29.

46. Zhang, H. 2004, The optimality of naive Bayes. *Proceedings of 17th International Florida Artificial Intelligence Research Society Conference (FLAIRS 2004)*, Miami Beach, FL: AAAI Press, pp. 562–567.

47. Koprowski, R. 2014, Quantitative assessment of the impact of biomedical image acquisition on the results obtained from image analysis and processing. *Biomedical Engineering*, **13**(1), 1–21.

48. Mandrekar, J. N. 2010, Receiver operating characteristic curve in diagnostic test assessment. *Journal of Thoracic Oncology*, **5**(9), 1315–1316.

49. Tang, X., Ding, H., Yuan, Y., and Wang, Q. 2008, Morphological measurement of localized temperature increase amplitudes in breast infrared thermograms and its clinical application. *Biomedical Signal Processing and Control*, **3**(4), 312–318.

50. Madhavi, V. and Bobby, C. 2017, Assessment of dynamic infrared images for breast cancer screening using BEMD and URLBP. *International Journal of Pure and Applied Mathematics*, **114**(10), 261–269.

8 Histopathological Image Analysis and Classification Techniques for Breast Cancer Detection

Gaurav Makwana, Ram Narayan Yadav, and Lalita Gupta
Maulana Azad National Institute of Technology

CONTENTS

8.1   INTRODUCTION

Malignant growth is portrayed by the uncontrolled development of the cells. It can be characterized by different kinds of influenced cells or organs. There are roughly 100 types of cancer, which indicate an intricate and diverse illness, as malignant development can happen in any part of the body. Breast malignancy is the most common disease in the women worldwide and the second most common reason for cancer-related death in women. It is because many times breast disease doesn't cause any agony or uneasiness until it has spread to close-by tissue. According to the World Health Organization report published in February 2020, lung cancer is the most diagnosed cancer (11.6% of all cases) followed by breast cancer (11.6%). Breast cancer growth is an uncontrolled development of epithelial cells in the breast. The statistical data show that 20,88,849 new cases were detected, and 6,26,679 female deaths occurred in the year 2018 [1]. North America, Sweden, and Japan have nearly 80% survival rate, while middle-income nations have around 60% and low-income nations have less than 40% survival rate in case of breast cancer [2]. A report on cancer in India by GLOBOCON in 2018 shows breast cancer is the leading cause of death in Indian women. Approximately 87,090 women died due to breast cancer, and 1,62,468 new cases were detected [3]. Report of E&Y Ficci 2015 informed that breast cancer contributes 19% of all cancer in women, and ~2000 new women are diagnosed with cancer every day and around 1200 cases are detected in the last stage. Breast conservation rates are low even for stage I & II in most Indian women and reflect the absence of access to current radiotherapy [4]. According to statistical data collected by the Union health ministry of India [5], it is estimated that 18 lakh new cases of breast cancer are possible by the end of the year 2020 and one out of 28 women has the possibility of breast cancer development. To improve breast cancer diagnosis and cure, early detection of the disease is very important to have more effective treatment which reduces the risk of death. Breast cancer can be classified according to its place of onset. Invasive ductal carcinoma in situ (IDC), infiltrating ductal carcinoma, lobular carcinoma in situ, and invasive lobular carcinoma are some types of breast cancer. IDC is the most commonly found breast cancer in the woman of all age; however, woman older than 55 years have a higher danger of being influenced [6], so early diagnosis may improve the survival rate.

Histopathological image analysis is the most common type of method for breast cancer diagnosis. Digital pathology has replaced conventional pathology with the development of computers and the advancement of its computational power. Digital pathology uses a camera-equipped microscope for the digital screening of tissue samples. This development of digital pathology increases the data-sharing capacity and storage power but still has some disadvantages like manual interpretation. Pathologists apply visual inspection on the histological image and make the decision based on their knowledge and experience for the presence of cancer. This method is time-consuming, costly, error-prone, and inconsistent. Thus, an efficient and reliable computer-based automatic diagnosis method can be implemented to solve this problem. Recent advancements in image processing and deep learning techniques can assist the pathologist in decision-making for breast cancer diagnosis. Computer-aided diagnosis (CAD) methods are more productive and consistent for the diagnosis of the patient's current situation accurately. It can identify the suspected area and classify the image with better diagnosis accuracy in less time. For automatic diagnosis, it is important to extract special diagnostic feature which can provide a specific clinical measure of the disease. This computer-aided diagnosis can reduce inter-observer discrepancy. Also it increases the consistency and better diagnosis efficiency. Therefore, a lot of research is going on to develop and improve the existing image classification algorithms.

8.2 METHODOLOGY

Computer-assisted diagnosis (CAD) alludes to the strategies in medication where computer algorithms and programs help doctors in the understanding of medical images. Computer-aided design is getting significant research option in therapeutic imaging and has been the inspiration for advancement in various domains including image processing, AI, and clinical frameworks integration. Histological images contain large quantities of cells and various structures that are appropriated and encompassed by a variety of tissues. Thus, the manual understanding of histological images is tedious and requires a lot of experience. Studies show that the elucidation and scoring of strained specimens that utilized the microscope don't just work exceptional yet additionally it is a profoundly visual and conceptual procedure [7,8]. Regardless of standardizing the scoring process, the inter-observer and intra-observer reproducibility by pathologists are not perfect [9]. Additionally, recognizing certain histological structures, such as tumors, cell membranes, or nuclei, is one of the per-requirements to malignancy evaluating in histological images. Quantitative and subjective information concerning the presence, degree, measure, and state of these structures is a significant pointer for treatment expectation and anticipation. The advancement in image analysis method and availability of high-quality digital cameras and whole slide scanners has permitted the development of many powerful computer-assisted approaches for histopathological image investigation [10]. The techniques can't just offer effective and robust quantification of cell expression, yet also objectivity and reproducibility. The utilization of computer-assisted diagnosis of histological images permits both automation and consistent interpretation of both malignant and benign tumors [11,12]. Utilization of CAD for histopathology can be extensively categorized into three significant groups, first is detection and segmentation of nuclei to analyze nuclear morphology for cancer metastases detection. Second is classification, i.e., grading and identification of lesion type in histopathology image, and the last is the disease diagnosis. Computerized tools may yield significant data for diagnosis, relying upon the evaluation of exquisite sub-visual changes in the patterns of significant structures in histopathological images that are imperceptible to or hard to see for human vision. This can conceivably prompt to early analysis of illness.

The prerequisite to structure computer-aided design frameworks for histopathology that can be utilized in clinical practice requires the improvement of vigorous and high throughput methods that can work at the whole slide images (WSI) level. Histological stains show a large variability in their color and intensity. Such varieties can conceivably hamper the viability of quantitative image analysis.

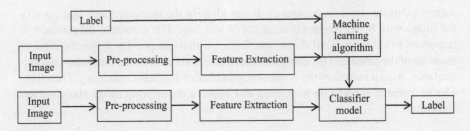

FIGURE 8.1   Block diagram representation of CAD system.

Furthermore, tissue arrangement and digitization normally create lots of artifacts that cause difficulties for automated investigation. All through in this chapter, we have concentrated on the improvement of CAD that can work at the WSI level by defeating these difficulties, thus empowering the frameworks to be used in a viable clinical setting.

Machine learning algorithms can be used to solve complex datasets. It can find different patterns and relationships between these datasets. This feature has made them popular to design computer-aided diagnostic systems. Figure 8.1 shows the functional block diagram of the CAD system. The stepwise process that makes this system for classification of breast cancer is explained in the following section.

8.2.1   IMAGE PRE-PROCESSING

Pre-processing is the main procedure in image analysis. It reduces the computational cost of the process. Before applying machine learning on histopathological images, some pre-processing methods are required like identifying the area of interest, filtering, enhance contrast, resize, and remove noise from the image. Image enhancement converts an image in a more meaningful form that is appropriate for a given application [13]. Although the contrast enhancement method increases the contrast of the image over a threshold [14] it has some drawbacks like it may cause some regions under enhanced or over enhanced, which may result in information lost [15]. Pixel modification may also be used for contrast enhancement [16,17]. This can be achieved by increasing the contrast ratio of the image. Another way for image enhancement is local histogram modification like histogram equalization, contrast stretching, and local histogram processing in the local area of an image [18,19]. This local enhancement method distorted the image quality to some extent because these transformations are not monotonic mapping. Tourassi et al. [20] utilized a template matching technique. A database is created by extracting ROI from the known image served as a template and mutual information between the test image and database decide the similarity to find the region of interest contained mass, although mass may be of different length, width, and density, etc., computer-aided diagnosis help to solve such problem. Morton et al. [21] and Brem et al. [22] proposed CAD for interpretation of low contrast mammograms with increased sensitivity of 7.62% and 21.2%, respectively. Image denoising is another challenge to remove noise and distortion from an image. Thakur et al. [23] shows different types of image noises and denoising dataset. They have also provided a comparative analysis of various CNN models for different types of Gaussian noise reduction techniques. In histopathological image analysis, background noise, blood vessels, and tissues cause false-positive detection. In this condition, median filtering can be a solution to minimize these problems under certain conditions; it preserves edges while removing noise.

8.2.2 IMAGE SEGMENTATION

Image segmentation extracts important information and attributes from the images. This method separates the suspicious regions from the background. The main objective of image segmentation is to identify suspicious areas to assist pathologists in diagnosis [24]. Image segmentation can be of two types, edge-based and region-based [25,26]. The edge-based segmentation method depends upon the relationship between neighboring pixels. Edge detection is performed by the convolution between the gradient operators like canny, Sobel, etc.; when this gradient value crosses the threshold limit, the edge is detected [27]. Edge detection algorithms are high-frequency phenomena [28] same as noise; therefore, it becomes very difficult to identify edge from noise or trivial geometry feature. However, for ultrasound, MRI images, it can be used [29].

Region-based segmentation predefines the boundaries in the selected region of interest, based on the dissimilarity like texture, intensity, color, etc. [30,31]. Structural, statistical, and spectral are some principal approaches for image analysis. The structure method depends upon the local properties and spatial organization of the image and therefore is not suitable for natural textures, whereas in the statistical method, the spatial distribution of pixel value in an image is analyzed indirectly. Local features of an image are calculated at each point in the image, deriving a set of local features. The local feature depends upon the number of pixels used in the calculation. Clustering is another method that classifies an object into a group of clusters. Fuzzy clustering can be used for the images which have been shaped with varying densities, so this method is appropriate for micro-calcification identification in digital mammogram [32]. Saha et al. [33] have implemented a fuzzy algorithm for segmenting dense regions from the fattest region in the mammogram. They have derived the feature of the segmented region from measured area and density. This feature linearly correlates between cranial candal (CC) and mediolateral oblique (MLO) view. This process organizes the group having similar characteristic like the intensity and texture of the object [34]. Some clustering methods such as k-means clustering [35,36], fuzzy c means clustering [37,38], and expected maximization [39] can be used to find the group of the same intensity. Malek et al. [40] used active contours for nuclei segmentation and fuzzy c-means algorithm for image classification. Their result shows 95% test accuracy. Dundar et al. [41] used clustering algorithms and the watershed-based segmentation algorithm to identify the region of individual cells based on their size, shape, and intensity-based feature. Santhos et al. [42] have segmented mammogram images using HAS, EMO, and MACSO multi-level thresholding techniques. The performance of these methods was analyzed using parameters like fitness, MSE, TIME, SSIM, and PSNR, etc. k-means clustering method classifies the image based on its features like texture and color. In the stage of clustering this method, first identify the cluster centroid, and then find the closest cluster to determine the distance between the sample and the cluster centroid [43]. After, all points are assigned; it updates the location of the centroid of the cluster. In this chapter, k-means clustering algorithm is applied on the microscopic images. This algorithm starts with the initialization of the k centroids which might be either randomly selected or selected from the data set. Clustering procedure will be as follows

$$\underset{c_i \in C}{\arg\min} \ \mathrm{Dis}(c_i, x) \tag{8.1}$$

Where c_i are the centroids, and $x = \{x_1, x_2, x_3, \ldots, x_n\}$ is the data point. Next step is to recomputed centroid by taking mean of all data point assigned to that centroid cluster. Let $S = \{S_1, S_2, S_3, \ldots, S_k\}$ is data point set, then centroid can be recomputed as

$$c_i = \frac{1}{|S_i|} \sum_{x_i \in S_i} x_i \tag{8.2}$$

This procedure will continue until no change in data point is achieved.

8.2.3 FEATURE EXTRACTION

Feature extraction is a very important aspect of the classification process. It converts data into a simpler form [44]. An image has a finite number of pixels, and this pixel relation can be represented using a set of features that contains the relevant information of an image. It is very difficult to use the full-dimensional training data directly to the machine learning algorithms; feature extraction can reduce this complexity. These extracted features can be used as an input to the classification system.

In digital image analysis feature can be described as a repetition of a certain pattern of the object like roughness, granularity, edge etc. of the surface. The texture feature becomes very important in object recognition. Texture analysis is divided into four class geometrical (structure), spectral, logical operation based, and statistical texture [45]. The structure method depends upon the local properties and spatial organization of the image. This method is not suitable for natural textures because there is no clear distinction between local properties (micro-texture) and spatial organization (macro-texture); both have variability. Spectral features like Fourier, wavelet, Gabor, and Haar transform, etc. can be used for image classification. They decompose the image into its constituent frequency and phase component. These features become very important for the classification of multiresolution and multichannel texture, whereas logical operation texture analysis can be used for image coding, spectral decomposition, cryptography, etc. The statistical feature analyzes the spatial distribution of pixel value in an image indirectly. Local features of an image are calculated at every point of the selected region. The local feature depends upon the number of pixels used in the calculation. First-order texture measure which is also known as the histogram-based approach depends on the individual pixel values, i.e., intensity distribution, and does not have neighboring pixel dependencies. In this chapter, various first order statistical measures have been evaluated as defined below.

The histogram of an image is defined as

$$P(i) = \frac{n_i}{\text{Total no. of pixel}} \tag{8.3}$$

where n_i is number of pixels of intensity 'i' in all observations.

8.2.3.1 Mean

Mean gives an average gray level to get an idea about intensity profile. It can be expressed as

$$\mu = \sum_{i=0}^{N-1} iP(i) \tag{8.4}$$

where 'N' is number of possible gray levels in an image.

8.2.3.2 Variance

Histogram width is the measure of the amount of grey level in an image. Variance measures the deviation of that gray level from the mean value. It can be given as

$$\sigma^2 = \sum_{i=0}^{N-1} (i-\mu)^2 P(i) \tag{8.5}$$

8.2.3.3 Kurtosis

Kurtosis provides the shape of the tail of the histogram to measure the histogram sharpness. It can be written as

$$m_4 = \sum_{i=0}^{N-1} (i-\mu)^4 P(i) \tag{8.6}$$

8.2.3.4 Entropy

Entropy of an image is a measure of the randomness of the image pixel. It can be expressed as

$$H = -\sum_{i=0}^{N-1} P(i) \log_2 [P(i)] \tag{8.7}$$

The higher-order statistics evaluate the pixel relationship of two or more pixel values. The gray-level co-occurrence matrix (GLCM) is a well-known method for texture analysis. It is similar to the local binary pattern analysis, which is created by calculating the existence of a pair of the pixel with a specific value and then extracting statistical measures from this matrix. In an image, GLCM tabulates the different combinations of gray-level that co-occurs in an image so it can give the statistics of variation in the intensity at the pixel of interest [46]. GLCM consists of the relative distance between the pixel pair and their relative orientation. Feature extraction using GLCM is a primitive type of pattern recognition. GLCM defines 14 higher order statistical features. These features provide image pixel information like coarseness, smoothness, and texture-related information. Contrast, correlation, energy, and homogeneity are some important features used in this chapter for image classification and are defined in the following section.

8.2.3.5 Contrast

Contrast of an image is the measure of the change in luminance in image regions, i.e., it determines the local level variation in an image. Image contrast can be expressed as

$$\text{Contrast} = \sum_{i,j=0}^{N-1} (i-j)^2 G(i,j) \tag{8.8}$$

where $G(i,j)$ is the relative frequency of two pixels i and j within a given neighborhood. Equal value of the image pixel i and j represents the diagonal pixel. Weight of the diagonal pixel is always zero which denotes no contrast and an exponential increase in the weight difference in $(i\text{-}j)$ pixel represents an increase in contrast.

8.2.3.6 Homogeneity

Homogeneity increases with less contrast so if weight difference $(i\text{-}j)$ decreases, the calculated texture measure will be large for the window with little contrast as shown in Equation (8.9).

$$\text{Homogeneity} = \sum_{i,j=0}^{N-1} \frac{G(i,j)}{1+(i-j)^2} \tag{8.9}$$

8.2.3.7 Correlation

It is a linear dependency measure between the neighboring gray-level pixels in an image. High correlation indicates high predictability of pixel relationships. It can be expressed as

$$\text{Correlation} = \sum_{i,j=0}^{N-1} \left[\frac{(i \times j)G(i,j) - \mu_x \mu_y}{\sqrt{\sigma_x^2 \sigma_y^2}} \right] \tag{8.10}$$

Where μ_x and μ_y are the mean and σ_x and σ_y are the standard deviation. x and y are the row and column coordinate of the matrix G.

8.2.3.8 Energy

Energy predicts uniformity in the gray level distribution in an image. It will be maximum for constant or periodic distributed gray level as shown below

$$\text{Energy} = \sum_{i,j=0}^{N-1} [G(i,j)]^2 \tag{8.11}$$

8.2.4 CLASSIFICATION METHOD

Machine learning techniques are pattern-based mechanism that can be used to model complex problems. Markov chain analysis, genetic algorithms, Support Vector Machine (SVM), artificial neural network (ANN), etc. are different types of supervised machine learning algorithms; they learn labeled dataset and give an answer key to assess its accuracy on training data. ANN is an advanced area of research for image classification, but it requires large computation, the proneness of over-fitting, and large data for training input class [47]. SVM is another type of supervised learning technique that classifies labeled data into two distinct classes [48,49]. It requires extracting the features of the labeled data before classifying images into its classes, whereas in unsupervised learning like in k-means, c-means, etc., the system is trained with the unknown sample. George et al. [50] uses nuclei segmentation of cytological images, after segmentation they have used neural network and support vector machine model for image classification with classification accuracy ranging from 76% to 94%. Huang and Lee [51] used multi-wavelet, Gabor filter, and statistical features like contrast, energy, correlation, homogeneity, and energy for texture analysis of pathological prostate image and Bayesian, KNN, and SVM classification algorithms are used for image classification. LBP is widely used in face recognition [52,53]. Linder et al. [54] use the LBP texture descriptor for epithelium and stoma identification. Fuchs et al. [55] use the LBP and random forest texture descriptors for renal cell carcinoma identification. Qi et al. [56] combined LBP and GLCM parameters for breast tissue classification. They have used adaptive boosting and perceptron least-square classifiers for breast cancer classification. Doyal et al. [57,58] uses a statistical descriptor (1st and 2nd order) and wavelet for prostate tissue diagnosis. Sertel et al. [59] designed a CAD system to identify lymphoid malignancy also known as centroblasts (CB). They have tested their method on H&E-stained tissue sections in two steps, in the first step they have detected non-CB cells by identifying their size and shape, and in the second step CB cells have detected by utilizing texture distribution of non-CB cells. Nahid et al. [60] have proposed two algorithms for image feature extraction for BreakHis dataset with magnification factor 40×, 100×, 200× and 400×. Algorithm one used to extract Tamura features directly whereas in algorithm two contrast enhancement is required before Tamura feature extraction. These extracted features fed to the Restricted Boltzmann Machine (RBM) for image classification. Algorithm 1 gave classification accuracy of 82.2%, 74.7%, 69%, and 81.7%, while Algorithm 2 gave 88.7%, 85.3%, 88.6% and 88.4% for different magnified image of the database. Reis et al. [61] classify the breast histological images by first extracting the multi-scale basic image features and LBP feature and then classify these extracted features using random decision trees classifier with 84% detection accuracy. Santhos et al. [62] utilized the GLCM feature for breast cancer histopathological image analysis. These extracted feature feed into fuzzy min-max with k highest (Kh-FMM) classifier. They have achieved the highest classification accuracy on the image with a magnification factor 200× in the BreakHis dataset. Bajaj et al. [63] utilized the bidimensional empirical mode decomposition method (BEMD) for mammogram image classification. First, they extracted the GLCM feature from the 2-D intrinsic mode function, and then these extracted features feed into the SVM classifier. Their experimental result shows 95% classification accuracy.

Cosatto et al. [64] used a Gaussian filter of size same as the size of the nucleus within the area of interest but due to the complex structure of H&E-stained image, it becomes very difficult to trace the outline of the nuclei. They used active learning method to extract the feature like shape (area, smoothness, symmetry) and texture (variance) between normal and malformed outlines then SVM classifier train for breast cancer classification and test the image with <5% error in detection. Fatakdawala et al. [65] use expectation-maximization algorithm for breast microscopic image segmentation to initialize the geodesic active contour. The proposed EMaGACOR model provides 86% sensitivity without using a supervised classifier model. Veta et al. [66] model uses unsupervised learning method for nuclei segmentation of H&E-stained image of breast biopsy. They preprocessed the image using color deconvolution and morphological operation to remove the irrelevant structure, then they used fast redial symmetry transform for detection of nuclei location followed by the watershed segmentation algorithm. A radial symmetry marker provides 81.5% correct segmentation per ROI, whereas the regional minima marker gives 81% exactness. Lado et al. [67] proposed a generalized additive model (GAM) for the detection of micro-calcification cluster. They have involved thresholding techniques for breast border detection to avoid artifacts. A wavelet transform is used to enhance the high-frequency component and remove the low-frequency background followed by local gray-level thresholding. A discriminant analysis applied on the extracted properties like the size of the cluster, the moment of microcalcification, difference in gray level value of the clustered mammogram. This method provides 83.12% sensitivity and 1.46 FP/image.

Niwas et al. [68] first filter images using Gabor wavelets statistical descriptor for classification of breast carcinoma using an SVM classifier. Zhang et al. [69] combines local binary pattern, curvelet transform, and statistical texture descriptor for the normal, and carcinoma in situ breast tumor classification. They have used fusion of SVM and multi-layer perceptron for classification. Bayramoglu et al. [70] employed deep learning for histopathological images with 83% accuracy. Wang et al. [71] use cancerous histopathological images to test the SVM classifier. Wan et al. [72] suggested a statistical method for automatic detection of cancer cells. Kowal et al. [73] test dataset of 500 images on different algorithms for nuclei segmentation with 96% accuracy. Erfankhah et al. [74] used homogeneity and variance of local neighborhoods to extend the local binary pattern histogram with heterogeneity information of histological image and SVM classifies the IDC and BreakHis dataset with accuracy 85.2% and 87%, respectively. Spanhol et al. [75] trained the model using the extracted patches of the histopathological image of BreakHis dataset and compute the mean image of all the extracted patches then subtract the mean image from each input patch before feeding into convolution neural network architecture for classification. Morphological image features are also a very important tool for detecting vessel like pattern [76]. Histograms and wavelet features are also used by many researchers for medical imaging [77,78]. Orlov et al. [79] have proposed features like texture information, image statistical parameters, and transform domain coefficients. These features are used for image classification using support vector machines (SVM) and neural network classifiers. Deniz et al. [80] utilized fine-tuned AlexNet and VGG16 model for histopathological image feature extraction. These extracted feature vectors are then classified by the SVM classifier. They have found the best accuracy of 93.78% with a 200x magnification factor on the BreakHis dataset. Zhu et al. [81] utilize assembling multiple compact CNN for image classification. First, they have created a hybrid CNN model; then a squeeze excitation pruning (SEP) block is used to reduce data overfitting. Nahid et al. [82] use two models for IDC breast cancer image classification. In the first model, they have directly fed image to the CNN architecture and in model 2 they have utilized local descriptive features like LBP, Histogram information, contourlet transform before training CNN architecture. Reza and Ma [83] use the oversampling and undersampling technique before CNN classifier to overcome class imbalance during training. Balazsi et al. [84] uses the random forest classifier for IDC image classification. They have used simple linear iterative clustering for image segmentation and extract color and texture features to train and test the random forest classifier performance. Doyle et al. [85] uses spectral clustering with architectural and textural image features for histopathological images to grade breast cancer. Cahoon et al. [86] uses fuzzy models for image segmentation and the crisp k-nearest neighbor (KNN) algorithm for the classification of breast cancer. They have shown that only the intensity feature will give higher misclassification rates whereas the intensity feature in combination with standard deviations can improve the classification accuracy for particular regions within the image. There are some classification algorithms used in this chapter for breast histological image analysis described in the following section.

8.2.4.1 Distance-Based Classifier

A minimum distance classifier (MDC) is the supervised classification technique. MDC inherently separates two classes by correlating the features of the input test image with a closely related feature from the training set. It calculates the prototype vector for each class and calculates the Euclidian distance between the unknown prototype vectors, and then allocates the pixel to that class with the shortest Euclidian distance. The image classifier performs the role of a discriminant. The discriminant value will be higher for one class and lower for the other class [87]. Let $f(y_l,x)$ is the discriminant function, and x and y_l are the feature vector and class, respectively. For two-class classification, discrimination function can be classified as

$$f(x)>0 \qquad \text{for} \quad x\varepsilon y_1$$
$$f(x)<0 \qquad \text{for} \quad x\varepsilon y_2$$

$$(8.12)$$

Let $X \subseteq R^d$ is a subset of the n-dimensional real feature space and the feature vector $\vec{x} = [x(1)\ x(2)...x(n)] \in X$ represents a given class of image and $y = [1,2,...,L]$ are the class label. These patterns are represented by pair $(\vec{x},y)$. Any unlabeled object assigns a label during the classification process by learning about the set of objects of known class. Computation of mean of each class is given as

$$\bar{\mu}_l = \frac{1}{N_l} \sum_{n=1}^{N_l} \vec{x}_n \quad l = 1,2,3,...,L$$

$$(8.13)$$

Where 'l' is the class label. Classification of the object is given by the minimum Euclidian distance d_E given as

$$d_E\left(\vec{x}_n,\vec{\mu}_l\right)=\sqrt{\sum_{n=1}^{N_l}\left(\vec{x}_n-\vec{\mu}_l\right)^2} \qquad (8.14)$$

The MDC uses mean vector of the ROI and calculates the d_E between unknown image feature vector and mean vector of each class. The unknown image is assigned to the class with minimum Euclidian distance.

8.2.4.2 Support Vector Machine

A SVM can be used in image pattern recognition and classification. SVM function projects nonlinear training data in the input space to higher-order feature space by using a kernel function. This process enables the classification of a cancerous image dataset which is highly nonlinear. Classification of higher dimensional feature space results in over-fitting in the input space [88,89]. The SVM separates training images into two classes. Suppose sample x_i have features $\left[\,x(1)\ x(2)\ldots x(n)\,\right]$ in R^d, in d dimensional feature space and y_i class label has two possible pattern either $y_i=1$ or $y_i=-1$.

SVM classifier separates the data objects into two classes in the feature space by optimal decision boundary as shown in Figure 8.2. The optimal decision boundary

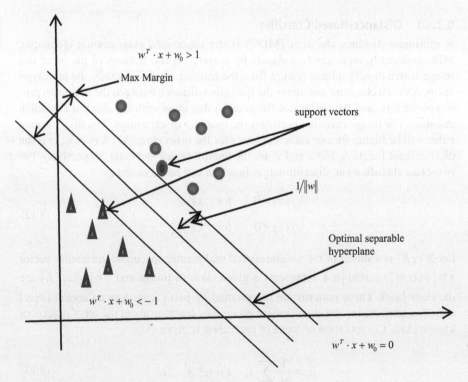

FIGURE 8.2 Optimal separating hyperplane.

or linear hyperplane has the maximum margin between the two classes. The SVMs can be Linear SVM or Non-Linear SVM.

Let S be a set of points $x_i \in R^d$ with $i = 1,\ldots,m$. Each point x_i belongs to either of two classes, with label $y_i \in \{-1,+1\}$. The set S is linear separable if $w \in R^d$ and $w_0 \in R$ such that

$$y_i\left(w^T \cdot x_i + w_0\right) \geq 1, \quad i = 1,\ldots,m \tag{8.15}$$

where the w represents an adjustable vector, and w_0 bias.

$$w^T \cdot x + w_0 = 0 \tag{8.16}$$

Equation (8.16) defines hyperplane also known as the separating hyperplane. The expression of a signed measure of distance d_i from the sample x_i to the optimal hyperplane is given by:

$$d_i = \frac{\left(w^T \cdot x_i + w_0\right)}{\|w\|} \tag{8.17}$$

From Equations (8.15) and (8.17)

$$y_i d_i \geq \frac{1}{\|w\|} \tag{8.18}$$

Equation (8.18) shows that the minimum value of the Euclidean norm 'w' will give maximum margin between the point x_i and the separating hyperplane.

In non-linear SVMs as shown in Figure 8.3, data points are transferred to higher dimensional space so that data points can be linearly separated which appear as dot products $x_i \cdot x_j$, in input space R^d. Transfer of data points from the input space R^d to the higher dimensional space $R^n (n > d)$ can be done by function $\Phi : R^d \rightarrow R^n$.

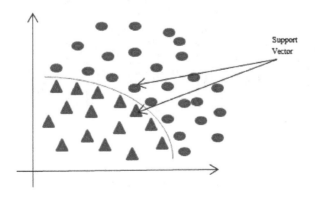

FIGURE 8.3 Decision surface by a polynomial classifier.

The data points appear in the form of dot products updated to $\Phi(x_i) \cdot \Phi(x_j)$ in the training phase. This updated function provides maximum margin in the higher-dimensional space to make nonlinear decision boundary in the input space.

A kernel function K: polynomial, RBF, Sigmoid can be used in the training algorithm such that

$$K(x_i, x_j) = \Phi(x_i) \cdot \Phi(x_j) \tag{8.19}$$

However, selection of the kernel function is a "trial and error" approach.

8.2.4.3 Convolutional Neural Network

Convolutional neural network (CNN) is a deep learning tunable architecture that is feed-forward neural network consists of multiple trainable nonlinear transformation stages. It extracts the feature vector in every stage, which represents the input and output vector, respectively. CNN includes the feature extractor in the training process, which modifies the weights during the training process. In CNN the structure of weight to a particular neuron is the same and their value is also the same. These are not the separate neuron applied in different locations and the number of inputs that it gets is not the entire previous layer but a small portion of the previous layer.

CNN consists of two networks one that extracts a feature of the input image and second classifies the extracted feature. CNN extracts patch-wise information, which provides better local information of the pixel for correct image classification. CNN has three types of network layers that are convolution, pooling, and fully connected layers as shown in Figure 8.4.

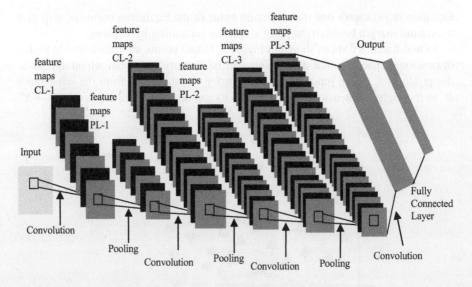

FIGURE 8.4 Convolution neural network architecture.

Convolution kernel $h(m,n)$ also known as convolution filter pass over the image $x(m,n)$ and feature map value are calculated as follows

$$G(m,n) = x(m,n) * h(m,n)$$

$$= \sum_j \sum_k h(i,j)x(m-j,n-k) \tag{8.20}$$

where 'm' and 'n' are the pixel coordinate.

Convolution is a proficient method of feature extraction with minimum data set redundancy. It also lowers the data dimension. Convolution kernel is a feature identifier that filters the important feature of the image and produces a convolution map about the distribution of these features. The neurons figure out how to assemble the data to gain a higher-order feature of the image in the next convolutional layers [44]. Since convolution operation shrinks the image, when the kernel moves through the image, impact of the outskirt pixels on feature map is very small than the center pixels, and it might decrease the spatial resolution of the image. The solution to this problem is image padding, which is defined as

$$p = \frac{d-1}{2} \tag{8.21}$$

where 'p' is padding and 'd' is the filter dimension. Another method to reduce the feature map size is stride. The stride function is just like the sliding window, which controls the amount of movement of the convolution kernel to the input image [48]. Usually, we shift the kernel by one pixel, but step size can be increased if a small feature map requires or less overlapping of the receptive field is required. The output matrix dimension can be calculated as

$$n_o = \left[\frac{n+2p-d}{s} + 1 \right] \tag{8.22}$$

Where 's' is stride.

In the convolution operation number of channels in the filter and image must be the same; multiple filters can be used on the same image for convolution. Generalized output dimension is given as follows

$$[m,n,n_c] * [d,d,n_c] = \left[\frac{m+2p-d}{s} + 1 \right] \left[\frac{n+2p-d}{s} + 1 \right], n_D \tag{8.23}$$

where n_c is the number of channel in the image filter and n_D is the number of filters.

The second layer i.e. pooling layer reduces the feature map size of the input matrix. There are two pooling mechanisms used in practice: max pooling and average pooling. Both pooling mechanisms give translation invariance, but max pooling is widely used [90]. It can be expressed as

$$G(m,n) = \max\{0, x(m,n)\}$$

$$\text{or} \tag{8.24}$$

$$\max\{x(m-j, n-k)\} \quad \forall\ 0 \le j \le m$$

$$\text{and} \quad 0 \le k \le n$$

The convolutional layer typically includes a non-linear activation function. Non-linearity is important to model the complex function. Rectified linear unit (ReLU) activation function (Equation 8.25) is generally used because it shows better convergence rate as compared to the other activation function. It gives zero output for the negative input value, which means it doesn't activate neurons with negative value, and for the positive input values, it gives output exactly the same as input.

$$\psi(x) = \max(0, x) \tag{8.25}$$

The last hidden layer of CNN is known as fully connected layers [24]. The output of this layer is a column vector, and each row of the column vector is corresponding to a different class. Classification of these different classes based on the probability estimation of each class. Practically, input to the fully connected layer must be as small as possible, pooling and stride reduce the size of the data matrix of fully connected layers. This fully connected layer and multinomial logistic regression layer with softmax function define the output class. Softmax is given as

$$f(x) = \frac{e^{x_k}}{\displaystyle\sum_{k=1}^{K} e^{x_k}} \tag{8.26}$$

It calculates the probability of the 'K' target class. The calculated probabilities are in the range between 0 and 1. In case of a multi-classification problem, it identifies the probabilities of all define classes and the highest probability among all will define as the detected target class.

8.3   IMAGE DATABASE

Image data were acquired from the examination of breast tissue biopsy. Breast Cancer Histopathological Database (BreakHis) and IDC database are used in this chapter to test the classifier performance.

8.3.1   BREAST CANCER HISTOPATHOLOGICAL DATABASE

The BreakHis database is freely available data for breast cancer research and contains hematoxylin and eosin (H&E)-stained microscopic images of benign and malignant breast tumors. The BreakHis database is created by the P&D Laboratory–Pathological

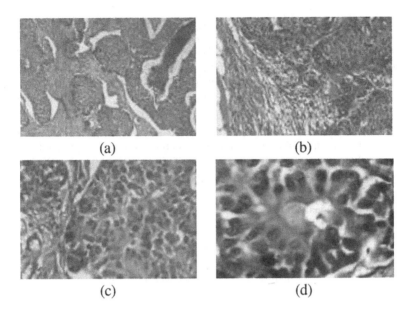

FIGURE 8.5   Sample BreakHis dataset breast malignant tumor with different magnification factors (a) 40×, (b)100×, (c) 200×, and (d) 400×.

Anatomy and Cytopathology, Parana, Brazil between January 2014 to December 2014 [91]. Surgical open biopsy (SOB) has prepared the sample using standard paraffin process for histological examination, each sample marked by experienced pathologists.

RGB images are acquired with different magnification factors as shown in Figure 8.5. The original images crop to remove black border and text annotation and saved in three-channel RGB form. This image is saved in portable network graphics (.png) format without compression. The dimension of the image is 700 × 460 pixels. The acquisition of images at different magnification is performed by the pathologist. The pathologist recognizes images at different magnifications and identified the tumor region. Until now this database consists of total 7909 microscopic images of 82 patients, out of which 2480 are identified as benign sample and remaining 5429 images are malignant.

8.3.2   INVASIVE DUCTAL CARCINOMA (IDC) DATABASE

IDC is the most common type of breast cancer in women, so it becomes a very challenging task for the pathologist to identify and differentiate the area corresponding to the healthy tissues and invasive tumor during microscopic image analysis. The University of Pennsylvania and the Cancer Institute of New Jersey created an IDC dataset using breast tissue surgical biopsy of 162 patients. This dataset is freely available for the researchers for histological study. The whole slid scanner scans the images at 40x magnification [92,93]. The dataset is composed of 2,77,524 patches of size 50 × 50 × 3 out of which 1,98,738 are benign and remaining 78,786 patches are malignant.

8.4   PERFORMANCE EVALUATION OF CAD SYSTEM

The performance measure of the system is a significant challenge to provide guarantee in accuracy. The importance of system validation is to approve the performance of the classification method. In the context of image analysis, the system should perceive and characterize the normality and irregularity of the cell. Commonly used performance indication for the normal and abnormal image detection is true positives (T_P), true negatives (T_N), false positives (F_P), and false negatives (F_N) values. T_P and T_N are indications of the right diagnosis, whereas F_P and F_N are the indications of the wrong diagnosis of suspected abnormality. This system performance is measured in terms of accuracy (A_{cc}), sensitivity (S), specificity (S_p), precision (P_{RE}), F-measure (F_{MEA}), true positive rate, false-positive rate, etc. To calculate the error of the system's performance i.e. Mean relative error (MRE), the comparison is made between the values obtained from the measured data and those from actual known data [75,94]. MRE is defined as

$$\text{MRE} = \frac{\text{Measured value} - \text{true value}}{\text{True value}} \tag{8.27}$$

Sensitivity is the measure of the true diagnosis of the positive case which can be calculated as

$$S = \frac{T_P}{T_P + F_N} \tag{8.28}$$

Equation (8.28) shows that high value of sensitivity which is the indication of minimum false negative detection and perfect sensitivity means the system effectively identifies the existence of the abnormalities.

Specificity is defined as the fraction of the true negative cases over the real negative cases as shown in Equation (8.29). It shows that false positive detection must be as low as possible to have high specificity. A perfect specificity indicates that the system can identify the normal condition efficiently.

$$S_p = \frac{T_N}{T_N + F_P} \tag{8.29}$$

Accuracy indicates that the system can correctly identified the data class out of all as defined in Equation (8.30). It is the measure of correct decision.

$$A_{CC} = \frac{T_P + T_N}{T_P + T_N + F_P + F_N} \tag{8.30}$$

Precision is defined as the ratio of T_P to all positive class as shown in Equation (8.31). It is the measure of the number of positive class, which actually belongs to positive class.

$$P_{RE} = \frac{T_P}{T_P + F_P} \tag{8.31}$$

Another performance criterion of the classifier is F-measure defined by Equation (8.32). It is combination of the precision and sensitivity corresponding to their harmonic mean.

$$F_{\text{MEA}} = 2 * \frac{P_{\text{RE}} \times S}{P_{\text{RE}} + S} \qquad (8.32)$$

8.5   RESULTS AND DISCUSSION

In this section, the performance of the three-image classification method: SVM, MDC, and CNN for breast histopathological images are analyzed. Histopathological image processing is very difficult because of muddled structure compared to other images like radiological images etc. The SVM and MDC classifier model was tested on the BreakHis dataset. These methodologies use a random dataset of the histopathological images. The dataset is divided into two parts: training set and testing set. A total of 1700 images (850 benign and 850 malignant) were used to train the classifier for breast cancer diagnosis. Breast histopathological image segmented into their region using the k-mean clustering algorithm. Figure 8.6 shows the result of applying k means clustering to the histopathological image. These images have four colors that indicate the different characteristics of the cell, like background represented by white color, nuclei by dark purple color, cytoplasm by dark pink color, and stroma represented by light pink color [95].

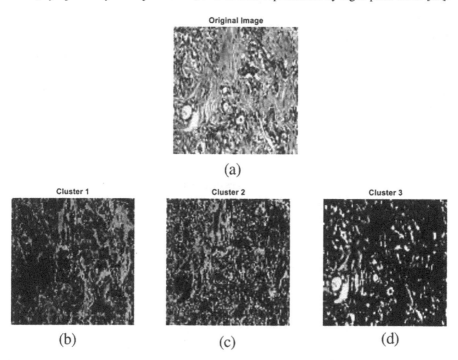

FIGURE 8.6   Results of image segmentation using k means clustering (a) original image, (b) cluster-1 (cytoplasm), (c) cluster-2 (stroma), (d) cluster-3 (background image).

First order statistical texture feature, i.e., mean, variance, Kurtosis, and Entropy and GLCM features, i.e., Contrast, Correlation, Homogeneity, and Energy as defined in the previous sections are used to train the classification model MDC and SVM. This trained model is tested with 200 benign and 200 malignant images.

Table 8.1 shows the testing data output of MDC and SVM classifier. Tables 8.2 and 8.3 is the confusion matrix that shows the performance of the classification models. A comparison of the performance of the SVM and MDC classifier with other classifier model including performance parameter accuracy, specificity, sensitivity, F_P rate, F_P per image are summarized in Tables 8.4 and 8.5. In Table 8.2, the confusion matrix of MDC classifier reveals the sensitivity (90.6%) and specificity (91.41%) of the classifier. Table 8.5 shows that MDC gives high precision (91.5%) and very low false-positive per image (0.085) that indicates the low number of F_p values. A high F-measure (0.9) also suggests the effectiveness of this classifier.

Performance evaluation of the SVM classifier gives 93% classification accuracy and 0.93 F-score as shown in Table 8.5. Table 8.3 the confusion matrix of SVM classifier shows 92.57% sensitivity and 93.43% specificity. Tables 8.4 and 8.5 show that SVM classifier gives better performance compared to the other classier model and MDC classifier with 93% classification accuracy, 6.5 FP rate, and 0.065 FP per image. This proposed CAD system has shown better diagnosis accuracy to support pathologists for breast cancer diagnosis.

TABLE 8.1

Testing Data Result of Classifier

Category	Testing Dataset	Images Correctly Detected by MDC	Detection Accuracy (%)	Images Correctly Detected by SVM	Detection Accuracy (%)
Benign	200	183	91.5	187	93.5
Malignant	200	181	90.5	185	92.5

TABLE 8.2

Confusion Matrix (Testing Set) for MDC

Benign	183	17	91.5%
Malignant	19	181	90.5%
	90.6%	91.41%	91%
	Benign	Malignant	

TABLE 8.3

Confusion Matrix (Testing Set) for SVM

Benign	187	13	93.5%
Malignant	15	185	92.5%
	92.57%	93.43%	93%
	Benign	Malignant	

TABLE 8.4
Comparative Analysis of MDC and SVM Classifier

Authors	Segmentation Method	Classification Method	Matrix
Dundar et al. [41]	Statistical features mean, standard deviation, median, and mode	Multiple instances (MIL) approach with asymmetric loss functions	Acc = 87.9%
Nahid et al. [60]	Algorithm-1: Tamura features Algorithm-1: Tamura features and contrast correlation	Restricted Boltzmann Machine (RBM)	Acc = 74.7% Acc = 85.3%
Reis et al. [61]	Multiscale image features and local binary patterns	Random decision Trees classifier	Acc = 84%
Cosatto et al. [64]	Hough transform & ACM	Morphology and texture with SVM	TPR = 0.92, FPR = 0.8, F-score = 0.86
Fatakdawala et al. [65]	GMM & EM and ACM	Intensity and k-means clustering	TPR = 0.8
Veta et al. [66]	RST and marker controlled watershed	–	Acc = 81%
MDC and SVM performance	First order statistical and second order (GLCM) texture feature	MDC SVM	Acc = 91% Acc = 93%

As we have seen that the existing CAD methods have shown possibility in the analysis of medical images, but it has some limitation such as the performance of the conventional classifier like SVM, MDC, KNN, wavelet, etc. depends upon the extracted features, i.e., data representation. If we change the hierarchy of the extracted feature, it will change the detection accuracy and also it requires a larger database to make the correct decision. To simplify these problems, another deep-learning classification technique known as convolution neural network is used. CNN is a very powerful tool in automated segmentation and classification of the histology image. It can automatically learn the given input data and predicts different classes.

The CNN system architecture is shown in Figure 8.7. This CNN architecture shows layer by layer transformation of the input image to the final output image. This CNN model is tested on IDC dataset. This dataset has input image of size $50 \times 50 \times 3$ with three color channel RGBs consisting of raw pixel value. The convolution layer consists of convolution filters; each of these filters will extract different relevant features from the input image for correct class prediction and pass it to the next layer. In every hidden layer ReLU function applies as an element-wise activation function. Pooling layers will perform the down sampling of the feature map that reduces the number of parameters. In this CNN architecture four convolution and pooling layers were added for more specific feature extraction before class prediction. The output layer (fully connected layer) with softmax function computes the output class score.

The input images are automatically divided into training and test set. Training of the dataset continues as long as the network keeps enhancing the validation set. This generated output is compared with the training dataset for error generation. A loss function is computing the mean square loss as shown in Figure 8.8. This generated

TABLE 8.5
Comparative Analysis of MDC and SVM Classifier with Other Classifier Model on BreakHis Database

Authors	Reza and Ma [83]	Nahid et al. [82]		Spanhol et al. [75]	Zhu et al. [81]	Erfankhah et al. [74]	Performance Evaluation of Classifier	
Classifier Model	Oversampling+ Undersampling+ CNN	CNN-I	CNN + LDF	AlexNet	Hybrid CNN + SEP	LBP Homogeneity and Variance + SVM	MDC	SVM
Performance								
Sensitivity (%)	90.9	95	88.78	89.6	94.5	83.2	90.6	92.57
Specificity (%)	80.34	67.42	81.87	88.6	–	88.4	91.41	93.43
FP rate (%)	–	32.5	18.18	–	–	11.57	8.5	6.5
FP/image	–	–	–	–	–	NA	0.085	0.065
Accuracy (%)	85.62	87.15	86.12	–	83.4	–	91	93
Precision (%)	–	88	93	–	83.7	–	91.5	93.5
F-measure	0.86	0.95	0.87	–	0.89	–	0.9	0.93

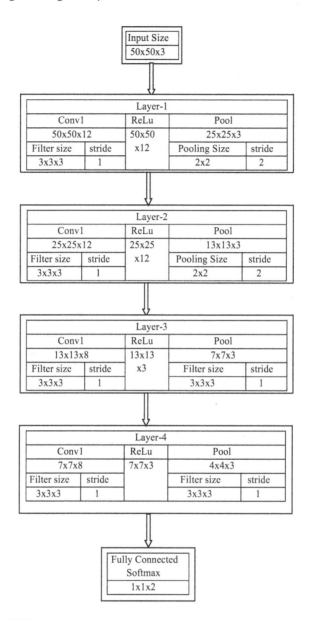

FIGURE 8.7   CNN system architecture.

error backpropagated to update the weights and bias values to minimize the error. This training continues until the right class prediction occurs.

The performance result of the CNN based CAD system is compared with the other competitive CAD systems are shown in Table 8.8. Tables 8.6 and 8.7 shows the confusion matrix of the training and testing dataset that shows the performance of the classification model. Table 8.8 shows that this classification model exhibits the highest classification accuracy of 94.11% and sensitivity of 0.9379 which means minimum

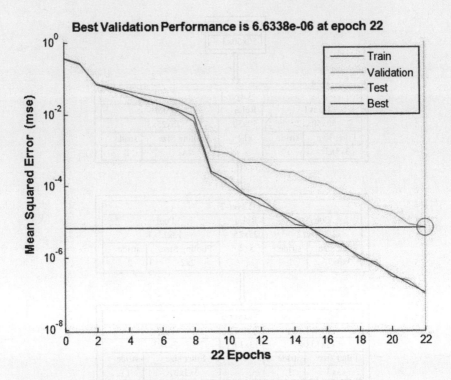

FIGURE 8.8   Performance measure of the CNN classifier.

TABLE 8.6
Confusion Matrix (Training Set)

Benign	15526	1228	92.67%
Malignant	1237	15517	92.28%
	92.62%	92.66%	92.54%
	Benign	**Malignant**	

TABLE 8.7
Confusion Matrix (Testing Set)

Benign	1997	117	94.47%
Malignant	132	1977	93.74%
	93.79%	94.4%	94.11%
	Benign	**Malignant**	

false-negative rate as compared to the other classification model. The specificity of the system is 94.4% which is high enough to indicate that this model is able to detect abnormalities with minimum false positive detection. High precision (0.9447) indicates the exactness of this classifier which also reveals low number of false positive values.

Histopathological Image Analysis

TABLE 8.8
Comparative Analysis of CNN Classifier with Other Classifier Model on IDC Database

Authors	Reza and Ma [83]	Balazsi et al. [84]	Cruz-Roa et al. [92]	Erfankhah et al. [74]	Classifier Output
Classifier Model					
Performance	Oversampling Undersampling + CNN	LIC, Color and Texture Feature + RFC	Handcrafted Feature + CNN	LBP Homogeneity and Variance + SVM	CNN
Accuracy (%)	85.48	88.7	84.23	85.2	94.11
Sensitivity (%)	80.85	88.4	79.6	84	93.79
Specificity (%)	90.12	83.8	88.86	88.4	94.4
Precision (%)	–	–	65.84	95	94.47
F-score	0.8478	0.661	0.7180	0.7665	0.9413
FP rate (%)	–	–	–	11.57	5.59

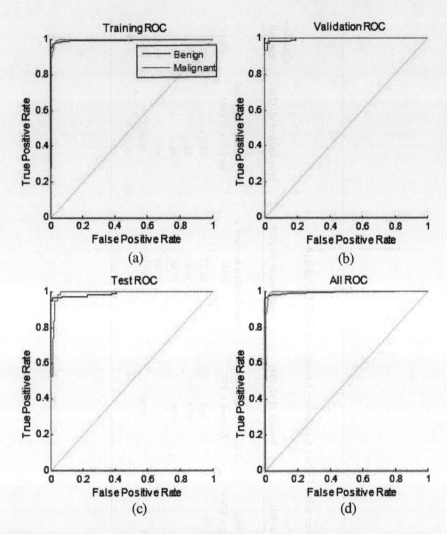

FIGURE 8.9 Receiver operating characteristic plot (a) training dataset, (b) validation dataset, (c) testing dataset, and (d) resulting output of all ROC curve.

A high F-measure (0.9413) also suggests the effectiveness of this classification method as high F-measure indicates the perfect evaluation of precision and sensitivity.

ROC is also investigated as shown in Figure 8.9 which illustrates that the proposed method increases the true positive rate and reduces the false-positive rate as output varies from 0 to 1. This proposed CAD system has shown better diagnosis accuracy to support pathologist for breast cancer diagnosis.

8.6 CONCLUSION

In this chapter, classification model for histopathological image investigation has validated with positive outcomes. The results obtained for image classification revealed that these classifiers can detect histological image morphology more accurately and efficiently. The classification methods categorize the histological image as benign and malignant. These methods involve two process training and testing. The SVM, MDC, and CNN classifiers have been used for breast cancer analysis. During the training process, the first order and GLCM statistical features were calculated. In the testing phase, statistical features of the testing image were calculated which is assigned to one of the class based on the SVM and MDC classification method. We have also revealed that the performance of the conventional machine learning method depends on the data representation and feature extraction. To simplify these problems we have used another deep-learning classification technique known as convolutional neural network. The convolutional neural network includes the feature extractor in the training process, which modifies the weights during the training process. Training of the CNN architecture with large input patches enables the learning of cellular detail and tissue structure. We have also compared the result of the CAD framework with the result of the other researchers on the same standard dataset provided by the pathologist concerning the detection and characterization of breast tumors. Based on the capacity of the CAD structure to mark retrospectively visible malignant growth, it is assessed that the potential benefit of CAD can be an extension in the breast disease recognition. This CAD framework can diagnose the image quickly which is currently being performed manually by the pathologist. These techniques show improvement in the existing CAD performance. The results of these classifiers show significant improvement in detection accuracy, specificity, F-measure, etc. that shows the effectiveness and proficiency of these classifiers.

REFERENCES

1. World Health Organization (2020) WHO report on cancer: Setting priorities, investing wisely and providing care for all, (February, 2020), pp. 17–36.
2. Freitas R Jr., Soares LR and Barrios CH (2015) Cancer survival: The CONCORD-2 study, *The Lancet*, vol. 386, 428–429.
3. Bray F, Ferlay J, Soerjomataram I, Siegel RL, Torre LA, Jemal A (2018) Global cancer statistics 2018: GLOBOCAN estimates of incidence and mortality worldwide for 36 cancers in 185 countries, *CA: A Cancer Journal for Clinicians*, vol. 68(6), 394–424.
4. http://cancerindia.org.in/wp-content/uploads/2017/11/Breast_Cancer.pdf (accessed August, 5, 2020).
5. http://cancerindia.org.in/india-still-low-breast-cancer-survival-rate-66-study/ (accessed August, 10, 2020).
6. https://seer.cancer.gov/csr/1975_2017/ (accessed August, 10, 2020).
7. Al-Kofahi Y, Lassoued W, Grama K, Nath SK, et al. (2011) Cell-based quantification of molecular biomarkers in histopathology specimens, *Histopathology*, vol. 59, 40–54.
8. Rizzardi AE, Johnson AT, Vogel RI, Pambuccian SE, et al. (2012) Quantitative comparison of immune-histochemical staining measured by digital image analysis versus pathologist visual scoring, *Diagnostic Pathology*, vol. 7(42), 1–10.
9. Wei B, Bu H, Zhu C, Guo L, Chen H, et al. (2004) Interobserver reproducibility in the pathologic diagnosis of borderline ductal proliferative breast diseases, *Journal of Sichuan University, Medical Science Edition*, vol. 35(6), 849–853.
10. Madabhushi A (2009) Digital pathology image analysis: Opportunities and challenges, *Imaging in Medicine*, vol. 1, 7–10.

11. Gavrielides MA, Gallas BD, Lenz P, Badan A and Hewitt SM (2011) Observer variability in the interpretation of HER2/neuroimmuno histochemical expression with unaided and computer-aided digital microscopy, *Archives of Pathology & Laboratory Medicine Online*, vol. 135(2), 233–242.

12. Mohammed ZMA, McMillan DC, Elsberger B, Going JJ, Orange C, Mallon E, Doughty JC and Edwards J (2012) Comparison of visual and automated assessment of ki-67 proliferative activity and their impact on outcome in primary operable invasive ductal breast cancer, *British Journal of Cancer*, vol. 106(2), 383–388.

13. Jayaraman S, Veerakumar T and Esakkirajan S (2010) *Digital Image Processing*, 3rd edn, New York: Tata McGraw-Hill Education, pp. 243–297.

14. Papadopoulos A, Fotiadis D and Costaridou L (2008) Improvement of microcalcification cluster detection in mammography utilizing image enhancement techniques, *Computers in Biology and Medicine*, vol. 38(10), 1045–1055.

15. Cheng HD, Cai X, Chen X, Hu L and Lou X (2003) Computer-aided detection and classification of microcalcifications in mammograms: A survey, *Pattern Recognition*, vol. 36(12), 2967–2991.

16. Chen ZK, Tao Y and Chen X (2001) Multi-resolution local contrast enhancement of X-ray images for poultry meat inspection, *Applied Optics*, vol. 40(8), 1195–1200.

17. Cheng HD, Xue M and Shi XJ (2003) Contrast enhancement based on a novel homogeneity measurement, *Pattern Recognition*, vol. 36(11), 2687–2697.

18. Stark JA (2000) Adaptive image contrast enhancement using generalizations of histogram equalization, *IEEE Transactions on Image Processing*, vol. 9(5), 889–896.

19. Kim JY, Kim LS and Hwang SH (2001) An advanced contrast enhancement using partially overlapped sub-block histogram equalization, *IEEE Transactions on Circuits and Systems for Video Technology*, vol. 11(4), 475–484.

20. Tourassi GD, VargasVoracek R, Catarious D Jr. and Floyd C Jr. (2003) Computer-assisted detection of mammographic masses: A template matching scheme based on mutual information, *Medical Physics*, vol. 30(8), 2123–2130.

21. Morton M, Whaley D, Brandt K and Amrami K (2006) Screening mammograms: Interpretation with computer-aided detection prospective evaluation, *Radiology*, vol. 239(2), 375–383.

22. Brem R, Baum J, Lechner M, Kaplan S, Souders S, Naul L and Hoffmeister J (2003) Improvement in sensitivity of screening mammography with computer-aided detection: A multi institutional trial, *American Journal of Roentgenology*, vol. 181(3), 687–693.

23. Thakur RS, Yadav RN and Gupta L (2019) State-of-art analysis of image denoising methods using convolutional neural networks, *IET Image Processing*, vol. 13(13), 2367–2380.

24. Cheng HD, Shan J, Ju W, Guo Y and Zhang L. (2010) Automated breast cancer detection and classification using ultrasound images: A survey, *Pattern Recognition*, vol. 43(1), 299–317.

25. Costin H and Rotariu C (2003) Knowledge based contour detection in medical image using fuzzy logic, *International Symposium on Signal Circuit and System*, Romania, July 2003, pp. 273–276.

26. Ho J and Hwang W (2008) Automatic microarray spot segmentation using snake fisher model, *IEEE Transactions on Medical Imaging*, vol. 27(6), 847–857.

27. Gonzales R and Woods R (2008) *Digital Image Processing,* 3rd edn, Upper Saddle River, NJ: Pearson Education Prentice Hall, pp 144–172.

28. Huertas A and Medioni G (1986) Detection of intensity changes with sub pixel accuracy using Laplacian-Gaussian masks, *IEEE Transaction on Pattern Analysis and Machine Intelligence*, vol. 8(5), 651–664.

29. Somkantha K, Theera-Umpon N and Auephanwiriyakul S (2011) Boundary detection in medical images using edge following algorithm based on intensity gradient and texture gradient features, *IEEE Transactions on Biomedical Engineering*, vol. 58(3), 567–573.

30. Bosnjak A, Montilla G and Torrealba V (1998) Medical images segmentation using Gabor filters applied to echocardiographic images, *International Conference on Computers in Cardiology*, Cleveland, September 1998, pp. 457–460.

31. Kachouie NN and Alirezaie A (2003) Texture segmentation using Gabor filter and multi-layer perceptron, *IEEE International Conference on Systems, Man and Cybernetics*, Washington, DC, October 2003, vol. 3, pp. 2897–2902.

32. Sampat MP and Bovik AC (2003) Detection of speculated lesions in mammograms, *IEEE International Conference on Engineering in Medicine and Biology Society*, Cancun, September 2003, pp. 810–813.

33. Saha PK, Udupa JK, Conant EF, Chakraborty DP and Sullivan D (2001) Breast tissue density quantification via digitized mammograms, *IEEE Transactions on Medical Imaging*, vol. 20(8), 792–803.

34. Demir C and Yener B (2005) Automated cancer diagnosis based on histopathological images: A systematic survey, Technical Report, Rensselaer Polytechnic Institute, Department of Computer Science, pp. 1–15.

35. Boykov Y and Kolmogorov V (2004) An experimental comparison of min-cut/max-flow algorithms for energy minimization in vision, *IEEE Transactions on PAMI*, vol. 26(9), 1124–1137.

36. Micusik B and Hanbury A (2006) Automatic image segmentation by positioning a seed, *European Conference on Computer Vision*, Graz, May 2006, pp. 468–480.

37. Wang Y, Turner R, Crookes D, Diamond J and Hamilton P (2007). Investigation of methodologies for the segmentation of squamous epithelium from cervical histological virtual slides, *International Machine Vision and Image Processing Conference*, Kildare, September 2007, pp. 83–90.

38. Esgiar AN, Naguib RNG, Sharif BS, Bennett MK and Murray A (1998) Microscopic image analysis for quantitative measurement and feature identification of normal and cancerous colonic mucosa, *IEEE Transactions on Information Technology in Biomedicine*, vol. 2(3), 197–203.

39. Guillaud M, Cox D, Malpica A, Staerkel G, Matisic J, Van Niekirk D, Adler-Storthz K, Poulin N, Follen M and MacAulay C (2004) Quantitative histopathological analysis of cervical intra-epithelial neoplasia sections: Methodological issues, *Cellular Oncology*, vol. 26, 31–43.

40. Malek J, Sebri A, Mabrouk S, Torki K and Tourki R (2009) Automated breast cancer diagnosis based on GVF-Snake segmentation, wavelet features extraction and fuzzy classification, *Journal of Signal Processing Systems*, vol. 55, 49–66.

41. Dundar MM, Badve S, Bilgin G, Raykar V, Jain R, Sertel O and Gurcan MN (2011) Computerized classification of intraductal breast lesions using histopathological images, *IEEE Transactions on Biomedical Engineering*, vol. 58(7), 1977–1984.

42. Santhos KA, Kumar A, Bajaj V and Singh GK (2020) McCulloch's algorithm inspired cuckoo search optimizer based mammographic image segmentation, *Multimedia Tools and Applications*, vol. 39, 30453–30488.

43. Ilea DE and Whelan PF (2006) Color image segmentation using a spatial k-means clustering algorithm, *International conference on Machine Vision and Image Processing*, Dublin, August to September 2006, pp. 1–8.

44. Bishop CM (2006) *Pattern Recognition and Machine Learning: Information Science and Statistics*, New York: Springer Science and Business Media, pp. 225–281.

45. Tuceryan M and Jain AK (1998) *The Handbook of Pattern Recognition and Computer Vision*, 2nd edn, Singapore: World Scientific Publishing, pp. 235–276.

46. Aggarwal N and Agrawal RK (2012) First and second order statistics features for classification of magnetic resonance brain images, *Journal of Signal and Information Processing*, vol. 3, 146–153.

47. Alpaydin E (2004) *Introduction to Machine Learning*, 2nd edn, Cambridge, MA: MIT Press.

48. Cristianini N and Shawe-Taylor J (1999) *An Introduction to Support Vector Machines and Other Kernel-Based Learning Methods*, Cambridge: Cambridge University Press.

49. Kim HI, Shin S, Wang W and And Jeon SI (2013) SVM-based Harris corner detection for breast mammogram image normal/abnormal classification, *Proceedings of the 2013 Research in Adaptive and Convergent Systems*, New York, NY, USA, pp. 187–191.

50. George YM, Zayed HL, Roushdy MI and Elbagoury BM (2014) Remote computer-aided breast cancer detection and diagnosis system based on cytological images, *IEEE Systems Journal*, vol. 8(3), 949–964.

51. Huang PW and Lee CH (2009) Automatic classification for pathological prostate images based on fractal analysis, *IEEE Transactions on Medical Imaging*, vol. 28(7), 1037–1050.

52. Marcel S, Rodriguez Y and Heusch G (2006) On the recent use of local binary patterns for face authentication, *International Journal of Image and Video Processing*, IDIAP–RR 06-34, 1–9.

53. Maturana D, Mery D and Soto A (2009) Face recognition with local binary patterns, spatial pyramid histograms and Naive Bayes nearest neighbor classification, *International Conference of the Chilean Computer Science Society*, Santiago, November 2009, pp. 125–132.

54. Linder N, Konsti J, Turkki R, et al. (2012) Identification of tumor epithelium and stroma in tissue microarrays using texture analysis, *Diagnostic Pathology*, vol. 7(22), 1–11.

55. Fuchs TJ, Wild PJ, Moch H and Buhmann JM (2008) Computational pathology analysis of tissue microarrays predicts survival of renal clear cell carcinoma patients, *Medical Image Computing and Computer Assisted Intervention*, vol. 11(2), 1–8.

56. Qi X, Cukierski W and Foran DJ (2010) A comparative performance study characterizing breast tissue microarrays using standard RGB and multispectral imaging, *Proceeding of SPIE, Multimodal Biomedical Imaging*, San Francisco, California, United States, vol. 75570Z.

57. Doyal JTS, Feldman M, Tomaszewski J and Madabhushi A (2007) Automatic grading of prostate cancer using architectural and textural image features, Proceeding of Biomedical Imaging: From Nano to Macro, pp. 1284–1287.

58. Doyal JTS, Feldman M and Madabhushi A (2010) A boosted Bayesian multi-resolution classifier for prostate cancer detection from digital needle biopsies, *Transaction on Biomedical Engineering*, vol. 59(5), 1205–1218.

59. Sertel O, Lozanski G, Shana'ah A and Gurcan MN (2010) Computer-aided detection of centroblasts for follicular lymphoma grading using adaptive likelihood-based cell segmentation, *IEEE Transaction Biomedical Engineering*, vol. 57(10), 2613–2616.

60. Nahid A, Mikaelian A and Kong Y (2018) Histopathological breast-image classification with restricted Boltzmann machine along with backpropagation, *Biomedical Research*, vol. 29(10), 2068–2077.

61. Reis S, Gazinska P, Hipwell JH, et al. (2017) Automated classification of breast cancer stroma maturity from histological images, *IEEE Transactions on Biomedical Engineering*, vol. 64(10), 2344–2352.

62. Santhos KA, Kumar A, Bajaj V and Singh GK (2019) K-highest Fuzzy min-max network to classify histopathological images, *International Conference on Communication and Signal Processing (ICCSP)*, Chennai, April 2019, pp. 240–244.

63. Bajaj V, Mayank Pawar M, Meena VK, Kumar M, Sengur A and Guo Y (2017) Computer-aided diagnosis of breast cancer using bi-dimensional empirical mode decomposition, *Neural Computing and Applications*, vol. 31, 3307–3315.

64. Cosatto E, Miller M, Graf HP and Meyer JS (2008) Grading nuclear pleomorphism on histological micrographs, *Proceedings of 19th International Conference Pattern Recognition*, Tampa, December 2008, pp 1–4.

65. Fatakdawala H, Xu J, Basavanhally A, Bhanot G, et al. (2010) Expectation maximization-driven geodesic active contour with overlap resolution (EMaGACOR): Application to lymphocyte segmentation on breast cancer histopathology, *IEEE Transaction Biomedical Engineering*, vol. 57(7), 1676–1689.

66. Veta M, Huisman A, Viergever MA, van Diest PJ and Pluim JPW (2011) Marker-controlled watershed segmentation of nuclei in H&E stained breast cancer biopsy images. *Proceedings of 8th IEEE International Symposium Biomedical Imaging: Nano Macro*, Chicago, IL, April 2011, pp. 618–621.

67. Lado MJ, Cadarso-Suarez C, Roca-Pardinas J, Tahoces PG (2006) Using generalized additive models for construction of nonlinear classifiers in computer-aided diagnosis systems, *IEEE Transactions on Information Technology in Biomedicine*, vol. 10(2), 246–253.

68. Niwas SI, Palaniwamy P, Zhang WJ, Isa NAM and Chibbar R (2011) Log-Gabor wavelets based breast carcinoma classification using least square support vector machine, *IEEE International Conference on Imaging Systems and Techniques*, Penang, May 2011, pp. 219–223.

69. Zhang G, Yin J, Li Z, Su X, Li G and Zhang H (2013) Automatic skin biopsy histopathological image annotation using multi-instance representation and learning, *BMC Medical Genomics*, vol. 6(3): S10.

70. Bayramoglu N, Kannala J, Heikkila J (2016) Deep learning for magnification independent breast cancer histopathology image classification, *International Conference on Pattern Recognition*, Cancun, December 2016, pp. 2441–2446.

71. Wang P, Hu X and Li Y (2016) Automatic cell nuclei segmentation and classification of breast cancer histopathology images, *Signal Processing*, vol. 122, 1–13.

72. Wan T, Liu X, Chen J and Qin Z (2014) Wavelet-based statistical features for distinguishing mitotic and non-mitotic cells in breast cancer histopathology, *IEEE International Conference on Image Processing*, Paris, October 2014, pp. 2290–2294.

73. Kowal M, Filipczuk P and Obuchowicz A (2013) Computer-aided diagnosis of breast cancer based on fine needle biopsy microscopic images, *Computers in Biology and Medicine*, vol. 43(10), 1563–1572.

74. Erfankhah H, Yazdi M, Babaie M and Tizhoosh HR (2019) Heterogeneity-aware local binary patterns for retrieval of histopathology images, *IEEE Access*, vol. 7, 18354–18367.

75. Spanhol FA, Oliveira LS, Petitjean C and Heutte L (2016) Breast cancer histopathological image classification using Convolutional Neural Networks, *International Joint Conference on Neural Networks*, Vancouver, July 2016, pp. 2560–2567.

76. Zana F and Klein JC (2001) Segmentation of vessel-like patterns using mathematical morphology and curvature evaluation, *IEEE Transaction on Image Processing*, vol. 10(7), 1010–1019.

77. Chapelle O, Haffner P and Vapnik VN (1999) Support vector machines for histogram-based image classification, *IEEE Transaction on Neural Network*, vol. 10(5), 1055–1064.

78. Unser M, Aldroubi A and Laine A (2003) Guest editorial: Wavelets in medical imaging, *Transaction on Medical Image Processing*, vol. 22(3), 285–288.

79. Orlov N, Shamir L, Macura T, Johnston J, Eckley DM and Goldberg IG (2008) WND-CHARM: Multi-purpose image classification using compound image transforms, *Pattern Recognition Letters*, vol. 29(11), 1684–1693.

80. Deniz E, Sengur A, Kadiroglu Z, Guo Y, Bajaj V and Budak U, (2018) Transfer learning based histopathologic image classification for breast cancer detection, *Health Information Science and Systems*, vol. 6, 1–7.

81. Zhu C, Song F, Wang Y Dong H, Guo Y and Liu J (2019) Breast cancer histopathology image classification through assembling multiple compact CNNs, *BMC Medical Informatics and Decision Making*, vol. 19, 1–17.

82. Nahid A and Kong Y (2018) Histopathological breast-image classification using local and frequency domains by convolutional neural network, *Information Centered Healthcare, MDPI*, vol. 9, 1–26.

83. Reza MS and Ma J (2018) Imbalanced histopathological breast cancer image classification with convolutional neural network, *Proceedings of ICSP 2018*, pp. 619–624.

84. Balazsi M, Blanco P, Zoroquiain P, Levine M and Burnier M Jr. (2016) Invasive ductal breast carcinoma detector that is robust to image magnification in whole digital slides, *Journal of Medical Imaging*, vol. 3(2), 275011–275019.

85. Doyle S, Agner S, Madabhushi A, Feldman M and Tomaszewski J (2008) Automated grading of breast cancer histopathology using spectral clustering with textural and architectural image features, *IEEE International Symposia Biomedical Imaging: From Nano to Macro*, Paris, France, pp. 496–499.

86. Cahoon TC, Sutton MA, Bezdek JC (2000) Breast cancer detection using image processing techniques, *Ninth IEEE International Conference on Fuzzy Systems*, San Antonio, May 2000, pp. 973–976.

87. Santucci E (2017) Quantum minimum distance classifier, *MDPI Entropy,* vol. 19(12), 1–14.

88. Foody MG and Mathur A (2004) A relative evaluation of multiclass image classification by support vector machines, *IEEE Transactions on Geoscience and Remote Sensing*, vol. 42, 1335–1343.

89. Foody MG and Mathur A (2004) Toward intelligent training of supervised image classifications: Directing training data acquisition for SVM classification, *Remote Sensing of Environment*, vol. 93, 107–117.

90. Goodfellow I, Bengio Y and Courville A (2016) *Deep Learning*, Camberidge, MA: MIT Press, pp. 326–365.

91. Spanhol FA, Oliveira LS, Petitjean C and Heutte L (2016) A dataset for breast cancer histopathological image classification, *IEEE Transactions on Biomedical Engineering*, vol. 63(7), 1455–62.

92. Cruz-Roa A, Basavanhally A, González FA, Gilmore H, Feldman M, Ganesan S, Shih N, Tomaszewski J and Madabhushi A (2014) Automatic detection of invasive ductal carcinoma in whole slide images with convolutional neural networks, *SPIE Medical Imaging, International Society for Optics and Photonics*, vol. 9041, 904103-1–904103-15.

93. Pantanowitz L, Farahani N and Parwani A (2015) Whole slide imaging in pathology: Advantages, limitations, and emerging perspectives, *Pathology and Laboratory Medical International*, vol. 7, 23–33.

94. Sankar D and Thomas T (2007) Fractal modeling of mammograms based on mean and variance for the detection of microcalcifications, *International Conference on Computational Intelligence and Multimedia Applications*, Sivakasi, December 2017, vol. 2, pp. 334–348.

95. Rahmadwati R, Naghdy G, Ros M and Todd C (2012) Computer aided decision support system for cervical cancer classification. *Proceedings of SPIE*, vol. 8499, 1–13.

9 Study of Emotional Intelligence and Neuro-Fuzzy System

Mohan Awasthy
Rungta College of Engineering and Technology

CONTENTS

9.1 EMOTIONAL INTELLIGENCE

A cognitive intelligent system would be one which will be capable of comprehension and communicating its own feelings, recognizing and empathizing emotions in others, regulating and expressing feelings to spur versatile behaviors. Recognizing emotions in others comprises reasoning about reflection of emotion in a likely situation, sense of understanding what is imperative to other individual, what are his objectives, inclinations and predispositions and so on. Directing own passionate condition and its layers is normal for a developed and acculturated individual. Limit of using feelings and emotions, both in self and in others, for higher sane objectives, for example, picking up, instructing, profitability and inventiveness, is an incredible expertise.

These components of emotional intelligence rely on the three abilities [1] of CIS presented above: having, recognizing, and expressing emotions. Computers which we are developing as cognitive intelligent system have to be aware of it and will be able to regulate and utilize it.

9.2   EMOTIONS AND COGNITIVE INTELLIGENT SYSTEMS (CIS)

Cognitive Intelligent Systems (CIS) or Affective Computers build around combination of four computing domains as machine learning, algorithms, cognitive vision, and big data [2]. Emotions can be expressed without actually having any of them. Is it correct to make computers emotionally intelligent? Emotional registering likewise has its expected concerns and good dangers, which will be examined finally.

In the chart, high explanation and discerning idea additionally add to elevated level social and good feelings to shape the particular part of dynamic [3] that is morals.

Propelled thinking works along these lines and alludes to the procedure by which enthusiastic musings increase extra essentialness through the utilization of objective proof and information. The other way, discerning proof can be forced upon specific sorts of enthusiastic idea to deliver the kind of programmed moral dynamic that underlies instinctive thoughts of good and insidiousness. For instance, in assessing the profound quality of interbreeding, trial proof proposes that individuals choose rapidly at the inner mind and natural level and later force impromptu balanced proof on their choice. Then again, complex good difficulties, for example, regardless of whether to send a country to war are (one expectations) educated by a plenitude of judicious proof.

On the left half of the Figure 9.1, the real parts of feeling are spoken to as a circle from passionate idea to the body and back. Here, enthusiastic contemplations [3], either cognizant or non-cognizant, can adjust the condition of the body in trademark ways, for example, by straining or loosening up the skeletal muscles or by changing the pulse. Thusly, the substantial impressions of these changes, either real or reenacted, contribute either intentionally or non-deliberately to sentiments, which would then be able to impact thought. (Reenacted body sensation alludes to the way that occasionally envisioning real changes is adequate; really straining the clench hands, for instance, isn't vital.) This is the course by which sound consultations over, state, a country's wartime choices can deliver significant level social feelings, for example, irateness, just as the real indications of these feelings, for example, strained clench hands, expanded pulse, or loss of hunger. The sentiment of these substantial sensations, either intentionally or not, would then be able to inclination intellectual procedures, for example, consideration and memory toward, for this situation, animosity. The final product might be a ridiculous contention with one's companion over a point

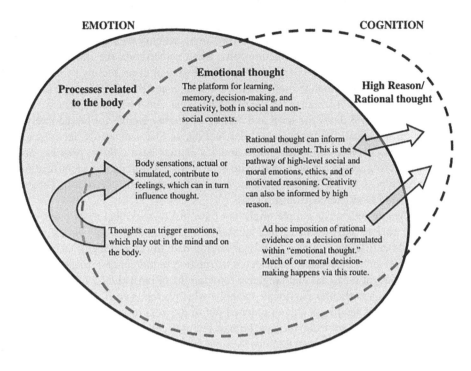

FIGURE 9.1   The emotion-cognition domain.

absolutely inconsequential to the war, the making of a somber and furious conceptual artistic creation, or for the most part tense state of mind.

Notwithstanding the proof talked about above, support for these connections
between the body, feeling, and insight comes mostly from neurobiological and
psychophysiological research [4], in which the acceptance of feeling, either legiti-
mately by a boost in the earth or in a roundabout way through musings or recol-
lections, causes mental changes just as physiological impacts on the body. Thus,
sentiments of feeling depend on the somatosensory frameworks of the mind. That
is, the mind territories related with interoception (the detecting of body states)
are especially dynamic as individuals feel feelings [5], for example, bliss, dread,
outrage, or pity.

9.2.1   RESEARCH CHALLENGES IN EMOTION RECOGNITION

To finish up, in introducing this model, we will likely comprehend forms that incorporate learning, memory, dynamic, and imagination, just as high explanation and
balanced reasoning. They likewise remember the impact of the psyche for the body
and of the body on the brain.

Feelings are mainly recognized by three factors in brain science: Valence
(Pleasure), Arousal and Dominance (Control) otherwise called PAD [6]. The main
reason is Valence or Pleasure, which shows the level of joy, regardless of whether it

is a positive or a negative feeling. The subsequent factor is Arousal which discusses the measure of vitality of feelings. In writing, extraordinary measure of vitality has been named, for example, satisfaction, pity, etc. Furthermore, the third reason is Dominance (Control), which depicts the quality degree of every feeling.

Right off the bat, human feelings are not consistent, and they don't happen remain in one single level. For instance, on the off chance that somebody feel upbeat for few days may be 2 or 3 days, the level and feeling of quality would not be the equivalent for each second that he has a similar feeling. In different words, feelings are more than the twofold estimations of 0 and 1. Feelings have a fluffy premise, and discrete figuring would not be a legitimate answer for assessment and identification. Figure has represented in most of the time by the three elements of Arousal, Valence (Pleasure) and Dominance (Control) [7,8].

Furthermore, clients as people might not have just a single feeling at once; they may have various feelings in various qualities. It is certain that our way of life is influenced by different feelings, and we are living with the mix of them.

Thirdly, mental exploration indicated a distinction in human enthusiastic change design dependent on social and language foundations of individuals. These distinctions state that a particular passionate example which is removed from an individual from a particular land region can't reached out to the next piece of the world. This examination endeavors to investigate human feelings dependent on a fluffy model which can perceive the human feelings.

9.2.2 RECOGNIZING EMOTIONS

Recognition of emotion might be intuitive and instantaneous; this does not preclude the involvement of much knowledge and experience. It may consist of phenomenally instant integrations in simple or well-known situations. Recognition of emotion states usually comprises of conscious hypotheses and continuous self-corrections mechanism. It often includes explicit inferential activities, utilizations of previous experiences [7] and reasoning by analogy.

The core of CIS is the study and advancement of frameworks that can decipher, perceive, measure, and reproduce human conduct. It is a mixed field of computer science and other discipline, cognitive science and psychology. One of the fundamental motivations of the research is to develop the ability in machines for having perception of emotional intelligence, including to simulate empathy [9]. Such a machine which cannot just decipher the passionate condition of people yet in addition be capable adjusts its conduct for giving a fitting reaction to those feelings.

For perceiving customary human feelings PC requires human detects like video and sound visuals, outward appearances and vocal pitches and so forth. In addition to that, it may recognize inputs like- electro thermal conductivity and infrared temperature etc. Sooner, sensing the inputs the system can use its knowledge about emotion it has gathered over learning to infer the underlying emotional state.

A computer is not necessarily being concerned or connected with human senses when it is recognizing emotions for human. Input in the form of Video and audio visuals could be enlarged with internal heat level and skin conductivity, pulse and breath, and so forth.

Having the option to just perceive client's feelings and adjust its conduct as needs be makes well-carrying on PC. NO one will like the computer with negatively emotional shouting and insulting the user. However, it's not practical in terms of necessity of making completely emotionally capable computer to the extent of computer's ability to understand human's negative behavior [10].

9.2.3 Recreating Emotions

The basic requirement for a cognitive intelligent system for expressing feelings is to have modes of correspondence, for example, voice, interactive avatars, and a sort of ability to mimic feelings using those mediums. This may include human like 1D imaging [10] for a better experience of virtual reality using knowledge learned from supervisor or user.

Data age empowered to flood with a great deal of data and examination improvements that may prompt psychological depletion and failure to handle new sources of info adequately. So, it is wise to process the information in parallel to relieve the cognitive load [11].

Here we have already quoted that a cognitive intelligent system can express emotions without really "having" it as when programmed to be happy (for example). On the other side, mostly there exists some form of emotions in human beings, whether or not they choose to express them (if they have ability to control the same). Passionate state and communicating feeling are additionally coupled. Giving a grin should cause an individual and encompassing to feel cheerful yet in the event that an individual inclination shocked can scarcely fulfill a persuading face, doesn't make a difference how earnestly he attempted. Having layers of emotions usually affects the ability to express it. This complex but usual, natural and common behavior is essential to consider for designing cognitive intelligent system.

9.2.4 Components of Emotions

Is there a chance that machines can feel? Absolutely this is the most significant inquiry in the field of CIS. Sentiments are considered as the division line between a human and a machine. The question is deeply related to the issue of computers having consciousness (The major concerns around the world). Awareness is likewise the fundamental capability for most human feelings, similar to disgrace and sew – on the off chance that somebody doesn't have cognizance, none should be embarrassed about anything. Fitting the bill to sort of machine having feelings [11].

The primary component is 'developing feelings', are those which are able to be the caring frameworks having perceptible enthusiastic conduct which spread towards the client.

Second component is 'fast primary emotions'. More precisely – reaction to precautionary or panic situation such as anger or fear. These essential feelings work through two imparting design acknowledgment frameworks: a harsh framework that demonstrates quick and gets seized, on the opposite side a refined component or structure that is more slow yet will in general be more exact.

Third component is 'cognitive emotions' that is significant or intense feeling to stay in memory for a long time such as Happiness or Satisfaction on achieving something.

The fourth component is 'emotional experience'. Derived through previous experiences in complex situations such as having awareness, physiological accompaniments or gut feeling about something good or bad, win or lose.

The fifth component is 'body-mind interactions' [12]. Emotions influence perception, interest, priorities, learning, creativity and ultimately the decision-making. Feelings impact perception and subsequently knowledge, particularly with regard to social dynamic and collaboration, which lead to impact vocal and outward appearances, stance and development. Needless to mention it also influence by biochemical processes like hormones and neurotransmitter triggers. This body-mind interaction is very crucial in terms of influence of emotion on cognitive processes.

9.3 METHODS FOR IMPLEMENTATION OF EMOTIONAL INTELLIGENCE

All in all, the brain research, intellectual science and neuroscience, proposes two primary methodologies as nonstop or straight out for unfurling that how people see and order feelings. The characterization is frequently characterized by utilizing measurements, for example, negative versus positive, quiet versus excited. On the opposite side, the unmitigated methodology will in general utilize discrete classes, for example, upbeat, dismal, furious, dread, shock, and sicken.

Various types of AI calculations and characterization models [13] are utilized for creating machines that produce consistent or discrete names. Here and there not many models are additionally proposing to construct blends over the classes, for example, a glad astonished or a dreadful amazed and so on.

The accompanying methods will be considered broadly for the undertaking of feeling acknowledgment.

9.3.1 EMOTIONAL SPEECH RECOGNITION

A few changes in the autonomic sensory system can in a roundabout way influence or modify an individual's discourse, and full of feeling advancements can impact these data to perceive feeling. For instance, discourse delivered in a condition of outrage, dread or delight reflects higher and more extensive territory in pitch with firm, uproarious and decisively voiced yield, though feelings, for example, trouble, disturb, sleepiness or fatigue grade to produce low-pitched, slow and now and again indistinct discourse.

Passionate discourse test handling advancements perceive the client's enthusiastic state utilizing computational investigation of discourse highlights. Verbal boundaries and prosodic highlights like pitch factors and discourse rate can be broke down through example acknowledgment procedures.

Discourse examination and combination is a compelling technique for distinguishing passionate states with a normal exactness of ~70% [14].

9.3.2 FACIAL EMOTION RECOGNITION

This can be accomplished through different strategies, for example, optical stream, shrouded Markov models, neural system handling or dynamic appearance models. Beyond this, modalities can be consolidated or combined to give an increasingly hearty estimation of the subject's passionate state [15].

9.3.2.1 Facial Expression Databases

To address this dull and tedious assignment, the specialists work with one kind of databases, for example, pictures of database of pinnacle articulation, picture succession database delineating a feeling from unbiased to the top, alongside concerned video cuts. Numerous outward appearance databases are accessible in open space for research. Two of the usually utilized databases are CK+ and JAFFE [16].

By Paul Ekman, it is the same as past segment as Happy, Anger, Disgust, Sad, Fear and Surprise yet during the 1990s Ekman extended his rundown of essential feelings, including a scope of positive and negative feelings. The new arrangement of feelings include: Pride in accomplishment, Guilt, Relief, Amusement, Contempt, Contentment, Embarrassment, Excitement, Shame, Satisfaction and Pleasure [17].

9.3.3 VISUAL AESTHETICS

Feel alludes to the standards of the nature and valuation for excellence. Making a decision about tasteful characteristics is an exceptionally emotional undertaking. Methods recommend [18] to remove certain visual highlights depending on the instinct that they can segregate between stylishly satisfying and disappointing pictures.

9.4 ARTIFICIAL NEURAL NETWORK AND FUZZY INFERENCE SYSTEM

9.4.1 ARTIFICIAL NEURAL NETWORK

In current several real-world applications, we need our machines and computers to be able to perform complex pattern recognition task. Since our conventional computers clearly do not cope-up with these types of problem, we therefore derive analogy from the biological neuron which has come to be known as Artificial Neural System (ANS) Technology [19,20] or simply Neural Networks.

Counterfeit Neural Networks are being touted as a flood of things to come in processing. This system makes them learn components that needn't bother with the conventional abilities of any software engineer.

Fake neural system is an intercommoned gathering of neuron which go about as are non-direct data (signal) preparing gadgets. The rudimentary preparing gadgets called neurons.

So by definition, when a lot of interconnected handling units offer a rich structure displaying a few highlights of the natural neural system, then such a structure is frequently called an Artificial Neural Network (ANN). Since ANNs are actualized on PCs, it merits contrasting the handling capacities of a PC with those of the brain.

9.4.2 Fuzzy Inference System

The articulation "fleecy method of reasoning" was given the 1965 recommendation of cushy set theory by Lotfi A. Zadeh. Cushioned method of reasoning is a kind of many-regarded justification or probabilistic basis; it oversees believing that is vague rather than fixed and positive. Diverged from standard combined sets (where components may take on evident or fake characteristics), feathery method of reasoning elements may have a reality regard that ranges in degree some place in the scope of 0 and 1 as shown in Figure 9.2. Soft method of reasoning has been connected with manage the possibility of fragmentary truth, where reality worth may run between absolutely evident and absolutely sham. Furthermore, when semantic elements are used, these degrees may be regulated by express limits. The investment limit of a cushioned set is a theory of the marker work in old style sets. In soft reason, it addresses the degree of truth as an enlargement of valuation.

Fuzzy Inference Systems take data sources and process them depending on the pre-indicated rules to deliver the yields as shown in Figure 9.3. Both the data sources and yields are genuinely esteemed, while the inner handling depends on fluffy guidelines and fluffy number-crunching. Let us study the handling of the fluffy derivation frameworks with a little model. To make things basic, let us think about a framework with just two sources of info and one yield. Think about the contributions as good ways from deterrent in front and vehicle at right path. Think about the yield as guiding. The principal thing to be done is to separate all sources of info and yields into participation capacities.

To make things clear, let the data great ways from impediment in front have three enlistment limits which are close, far and far. The underlying two support limits are Gaussian, while the third is sigmoidal. Further, let the data vehicle at right way have only two enlistment limits, which are close and far. Both the interest limits are taken as Gaussian. Let the yield coordinating have five investment limits

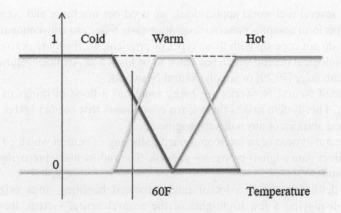

FIGURE 9.2   Illustration of Fuzzy membership function.

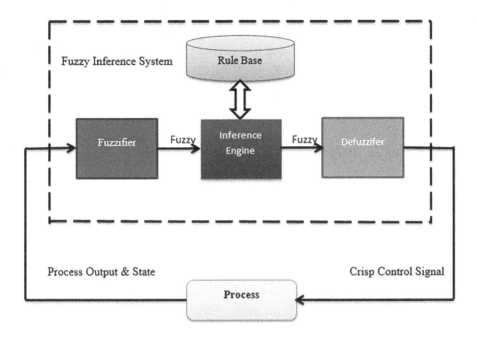

FIGURE 9.3   The Fuzzy logic controller.

steep left, left, no controlling, right and steep right. It must be centered around that the certifiable systems may have endless enlistment limits depending on the multifaceted nature.

9.5   THE INTEGRATED NEURO FUZZY APPROACH

In the field of electronic thinking, neuro-feathery insinuates blends of phony neural frameworks and cushy method of reasoning. Neuro-fleecy hybridization realizes a blend sharp structure that synergizes these two techniques by combining the human-like considering style cushioned systems with the learning and connectionist structure of neural frameworks. Neuro-fleecy hybridization is for the most part named Fuzzy Neural Network (FNN) or Neuro-Fuzzy System (NFS) [21] in the composition. Neuro-cushioned structure (the more notable term is used from this time forward) wires the human-like considering style soft systems utilizing cushy sets and a phonetic modeling including a ton of IF-THEN fleecy rules. The standard idea of neuro-woolen frameworks is that they are limitless approximators with the capacity to request interpretable IF-THEN measures. Following are the assortment to discuss around there – Adaptive neuro-fluffy surmising frameworks (ANFIS), which utilize neural systems to learn fluffy frameworks.

- Fuzzy neural systems, which are ANNs that take fluffy data sources.
- Rule extraction in ANNs, which separate fluffy guidelines from a prepared ANN.

9.5.1 ADAPTIVE NEURO FUZZY INFERENCE SYSTEM (ANFIS)

The versatile neuro-fluffy surmising framework (ANFIS) has a place with the class of frameworks usually known as neuro-fluffy frameworks. Neuro-fluffy frameworks consolidate the forces of ANN with those of fluffy frameworks. As examined before, the fundamental thought process of such a blend is to take the best highlights of both to make an increasingly strong framework as shown in Figure 9.4.

9.5.2 TYPES OF ANFIS

ANFIS might be grouped into different kinds, contingent upon the standards or rationale utilized. We briefly examine the various orders in this area. In view of the rationale in which the standards are framed or associated, the ANFIS is partitioned into two significant sorts,

- **OR type:** This framework uses OR rationale for the different standards that are encircled.
- **AND type:** This framework associates the principles with the assistance of the AND administrator.
- Besides the ANFIS might be partitioned based on the overseeing rationale.
- **Mamdani approach:** This framework depends on the Mamdani rationale.
- **Takagi and Sugeno's approach:** Here the FIS utilizes a sensible working of the principles.

9.5.3 ADVANTAGES OF ANFIS

- To characterize the conduct of a perplexing framework. ANFIS improves fluffy on the off chance that rules. It doesn't require earlier human aptitude.
- Larger decision of participation capacities to utilize.
- It utilizes participation capacities.
- It has extremely quick combination time.
- Desired dataset to estimated.

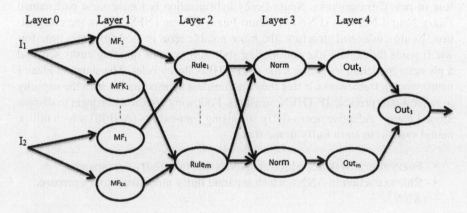

FIGURE 9.4 The ANFIS structure.

9.5.3.1 NFS through Mamdani Approach

Total five layers having two input and one output shown in Figure 9.5. This approach is used to implement fuzzy logic tool or controller using the structure of an ANN and shows that we can develop a neuro fuzzy system.

- Membership Function Distribution

    Reason behind using this network – so that we can modify this network in order to train the mamdani approach of fuzzy reasoning tool using the principle of backpropagation algorithm or we can use some nature inspired optimization tool like GA and other (Figures 9.6 and 9.7).

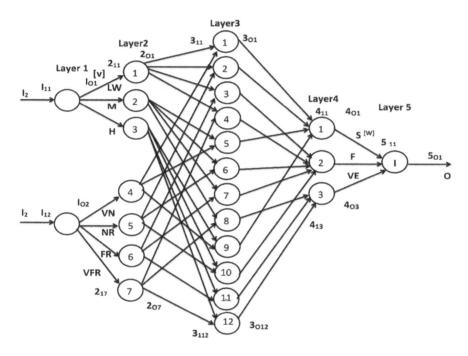

FIGURE 9.5 The NFS architecture (Mamdani approach).

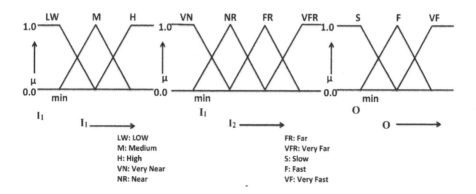

FIGURE 9.6 Membership function (Mamdani approach).

	VN	NR	FR	VFR
LW	S	S	F	F
M	S	F	F	VF
H	S	F	VF	VF

FIGURE 9.7   Rule base (Mamdani approach).

Layer 2: Fuzzification of the input
Layer 1: Performs logical AND operation
Layer 4: Carries out the task of fuzzy inference
Layer 5: Defuzzification – through center of sums method
• Tuning of NFS
  • Batch mode of training
  • BP algorithm
  • Nature inspired optimization tool (such as genetic algorithm)

9.5.3.2   NFS through Takagi & Sugeno's Approach

It processes with two inputs and one output. According to this approach:

$$Y^i = a_i I_1 + b_i I_2 + c_i$$

where $I = 1, 2, 1,\dots, 9$
a_i, b_i, c_i are coefficients

• Membership Function Distribution

9.5.4   Convergence in ANFIS

The preparation diagram of the ANFIS is shown in Figures 9.8 and 9.9, in which we plot the mistakes against preparing emphases, which carries on along these lines to the preparation chart of the ANN. While building in ANFIS, we consider that to be the emphasess develop, the blunder gets littler, and the yields of the data sources given to the ANFIS [22] intently follow real qualities. This is to a great extent a direct result of the angle drop approach that is utilized in both these calculations. Thus, the preparation bend of the ANFIS really combines. This combination, both as far as number of cycles and time, relies upon numerous boundaries, including the multifaceted nature of the FIS, the learning rate, etc. We again review that for the most part, three informational indexes are utilized. The first is for preparing, the second for approval, and the third for testing. Along these lines, we had the option to instigate ANN over FIS and to advance the FIS boundaries. Doing so diminished the need of utilizing experimentation to make a FIS [23]. Besides the subsequent FIS gives elite while additionally following the properties of the FIS.

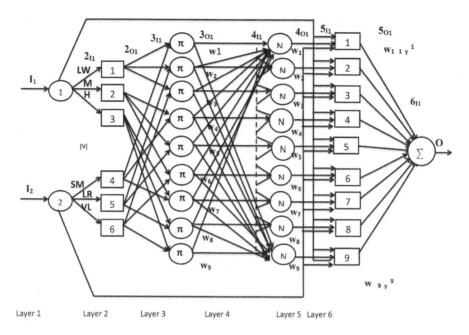

FIGURE 9.8 The ANFIS architecture (Takagi & Sugeno's approach).

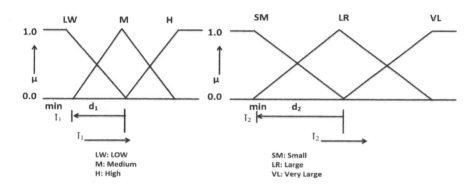

FIGURE 9.9 The membership distribution (Takagi & Sugeno's approach).

9.6 SUMMARY

All studies and experiments clearly indicate that ANFIS reaches the target faster than traditional neural network. When a more complex and sophisticated system with a large dataset is considered, the use of ANFIS in contrast of neural network results as more useful mechanism to achieve faster and accurate results in order to solve the complex problems. Also, in terms of training of the system with specific data-set, ANFIS delivers results with the minimum total error compared to other methods. The complete scenario and studies show that the ANFIS is the best learning methodology among the others, because it combines the advantages of both neural network and fuzzy logic, which offers good results.

However, when the trained parameters should be applied for testing the data than ANN reflects smaller total error than that of ANFIS. Even though it looks like a contradiction, but the core reason expected behind the phenomena should be the size of data that seems too short and possibly not enough for good learning.

The discussion led to motivate us for considering ANFIS approach mainly for carrying our experiment to expect better results. However, we tested traditional ANN approach as well with one dataset.

REFERENCES

1. Pandey, M., et al. (2020) Study and scheme of Neuro-Fuzzy system with various databases for emotional states recognition. *International Journal of Advanced Science and Technology*, 29(7), 804–810.
2. Diwan, R., et al. (2019) Power transmission through solar power satellite to earth surface with minimum power loss. *International Journal of Engineering and Advanced Technology (IJEAT)*, 8(6), 845–850.
3. Awasthy, M. (2020) Mobile robot path planning using weight grid algorithm. *International Journal of Control and Automation*, 13(1), 477–484.
4. Awasthy, M., et al. (2020) Design and analysis of mutual inductance based Wireless Power Transfer (WPT) system suitable for powering Solar Power Satellite (SPS). *International Journal of Advanced Science and Technology*, 29(12s), 941–952.
5. Tiwari, R., et al. (2020) Dynamic load distribution to improve speedup of multi-core system using MPI with virtualization. *International Journal of Advanced Science and Technology*, 29(12s), 931–940.
6. Awasthy, M., et al. (2020) Critical review and assessment of Space Based Solar Power Satellite (SBSPS). *Test Engineering and Management*, 8(6), 11322–11336.
7. Pandey, M. et al. (2020) Review, study and modelling of Neuro-Fuzzy system, it's performance comparision using various speech databases for recognition of emotional states. *Journal of Critical Reviews*, 7(12), 1231–1240.
8. Damasio, A. (1994) *Descarte's Error: Emotion, Reason, and the Human Brain*. New York: Gosset/Putnam Press, pp. 37–45.
9. Elliott, C. (1997) A multimodal approach to expressivity to 'emotionally intelligent' agents. *In Proceedings of the First International Conference on Autonomous Agents*, Marina Del Rey, CA, pp. 451–457.
10. Fogg, J. (1999) Persuasive technologies. *Communications of the ACM*, 42(5), 26–29.
11. MacLean, P. (1970) The triune brain, emotion, and scientific bias. In Schmitt, F. (editor): *The Neurosciences: Second Study Program*. New York: Rockefeller University Press, pp 336–349.
12. MIT (2001) *MIT Media Lab, Affective Computing Web Pages*. Cambridge, MA: Massachusetts Institute of Technology, pp. 79–83.
13. Mowrer, O. (1960) *Learning Theory and Behavior*. New York: John Wiley & Sons, Inc, pp. 39–41.
14. Picard, R. (1998) Affective computing. In Stork, D. (editor): *HAL's Legacy: 2001's Computer as Dream and Reality*, Cambridge, MA: The MIT Press, pp. 125–131.
15. Tao, J., Tan, T., and Pichard, R.W. (eds.) (2005) Affective computing: A review. In: Affective Computing and Intelligent Interaction, Lecture Notes in Computer Science 3784, Berlin Heidelberg: Springer, pp. 981–995.
16. Picard, R., Garay, N., Cearreta, I., and López, J.M. (1997) Inmaculada Fajardo (April 2006) Assistive technology and affective mediation. In Monica, D. (editor): *Affective Computing*. Cambridge, MA: MIT Press, pp. 1–12.

17. Pandey, M. and Awasthy, M. (2019). A comprehensive study of neuro Fuzzy system and emotional states recognition techniques: All India. *Conference on "New Age Skills"*, April 26th-27st, Chhatrapati Shivaji Institute of Technology, Durg (CG), India, pp. 7–14.
18. Breazeal, C. and Aryananda, L. (2002) Recognition of affective communicative intent in robot-directed speech. *Autonomous Robots*, 12(1), pp. 83–104.
19. Dellaert, F., Polizin, T., and Waibel, A. (1996) Recognizing emotion in speech. *In Proceedings of ICSLP 1996*, Philadelphia, PA, pp. 1970–1973.
20. Roy, D. and Pentland, A. (1996) Automatic spoken affect classification and analysis. *Proceedings of the Second International Conference on Automatic Face and Gesture Recognition*, Killington, Vermon, pp. 363–367.
21. Lee, C.M., Narayanan, S., and Pieraccini, R. (2001) Recognition of negative emotion in the human speech signals. *Workshop on Automatic Speech Recognition and Understanding*, Madonna di Campiglio, Italy, pp. 69–78.
22. Yacoub, S., Simske, S., Lin, X., and Burns, J. (2003) Recognition of emotions in interactive voice response systems. Proceedings of Eurospeech, Geneva, Switzerland, pp. 729–732.
23. Khoruzhnikov, S.E., et al. (2014) Extended speech emotion recognition and prediction. *Scientific and Technical Journal of Information Technologies, Mechanics and Optics*, 14(6): 137.

10 Essential Statistical Tools for Analysis of Brain Computer Interface

K. A. Venkatesh
Myanmar Institute of Information Technology (MIIT)

K. Mohanasundaram
AP, School of Business, Alliance University

CONTENTS

10.1 INTRODUCTION TO TESTING OF HYPOTHESIS

A hypothesis in statistics is an open statement or a conjecture with reference to one or more populations. This hypothesis can be verified that is true or false when we have the absolute knowledge about the concerning populations. Having the entire

209

population is a tedious and time-consuming job. In such scenario, one of the tools available to us is hypothesis testing. In hypothesis testing, we use random sample to judge the available evidence to support or reject the hypothesis. Hypothesis testing is a statistical approach to test the claims about a population.

Let us consider an example, in the Bangalore city, the children in the age group of 10–15 spent on an average of 4 hours in playing mobile games per week. To verify this statement, we will collect the data from a sample of 20 children from across the city. The mean from these 20 children, known as sample mean, and the given average is the population mean.

10.1.1 Testing of Hypothesis

A hypothesis in statistics is an open statement or a conjuncture with reference to one or more populations [1]. This hypothesis can be verified that is true or false when we have the absolute knowledge about the concerning populations. Having the entire population is a tedious and time-consuming job. In such scenario, one of the tools available to us is hypothesis testing. In hypothesis testing, we use random sample to judge the available evidence to support or reject the hypothesis. Hypothesis testing is a statistical approach to test the claims about a population.

Let us consider an example, in the Bangalore city, the children in the age group of 10–15 spent on an average of 4 hours in playing mobile games per week. To verify this statement, we will collect the data from a sample of 20 children from across the city. The mean from these 20 children, known as sample mean, and the given average is the population mean.

10.1.2 Steps in the Testing of Hypothesis

a. Formulate the claim/hypothesis (null & alternate hypothesis)
b. State the appropriate plan of analysis (set the level of significance and appropriate statistical test)
c. Calculate the test statistic
d. Based on the evidence, make a decision that is to accept H_0.
e. The statement made in the step (a), called null hypothesis, generally about the population and denoted by H_0 and the statement that contradicts the null hypothesis, known as alternate hypothesis H_1

In our example, the null hypothesis is the population mean and we assumed this statement is true that is children in Bangalore City spent 4 hours on mobile games per week and the alternate hypothesis is that children in Bangalore City spending either more than 4 hours or less than 4 hours in mobile games.

In general, "State the criteria for the decision" is the level of significance, for us, we normally set the level of significance to 5%. We compute the sample outcome, assuming that the null hypothesis was to be true. Using the test statistic value, one could decide to accept or reject the hypothesis. Importantly, based on the value of p, the value of p is the probability of obtaining the sample mean, assuming that the value given in the null hypothesis is true. Based on the value of p, one has to accept or reject the null hypothesis. If p-value is <5%, then reject else accept the null hypothesis. One can choose any one of the appropriate tests such as one-sample t test, one-sample z test, and so on. Importantly, accepting the null hypothesis not necessarily mean that it is true, and it indicates that there is a chance that not having enough evidence to support the alternate hypothesis [1].

Illustration 10.1

A firm producing electric motor informs that the defective rate of the entire production in a cycle is 5%. Let us verify this statement.

Solution: Let p be the probability of the defective item and let us formulate the null and alternate hypothesis as $H_0 : p = 0.05$ and $H_1 : p > 0.05$. Let us consider of a sample of 100 electric motors manufactured by this firm and X be the number of defective electric motors in this sample. If $X = 6$, it is tough to reject the null hypothesis; however, $X \geq 12$ then we must reject the alternate (null) hypothesis. Based on X, we are deciding to reject or accept the alternate (null) hypothesis, and hence X is called as test statistic. If we use the Bernoulli process, then the expected number of defectives in this sample is $np = 5$, and $X \geq 12$ is much more than expected number and hence the rejection.

Note (1): Based on the sample size, the information about the population, we make decision, there are possibilities committing mistakes in this process, the possible outcomes are shown in Table 10.1:

Note (2): α denotes the probability of selecting Type I error and it is known as the level of significance

Note (3): In our example, $\alpha = \Pr(X \geq 12, \text{ when } p = 0.05)$

Note (4): β denotes the probability of making Type II error

Note (5): Our example is a one-tailed test with critical region in the right side of the test statistic X

Illustration 10.2

A production line produces on an average of 100 units in a production cycle with a variation of 64.

Solution: Let $H_0 : \mu = 100$ and $H_1 : \mu \neq 100$. Consider a sample of 100 units

Let $\bar{X}$ be the sample mean. Let us say that we would reject the null hypothesis if the sample mean is ≤ 98 or ≥ 102.

TABLE 10.1
Possible Outcomes

Action/Decision	H_0 the Null Hypothesis Is True	H_1 the Alternate Hypothesis Is true
H_0 is accepted	Appropriate choice	Type II - error
H_0 is rejected	Type I - error	Appropriate choice

$$\alpha = \Pr\left(\bar{X} \geq 102, \text{ when } \mu = 100\right) + \Pr\left(\bar{X} \leq 98, \text{ when } \mu = 100\right)$$

$$= \Pr\left(Z < (\bar{X} - \mu)/\sigma/\sqrt{n}\right) + \Pr\left(Z > (\bar{X} - \mu)/\sigma/\sqrt{n}\right) = 0.0124$$

The significance level is given by $(1 - \alpha)100 = 98.76\%$ for the assumed confidence interval [98,102]

Note 1: The example 2 is a two-tailed test

Note 2: We were testing the population mean is equal to the sample mean or not, with the known variance of population

Note 3: When confidence interval is given,

$$1 - \alpha = \Pr\left(-Z_{\frac{\alpha}{2}} < \frac{\bar{X} - \mu_0}{\frac{\sigma}{\sqrt{n}}} < Z_{\frac{\alpha}{2}}\right)$$

Note (4): The critical values are the left, and right can be computed for two-tailed test at α as: left $= \mu_0 - Z_{\frac{\alpha}{2}}\frac{\sigma}{\sqrt{n}}$ and right $= \mu_0 + Z_{\frac{\alpha}{2}}\frac{\sigma}{\sqrt{n}}$. We reject the null hypothesis if $\bar{X} \leq$ left and $\bar{X} \geq$ right

Now we revisit the one-tailed test because we have to consider two cases with $H_0 : \mu = \mu_0$ and (i) $H_1 : \mu > \mu_0$ and (ii) $H_1 : \mu < \mu_0$ (one-tailed sample mean test with variance known).

Illustration 10.3

A white board marker manufacturing company says that a sample of 100 markers can write 470 boards (std size) on an average and the entire production cycle has a variance of 25 boards & the population mean is 480 boards. Apply the testing of hypothesis to verify the population mean $\mu = 480$ versus $\mu < 480$ at $\alpha = 0.05$ level.

Solution: Set up the null and alternate hypothesis

$$H_0 : \mu = 480 \text{ and } H_1 : \mu < 480$$

Step (a): $\alpha = 0.05$

Step (b): here the test-statistic is $\bar{X}$, the sample mean

$$\text{Now } \mu_0 - Z_\alpha\left(\frac{\sigma}{\sqrt{n}}\right) = 475.87$$

Step (c): Compare the sample mean with the test-statistic value

The sample mean is smaller than the test-statistic value, and hence we reject the null hypothesis.

In the previous problem, we tested the sample mean with known variance, similarly, one could test the sample mean with unknown variance. Note that s^2

denotes the sample mean. In this case, we use t-distribution, and instead of z-test, we will be using t-test with $(n-1)$ degrees of freedom.

10.1.3 PARAMETERS OF TWO INDEPENDENT POPULATIONS

Let $(x_{i1}, x_{i2}, \ldots, x_{in})(i = 1, 2)$ be two random samples of size n from the normal population with mean μ_i and the variance σ_i^2. We discuss three cases based on the following information [1]:

Case (1): When variances of both populations are known but unknown means

We first find the confidence interval for the means $\mu_1 = \mu_2$ (or the difference between the means is small positive number, <1)

Define $Z = \dfrac{(\bar{X}_1 - \bar{X}_2) - (\mu_1 - \mu_2)}{\sqrt{(n_2\sigma_1^2 + n_1\sigma_2^2)}} \sqrt{(n_1 n_2)}$ Clearly Z follows the normal distribution

The confidence interval for $\mu_1 - \mu_2 = \epsilon$ is given by

$$\left[(\bar{X}_1 - \bar{X}_2) - Z_{\frac{\alpha}{2}} \sqrt{\frac{n_2\sigma_1^2 + n_1\sigma_2^2}{n_1 n_2}}, \ (\bar{X}_1 - \bar{X}_2) = Z_{\frac{\alpha}{2}} \sqrt{\frac{n_2\sigma_1^2 + n_1\sigma_2^2}{n_1 n_2}} \right]$$

On the basis of available information, we have to test the following

$$H_0 : \mu_1 - \mu_2 = \varepsilon \simeq 0$$

$$\text{Now the test-statistic}, Z = \frac{(\bar{X}_1 - \bar{X}_2) - \epsilon}{\sqrt{\dfrac{n_2\sigma_1^2 + n_1\sigma_2^2}{n_1 n_2}}}$$

Possible alternate hypothesis is

$$H_1 : \mu_1 - \mu_2 > \epsilon, \text{ if } Z < Z_\alpha \text{ then accept } H_0, \quad \text{else reject } H_0$$

$$H_1 : \mu_1 - \mu_2 < \epsilon, \text{ if } Z < Z_{1-\alpha} \text{ or } Z < -Z_\alpha \text{ then reject } H_0, \quad \text{else acept } H_0$$

$$H_1 : \mu_1 - \mu_2 \neq \epsilon, \text{ if } |Z| > |Z_\alpha| \text{ then reject } H_0, \quad \text{else accept } H_0$$

Case (2): When means of both population is known but unknown variances

We first find the confidence interval for σ_1^2 / σ_2^2

Define $F_{m_1, n_2} = \dfrac{S_1^2 / S_2^2}{\sigma_1^2 / \sigma_2^2}$. Clearly F_{m_1, n_2} follows F-distribution with n_1 and n_2 degress of freedom $\left(\text{here } n_1 = n_2 = n \right)$

$$1 - \alpha = \Pr\left(F_{1-\frac{\alpha}{2}, m_1, n_2} \leq \frac{S_1^2 / S_2^2}{\sigma_1^2 / \sigma_2^2} \leq F_{\frac{\alpha}{2}, m_1, n_2} \right)$$

The confidence interval at level $100(1-\alpha)\%$ for $\dfrac{\sigma_1^2}{\sigma_2^2}$ is

$$\frac{S_1^2/S_2^2}{F_{\frac{\alpha}{2},m_1,n_2}}, \; S_1^2/S_2^2 \times F_{\frac{\alpha}{2},m_1,n_2}$$

On the basis of available information, we have to test the following

$$H_0 : \frac{\sigma_1}{\sigma_2} = \epsilon \cong 1$$

Now the test statistic, $F_{m_1,n_2} = \dfrac{S_1^2/S_2^2}{\epsilon^2}$

Possible alternate hypotheses are

$$H_1 : \frac{\sigma_1}{\sigma_2} > \epsilon, \; \text{if} \; F < F_{\alpha,m_1,n_2} \; \text{then accept} \; H_0, \; \text{else reject} \; H_0$$

$$H_1 : \frac{\sigma_1}{\sigma_2} > \epsilon, \; \text{if} \; F > F_{1-\alpha,m_1,n_2} \; \text{then accept} \; H_0, \; \text{else reject} \; H_0$$

$$H_1 : \frac{\sigma_1}{\sigma_2} \langle \epsilon, \; \text{if} \; F \rangle F_{\frac{\alpha}{2},m_1,n_2} \; \text{or} \; F < F_{1-\frac{\alpha}{2},m_1,n_2} \; \text{then accept} \; H_0, \; \text{else reject} \; H_0$$

Case (3): when means and variances of both populations are known

We first find the confidence interval for $\mu_1 - \mu_2$, given that $\sigma_1 = \sigma_2$

$$\text{Define} \; t_{m_1+n_2-2} = \frac{\left(\bar{X}_1 - \bar{X}_2\right) - \left(\mu_1 - \mu_2\right)}{S_k\sqrt{\dfrac{n_1+n_2}{n_1 n_2}}}, \; \text{where} \; s_k^2 = \frac{(n_1-1)S_1^2 + (n_2-1)S_2^2}{n_1+n_2-2}$$

The confidence interval at level $100(1-\alpha)\%$ for $\mu_1 = \mu_2$ is

$$\left[\left(\bar{X}_1 - \bar{X}_2\right) - t_{m_1+n_2-2} \, S_k\sqrt{\frac{n_1+n_2}{n_1 n_2}}, \; \left(\bar{X}_1 - \bar{X}_2\right) + t_{m_1+n_2-2} \, S_k\sqrt{\frac{n_1+n_2}{n_1 n_2}} \right]$$

On the basis of available information, we have to test the following

$$H_0 : \mu_1 - \mu_2 = \epsilon, \; \text{given that} \; \sigma_1 = \sigma_2$$

Now the test-statistic is $t_{m_1+n_2-2} = \dfrac{\left(\bar{X}_1 - \bar{X}_2\right) - \epsilon}{S_k\sqrt{\dfrac{n_1+n_2}{n_1 n_2}}}$

Possible alternate hypothesis are

$$H_1 : \mu_1 - \mu_2 > \epsilon, \text{if} \; t_{m_1+n_2-2} > t_{\alpha,\,m_1+n_2-2} \; \text{then accept} \; H_0, \text{else reject} \; H_0$$

$$H_1 : \mu_1 - \mu_2 < \epsilon, \text{ if } t_{m+n_2-2} < t_{1-\alpha,m+n_2-2} \text{ then reject } H_0, \text{ else accept } H_0$$

$$H_1 : \mu_1 - \mu_2 \neq \epsilon, \text{ if } \left| t_{m+n_2-2} \right| > \left| t_{\frac{\alpha}{2},m+n_2-2} \right| \text{ then reject } H_0, \text{ else accept } H_0$$

Illustration 10.4

Price of 10 essential items during covid19 period are given as: (in INR) 36, 42, 43, 55, 45, 49, 53, 46, 50, 51. Test the variance of the distribution of prices of all essential items during the normal time, from which the sample was drawn, is equal to 20 Rs (given that $\chi^2 = 16.92$ for $df = 9$).

Solution: This falls under the case, variance known, but the mean is unknown.

$$H_0 : \sigma^2 = 20 \text{ and } H_1 : \sigma^2 > 20$$

Test statistic: $\chi^2 = \dfrac{\Sigma_1^{10}(x_i - \bar{x})}{\sigma^2} = 14$. Since there are 10 observations, degrees of freedom is 9.

Now $\chi^2 = 14 < \chi^2_{0.05} = 16.92$. Hence, we have to accept the null hypothesis.

Illustration 10.5

Two random samples of size 9 and 13 are drawn from a population. The variances of both samples are 4.96 and 3.51, respectively. Verify that standard deviations of both samples are equal (given that $F_{0.05} = 2.85$, with df = 8 and 13)

Solution: From the given information, $n_1 = 9; n_2 = 13$ and $S_1^2 = 4.96$ and $S_2^2 = 3.51$
Set the hypothesis $H_0 : \sigma_1 = \sigma_2; \ H_1 : \sigma_1 > \sigma_2$

Test statistic: $F = \dfrac{S_1^2}{S_2^2} = 1.41 < 2.85 = F_{0.05}$ with $df = 9$ and $df = 12$. Hence, we conclude that the standard deviation of two samples may be equal.

Illustration 10.6

A weather lab gathers the temperature information every day at 5 p.m. For a particular study for five consecutive days, temperature is obtained from this lab, the temperatures are 36, 40, 32, 37, 40. Can we say that the average temperature is 38?

Solution: Form the given data, $n = 5, \bar{X} = 37$ and $S^2 = \dfrac{\Sigma_1^5(x_i - \bar{X})^2}{n-1} = 11, m = 38$

$$H_0 : m = 38 \text{ and } H_1 : m \langle 38 \text{ or } m \rangle 38$$

$$S = 3.3166$$

$$t = \frac{\sqrt{n}\left(\overline{X} - m\right)}{S} = -0.6742$$

From the table, $t_{0.05} = 2.78 > 0.6742 = t$. Hence, we have accepted the null hypothesis.

10.2 DESIGN OF EXPERIMENT

EXPERIMENTATION and *MAKING INFERENCES* are the twin essential features of general scientific methodology. To attain these goals, statistics is mainly viewed as the most appropriate tool. Always, it is based on problem of inductive inferences in conjunction with stochastic model describing random phenomena; the researcher may not have entire picture of the true variant of the phenomena under study. Whenever one is interested in studying the specific behavior of the unknown variant of the phenomena, statistics is the way to address.

After setting up the statistical problem, then one has toper form experiments in order to collect the necessary information on the basis of which inferences can be made in the most suitable way.

The *three* main aspects of making inferences are:

1. It *devices the ways for obtaining inference from observation* when these are not exact but subject to variation as such, the inferences are not exact but probabilistic in nature.
2. It *describes the ways for collection of data in an appropriate manner* in such a way that underlying assumptions of appropriate statistical methods for the application are satisfied.
3. *Possible methods for proper interpretation* of results are derived.

Design and analysis of experiments fall in the sphere of data collection and interpretation of the techniques.

10.2.1 Terminologies Used in DOE

An experiment starts with a problem, an answer for which is obtained from the interpretation of the set of observations collected suitably. For this purpose, a set of *experimental units* and adequate experimental materials are required. In agricultural experiments, equal sized plot of lands, a single or a group of plants etc. are used as experimental units. For animal husbandry experiments, animals, animal organ, etc. form the experimental units. For industrial experimentation, machines, nuts, bolts, etc. (all the industrial components) form the experimental units [2].

The problems are usually in the form of comparisons among a set of treatments with respect to some of their effects which are produced when they are applied to the experimental units.

A general name "TREATMENT" is used to denote experimental material among which comparison is desired by utilizing the effects which are produced when the experimental materials is applied to the experimental units. For example: in agricultural experiments, different varieties of food crop, different fertilizer doses, different levels of irrigation, different combinations of levels of two or more of the above factors namely variety, irrigation, nitrogen fertilizers, etc. may form the treatment.

A *design* for an experiment defines the size and number of the experimental units, the manner in which the treatment is allotted to the units along with grouping of the experimental units. It ensures validity and accuracy of the results obtained from an analysis of the observations.

10.2.2 PRINCIPLE OF DESIGN OF EXPERIMENT

For this purpose, we use the following three principles of experiments:

1. **Randomization**: Manner in which allocation of the treatments to experimental units, essentially to minimize the bias in the result. It is a simple method to achieve the independence of errors

   Suppose we want to test the effectiveness of a drug in controlling certain disease, we take two groups of 100 persons each and to one group we can administer the drug and to the other group we can administer the similar looking "placebo" (empty content of the drug) and then we can compare the rate of cures in the two groups. But there are certain difficulties, in general, the comparison may not be valid for the first group which may contain physically stronger persons and the second group may contain physically weaker person, then even if the rate of cure is higher in the first group, it cannot be attributed to the drug. To counter this argument, we can select two groups similar in physical condition. But, then the first group now may contain richer people and the second group may contain poorer people and again the comparison may not be valid. Then we can select the two groups now in such a way that they are similar w.r.t. physical and financial conditions are concerned. But even then the two groups may not be identical w.r.t. all possible causes affecting the rate of cure. Hence deliberate balancing of groups can be done only to certain extent. Hence, bias enters into the problem; to remove this bias (Statistical bias), the two groups may be selected at random from population, then there will not be any bias. After selecting the groups at random, if there is any difference in the rate of cure then that can be attributed to the effect of drug.

2. **Replication**: This specifies the number of units to be provided for each of the treatment. A treatment is repeated a number of times in order to get more reliable estimates than it is possible from a single observation. It is necessary to increase the accuracy of estimate of treatment effect. It also provides an estimate of the error variance which is the functions of different among observations from experimental units and their identical treatments. Though, more the number of replications, better it will be so, far as the precision of estimators is concerned, but is cannot be increased indefinitely as it increases the cost of the experimentation. Moreover, due to limited availability of the experimental resources, too many replications cannot be taken.

3. **Local control**: The term local control applies to any non-statistical method of increasing the efficiency and precision of the design. In animal experiments, it is better to use animals from the same litter for they are likely to be homogeneous and any difference in the effect seen may be attributed to relevant causes.

In a classical experiment of "lady tasting tea," R.A. Fisher gives a proper form of local control. A lady claims that she can distinguish tea prepared by adding tea infusion to milk from tea prepared by adding to milk to tea infusion. To test her claim, she is given eight cups of tea, four prepared in one way and another four prepared in another way. The local control here is that eight cups are similar in all respects, the temperature, a color, etc. of a tea are same in all respects [2].

10.3 COMPLETELY RANDOMIZED DESIGN (ONE WAY ANOVA)

10.3.1 ONE WAY CLASSIFICATION: COMPLETELY RANDOMIZED DESIGN

A *one–independent variable* experiment is called *one-way ANOVA (CRD-Completely Randomized Design)*. ANOVA is the acronym for Analysis of Variance, the techniques utilized for studying the cause-and-effect of one or more factors on a single dependent variable. Example: Marketing Manager wants to investigate the impact of different discount schemes on the sale of three major brands of edible oil. When the independent variables are of nominal scale (categorical) and the dependent variable is metric (continuous), ANOVA is applied. Example for dependent variable might be: sales, attitudes, preferences, and independent variables may be number of packages design, number of regions, and number of discount schemes. CRD is deployed when there is only one categorical independent variable and one dependent (metric) variable [3].

10.3.2 DESIGN FOR ONE WAY ANOVA

In Completely Randomized Design (CRD), the problem is to compare "v" treatments; we take "vn" plots homogeneous in all respects. We apply each treatment to n-plots chosen at *random* from "vn" plots, and then we observe the yield.

Let Y_{ij} be the yield in the jth plot of the ith treatment, $i = 1,2,...,v$, $j = 1,2,...,n$
We use the following linear model

$$Y_{ij} = \mu + t_i + \varepsilon_{ij}$$

Where,

μ: the overall average effect
t_i: effect of the ith treatment
ε_{ij}: the random component or error term.

TABLE 10.2

ANOVA

Source	df	SS	MSS	$F_{\text{calculated}}$	$F_{\text{tabulated}}$
Treatment	$v - 1(a_1)$	$T_r.S.S(a_2)$	$T_r\text{MSS} = (a_2)/(a_1)$	TrMSS/EMSS	$F_{v-1, (n-1)v}\text{df}$
Error	$(n - 1)v\,(b_1)$	ESS (b_2)	$\text{EMSS} = (b_2)/(b_1)$		
Total	$vn - 1$	TSS			

Inference: If the calculated F value exceeds the tabular value with $[(v - 1), (n - 1)v]$ at defined level of significance, we reject the null hypothesis, otherwise we accept.

Assumptions about parameter:

- μ and t_i are unknown parameters
- ϵ_{ij} is the random variable, ϵ_{ij}'s are iid random variables which are normally distributed with $N\left(0, \sigma^2\right)$

Estimating the unknown parameters using least square methods, we get the ANOVA Table 10.2.

Testing the null hypothesis that all t_i's are equal. $H_0 : \mu_1 = \mu_2 = \cdots = \mu_v$ at set a level of significance.

We use the test statistics as

$$F_{\text{cal}} = \frac{\text{TrSS}/(v-1)}{\text{ESS}/(n-1)v} \sim Fv - 1, \; (n-1)v\text{df}$$

Note 1: In our linear model, we assume that each treatment is applied to equal number of plots(n plot) but sometimes, we may have unequal number of plots to each treatment. In such cases, we can assume that i^{th} treatment is applied to n_i plots.

Note 2: Apply post-hoc tests (paired comparison between groups), if ANOVA is significant

10.3.3 ADVANTAGES AND DISADVANTAGES OF COMPLETELY RANDOMIZED DESIGN

Advantages:
- It results in the maximum use of the experimental units.
- Design is very flexible
- The statistical analysis remains simple if some or all the observations for any treatment are rejected.
- It provides maximum number of degrees of freedom for the estimation of the error variance, which increases the sensitivity or the precision of the experiment for small experiments, i.e., for experiments with small number of treatments.

Disadvantages:

In certain circumstances, the design suffers from the disadvantage of being inherently less informative than other more sophisticated layouts, if experimental units are heterogeneous.

Local control has not been used in CRD.

Illustration 10.7

Table 10.3 shows the number of claims processed per day for a group of four insurance companies' employees observed for a number of days. Test the hypothesis that the employees' mean claims processed per day are the same. Use the 0.05 level of significance.

Solution:

Step 1: Set up the null and alternative hypothesis:

H_0: Employee's mean claims processed per day are the same.

$$v / s$$

H_1: Employee's mean claims processed per day are not same.

Step 2: Level of significance $\alpha = 0.05$

Step 3: Test statistics

$$F = \frac{S_1^2}{S_2^2}$$

Where, $S_1^2 =$ variance between samples

$S_2^2 =$ variance within samples

Step 4: Calculations are shown in Table 10.4

TABLE 10.3

No of Claims Processed by Employee

Employee 1	15	17	14	12		
Employee 2	12	10	13	17		
Employee 3	11	14	13	15	12	9
Employee 4	13	12	12	14	10	

TABLE 10.4

Calculation

Employee 1	15	17	14	12			58
Employee 2	12	10	13	17			52
Employee 3	11	14	13	15	12	9	74
Employee 4	13	12	12	14	10		61

TABLE 10.5

ANOVA

Source	Degree of Freedom (df)	Sum of Squares	Mean Sum of Squares	$F_{calculated}$	$F_{tablulated}$
Treatment (employee)	$4 - 1 = 3$	14.65613	4.88537	$\dfrac{4.88537}{4.7429} = 1.0249$	$F_{(3,15)}df = 3.29$
Error	15	71.14389	4.7429		
Total	$19 - 1 = 18$	85.8	4.76666		

$$\text{Correction factor}(CF) = \frac{\left(\text{Grand total}\right)^2}{n} = \frac{(245)^2}{19} = 3159.2$$

$$= 3245 - 3159.2 = 85.8$$

$$\text{Treatment sum of square}(TrSS) = \sum \frac{(T_i)^2}{n_i} - CF$$

$$= \frac{(58)^2}{4} + \frac{(52)^2}{4} + \frac{(74)^2}{6} + \frac{(61)^2}{5} - 3159.2$$

$$= 14.65613$$

Step 5: Inference

Since the table value is greater than calculated value in Table 10.5, there is no sample evidence to reject H_0.

Therefore, the employee's mean claims processed per day are the same.

Illustration 10.8

In class 60 students, the teacher identified that 14 students require additional help in learning. In addition, the school counselor confirmed that these 14 students must be taught in three different methods, and she grouped them in three groups, which consisting of 5, 4, 5 students based on the methods to be adopted. This processes continued until the end of particular term. At the end of the academic term, results of the administrated test are shown in Table 10.6. Apply ANOVA to verify that if there is a difference in means of these three methods.

TABLE 10.6

Test Scores by Four Methods

Method	Test Scores					Total
Method 1	75	56	45	70	75	321
Method 2	60	65	70	62		257
Method 3	80	75	70	65	70	360
Total						938

TABLE 10.7
Computation

	Sum of Squares	Mean Square	Degrees of Freedom	F Test
Between the samples	107.3	53.63	2	0.729
Within the samples	809.3	73.65	11	
Total	916.6		13	Pr (>F) = 0.504

Solution:

H_0: means of all groups are equal and H_1: means of all groups are not equal

(here the alternate hypothesis is that at least one group is not equal to the others)
The computation are given in Table 10.7

From Table 10.7, $F_{2,11} = 0.729$ and from the F-table, the value of F at 5% level with df = 2 and df = 11 is 3.98, which is more the obtained F-value, hence accept the null hypothesis.

Note: ANOVA never indicates that which mean differs from the rest. In order to identify the differing group mean, we have to perform one more test called Scheffe Test, when our groups are of different size, or Tukey Test when our groups are of equal size.

10.4 TWO-WAY CLASSIFICATION OR RANDOMIZED BLOCK DESIGN

In one-way classification, we assume that all the plots are more or less homogeneous in all aspects, but in many cases, it is rather difficult to have a large number of plots to be homogeneous in all aspects, in such cases, a CRD cannot give efficient result and hence the differences among the treatments cannot be properly inferred. In such cases, the plots are divided into blocks such that plots occurring in any block are more or less homogeneous even though between any two blocks, there can be considerable variation. Each treatment may be considered as one-way classification as far as the application of treatment is concerned [1,2].

One important point to note is that the randomization for the applications of treatment within each block must be *independent* for the different blocks.

The primary interest in Randomized Block Design is one factor. It may also have several other nuisance factors which may affect the measured result. The way to control or eliminate the contribution of nuisance factor is by blocking.

10.4.1 DESIGN OF TWO WAY ANOVA (RANDOMIZED BLOCK DESIGN WITHOUT INTERACTION)

Suppose there are "v" treatments, we can have "b" blocks, containing "vn- plots". *Here, the combination of block and treatment is called the* "cell." Thus, we have "bv" cells, each cell will then contain "n" treatments. This is usually termed as RBD with *more than one observation per cell.*

Consider the case when $n = 1$

This is said to be RBD with one observation per cell and is called as *RBD with no interaction.*

Let Y_{ij} be the yield in the jth block of the ith treatment, $i = 1, 2, ..., v$, $j = 1, 2, ..., b$

We use the following linear model

$$Y_{ij} = \mu + t_i + b_j + \varepsilon_{ij}$$

Where,

μ: the overall average effect

t_i: effect of the ith treatment

b_j: effect of the jth block

ε_{ij}: the random component or error term.

Assumptions about parameter:

- μ, t_i, and b_j are unknown parameters
- ε_{ij} is the random variable, ε_{ij} - s are iid random variables which are normally distributed with mean zero and variance as σ^2 unknown. This property is called homoscedasticity property.

Estimating the unknown parameters using least square methods, we get the ANOVA-RBD Table 10.8.

Testing the null hypothesis that all t_i's are equal and all b_j's are equal.

$$H_{01} : T_1 = T_2 = \cdots = T_v \text{ and}$$

$$H_{02} : B_1 = B_2 = \cdots = B_b \text{ at a set level of significance.}$$

TABLE 10.8
ANOVA-RBD

Source	df	SS	MSS	$F_{calculated}$	$F_{tabulated}$
Treatment	$v - 1$	Tr.S.S	TrMSS	TrMSS/EMSS	$F_{v-1,\,[(v-1)(b-1)]}$df
Block	$b - 1$	B.S.S	BMSS	BMSS/EMSS	$F_{b-1,\,[(v-1)(b-1)]}$df
Error	$(v-1)(b-1)$	ESS	EMSS = ESS/$(v-1)(b-1)$		
Total	$vb - 1$	TSS			

Inference: If the calculated F value exceeds the tabular value with $[(v-1),(v-1)(b-1)]$ and $[(b-1),(v-1)(b-1)]$ at defined α, we reject the null hypothesis with treatment and block respectively, otherwise we accept.

Advantages of RBD

1. **Accuracy**: This design has been shown to be more efficient or accurate than CRD
2. **Flexibility**: There is no restrictions on the number of treatments or the number of replicates.
3 **Easy ways of analysis**: Statistical analysis is simple and rapid. Moreover, the error of any treatment can be isolated and any number of treatments may be omitted from the analysis without complicating it.

Disadvantages of RBD

It is not suitable for large number of treatments.

Illustration 10.9

The data on production rate by five workmen on four machines are given in Table 10.9. Test whether the rate is significantly different due to workers and machines. Test at 5% level of significance.

Solution:

Step 1:

H_{01}: There is no change in the workmen Production rate.

H_{02}: There is no change in the production rate due to machines

Step 2: $\alpha = 0.05$

Step 3: Test-statistic

$$F = \frac{S_1^2}{S_2^2}$$

Where, S_1^2 = variance between samples

S_2^2 = variance within samples

Step 4: Calculation

The calculated values are shown in Table 10.10

TABLE 10.9

Production Rate

Machines			Workmen		
1	46	48	36	35	40
2	40	42	38	40	44
3	49	54	46	48	51
4	38	45	34	35	41

TABLE 10.10
Calculation

Machines			Workmen			Total
1	46	48	36	35	40	205
2	40	42	38	40	44	204
3	49	54	46	48	51	248
4	38	45	34	35	41	193
Total	173	189	154	158	176	850

$$\text{Correction factor}(CF) = \frac{\left(\text{Grand total}\right)^2}{n} = \frac{(850)^2}{20} = 36125$$

$$TSS = 36754 - 36125 = 629$$

$$\text{Treatment Sum of square}(TrSS) = \sum \frac{(T_i)^2}{n_i} - CF$$

$$(\text{Machine}) = \frac{(205)^2}{5} + \frac{(204)^2}{5} + \frac{(248)^2}{5} + \frac{(193)^2}{5} - 36125 = 353.8$$

$$\text{Workmen sum of squares} = \frac{(173)^2}{4} + \frac{(189)^2}{4} + \frac{(154)^2}{4} + \frac{(158)^2}{4} + \frac{(176)^2}{4} - 36125$$

$$= 201.5$$

Step 5: Inference

As the table value is lesser than the calculated value in Table 10.11, both the null hypotheses are rejected. There is significant difference between the production rate of workers and between the machines.

TABLE 10.11
ANOVA

Source	df	Sum of Squares	Mean Sum of Squares	$F_{\text{calculated}}$	$F_{\text{tabulated}}$
Machine	4 − 1 = 3	353.8	117.9333	$\dfrac{117.9333}{6.1466} = 19.202$	$F_{3,12} = 3.49$
Workmen	5 − 1 = 4	201.5	50.375	$\dfrac{50.375}{6.1466} = 8.19558$	$F_{4,12} = 3.26$
Error	19 − 7 = 12	73.7	6.1466		
Total	20 − 1 = 19	629			

10.4.2   RANDOMIZED BLOCK DESIGN WITH INTERACTION

In the linear model, we assume that the effect of the ith treatment and j^{th} block is $t_i + b_j$, i.e., the sum of the effects due to the ith treatment and j^{th} block, but it may happen that apart from these effects, the combination of block and treatment may have its own effect over and above the effects of block and treatment. This is called as the interaction effect.

Suppose Y_{ijk}: yield in the kth plot of the ith treatment and j^{th} block then we can use the following linear model.

$$Y_{ijk} = \mu + t_i + b_j + x_{ij} + \varepsilon_{ijk}$$

Where,

μ: overall average effect
t_i: effect due to the ith treatment
b_j: effect due to the jth block
x_{ij}: the interaction between ith treatment and jth block
ε_{ijk}: the random component

Estimating the unknown parameters using least square methods, we get the ANOVA-RBDI Table 10.12.

Testing the null hypothesis that all t_i's are not significant, all b_j's are not significant, and there is no interaction between blocks and treatments.

$H_{01} : T_1 = T_2 = \cdots = T_v;$ $H_{02} : B_1 = B_2 = \cdots = B_b$ and H_{03}: no interaction between blocks and treatments at a set level of significance.

TABLE 10.12

ANOVA-RBDI

Predictor	df	SS	MSS	$F_{calulated}$	$F_{tabulated}$
Treatment	$v-1$	$T_rS.S$	$T_rMSS = T_rS.S/(v-1)$	TrMSS/EMSS	$F_{v-1,\ [bv(n-1)]}$df
Block	$b-1$	B.S.S	BMSS = BSS/(b-1)	BMSS/ EMSS	$F_{b-1,\ [bv(n-1)]}$df
Interaction	$(v-1)(b-1)$	ISS[a]	IMSS = ISS/(v-1)(b-1)	IMSS/EMSS	$F_{[(b-1)(v-1),\ [bv(n-1)]]}$df
Cell	$(bv-1)$	CSS[b]	CMSS = CSS/(bv-1)	CMSS/EMSS	
Error	$bv(n-1)$	ESS	EMSS = ESS/(v-1)(b-1)		
Total	$nvb-1$	TSS			

[a]   ISS = CSS − $T_rS.S$ − B.S.S.

[b] $CSS = \dfrac{\sum_i \sum_j c_{ij}^2}{n}$ − correction factor

Inference: If the calculated F value exceeds the tabular value with $\left[F(v-1),\, bv(n-1) \right]$, and $F\left[(b-1),\, bv(n-1) \right]$ and $F\left[(b-1)(v-1),\, bv(n-1) \right]$ at prescribed level of significance, we reject the null hypothesis with treatment, block and interaction, respectively, otherwise we accept the null hypothesis.

Illustration 10.10

Suppose that the following data represents the units of production turned out each day by three different machinists each working on the same machine for three different days. Test at 5% level whether the machines, machinists, and the interaction are significant.

Data are given as a four-tuple: (M_i, A, B, C), where A, B, C are three-tuples and $M_i, i = 1, 2, 3, 4$ are machines:

$(M_1, (15,15,17), (19,19,16), (16,18,21))$
$(M_2, (17,17,17), (15,15,15), (19,22,22))$
$(M_3, (15,16,17), (18,17,16), (18,18,18))$
$(M_4, (18,20,22), (15,16,17), (17,17,17))$

Solution

Step 1:

H_{01}: There is no significant difference between the machinists.

v/s

H_{11}: There is significant difference between the machinists.
H_{02}: There is no significant difference between the machines.

v/s

H_{12}: There is significant difference between the machines.
H_{03}: There is no significant interaction effect.

v/s

H_{13}: There is signification interaction effect.

Step 2: $\alpha = 0.05$

Step 3: Test-statistic

$$F = \frac{S_1^2}{S_2^2}$$

Where, S_1^2 = variance between samples
S_2^2 = variance within samples

Step 4: Calculations are given as M_i (row) total and Machine (column) total for A, B, C in Table 10.13.

$M_1 - 156, M_2 - 159, M_3 - 135, M_4 - 189$ (row-wise sum)

$A - 206, B - 198, C - 223$ (column-wise total)

Grand total $- 627$

$v = 3, b = 4, n = 3$

TABLE 10.13
ANOVA

Source	df	Sum of Squares	Mean Sum of Squares	$F_{calculated}$	$F_{tabulated}$
Machine	$4 - 1 = 3$	27.66	9.22	$9.22/1.702 = 5.42$	$F_{3,24} = 3.0088$
Machinists	$3 - 1 = 2$	2.75	1.375	$1.375/1.702 = 0.808$	$F_{2,24} = 3.4028$
Interaction	6	73.50	12.25	$12.25/1.702 = 7.197$	$F_{6,24} = 2.5082$
Error	24	40.84	1.702		
Total	$36 - 1 = 35$	144.75			

$$\text{Correction factor}(CF) = \frac{(\text{Grand Total})^2}{nvb} = \frac{(627)^2}{36} = 10920.25$$

TSS = Total Sum of Squares = 144.75

  Machine Sum of Square (MSS) = $10947.416 - 10920.25 = 27.166$

  Machinist Sum of Squares (McSS) = $10923 - 10920.25 = 2.75$

  Interaction Sum of Squares (ISS) = 73.50

Step 5: Inference

  H_{01} and H_{03} are rejected, whereas H_{02} is accepted.

10.5   LATIN SQUARE DESIGN

10.5.1   INTRODUCTION

Randomized Block Designs are improvements carried on completely Randomized Design. In the sense that they provide error control measures for elimination of block variation. This principle can be extended further to improve RBDs by eliminating more sources of variation. Latin Square Design is one such improved design with provision for the elimination of two sources of variation.

Suppose there are fertility gradient in two different directions among the plots, then unless the treatments are allotted to the various plots in a proper manner, some treatments may receive more favorable plots. One way of balancing this fertility is done by using LATIN SQUARE.

An $n \times n$ Latin square is an arrangement of n symbols (usually Latin letters) arranged in an $n \times n$ square array, such that each row contains all the symbols and each column contains all the symbols only once. Latin Square Design will have equal number of treatments and replications with n^2 experimental units. The experimental area is divided into n^2 experimental units (plots) arranged in a square so that each row as well as column contains n units. The n treatments are then allotted at random to these rows and columns in such a way that every treatment occurs once and only once in each row and in each column such a layout is known as $n \times n$ LSD and a particular lay out must be decided randomly.

For example with a treatment α, β, ζ, δ, one arrangement of 4×4 LSD is given by

α	β	ζ	δ
β	ζ	δ	α
ζ	δ	α	β
δ	α	β	ζ

Let Y_{ijk} denote the response from a unit in the ith row, j^{th} column, and receiving the kth treatment. The (i, j, k) triplet can assume only m^2 values if S represents the set of m^2 values, then symbolically $(i, j, k) \in S$. The most appropriate model for LSD is

$$Y_{ijk} = \mu + t_i + r_j + c_k + \varepsilon_{ijk} \quad 1 \le , i, j, k \le m$$

Where

μ: overall average effect
t_i: effect due to the ith treatment
r_j: j^{th} row effect
c_k: k^{th} column effect
ε_{ijk}: random component

Estimating the unknown parameters using least square methods, we get the ANOVA Table 10.14.

Testing the null hypothesis that all t_i's are not significant, all rows (b_j's) and all columns (c_k's) are not significant.

$$H_{01} : T_1 = T_2 = \cdots = T_m; \quad H_{02} : R_1 = R_2 = \cdots = R_m \text{ and}$$

$$H_{03} : C_1 = C_2 = C_3 = \cdots = C_m$$

at a set level of significance.

TABLE 10.14
ANOVA

Source	df	SS	MSS	$F_{calculated}$	$F_{tabulated}$
Treatment	$m-1$	T_rS.S	T_rMSS = T_rS.S/$(m-1)$	TrMSS/EMSS	$F_{m-1,\,[(m-1)(m-2)]}$df
Row	$m-1$	R.S.S	BMSS = BSS/$(m-1)$	RMSS/EMSS	$F_{m-1,\,[(m-1)(m-2)]}$df
Column	$m-1$	CSS	IMSS = ISS/$(m-1)$	CSS/EMSS	$F_{m-1,\,[(m-1)(m-2)]}$df
Error	$(m-1)(m-2)$	ESS	EMSS = ESS/$(m-1)(m-2)$		
Total	m^2-1	TSS			

Advantages:

- Latin Square Design controls more of the variation.
- Statistical Analysis is simple.
- More than one factor can be investigated simultaneously and with fewer trials

Disadvantages:

1. In general, the factors may not be independent which results in interaction effect. The fundamental assumption that there is no interaction between different factors (i.e., the factors are independent) may not be true in general.
2. In LSD, the number of treatments is restricted to the number of replications which limits its application.
3. Analysis will be complex if several missing plots are present.
4. Field layout should be square.

Illustration 10.11

The data in Table 10.15 relates to the purchase of apples in four different stores in the first 4 days of a chosen week. Analyze using Latin Square Design at 5% level of significance and interpret.

Solution:

H_{01}: There is no significant difference in stores ($A = B = C = D$)

H_{02}: There is no significant difference among the days.

H_{03}: There is no significant difference in purchase of apples in different stores

Level of significance $= 0.05$

Test statistics: F test

The calculated values are depicted in Table 10.16

TABLE 10.15
Apple Purchase-Data

Days		Stores		
1	A14	B8	C40	D48
2	B20	A22	D48	C25
3	D24	C7	B12	A27
4	C31	D16	A32	B22

TABLE 10.16
ANOVA

Source	df	SS	MSS	$F_{calculated}$	$F_{tabulated}$
Stores	$m - 1 = 4 - 1 = 3$	692.5	230.83	2.608	$F(3,6) = 4.7571$
Row (days)	$m - 1 = 4 - 1 = 3$	305.5	101.83	1.1506	$F(3,6) = 4.7571$
Column (stores)	$m - 1 = 4 - 1 = 3$	958.5	319.5	3.610	$F(3,6) = 4.7571$
Error	$(m - 1)(m - 2) = 6$	442.5	88.5		
Total	$m^2 - 1 = 15$	2399			

Inference: As the calculated F value is less than the F tabulated value in Table 10.16, there is no sample evidence to reject the null hypothesis.

10.6 FACTORIAL EXPERIMENTS

There are several types of experiments, which require statistical investigation. These are featured by the nature of independent variables and comparison required among them so as to meet the objectives of the experiment [2]. There are three main types of experiments

1. Varietal experiments
2. Factorial
3. Bio-assays

In varietal trials, treatment like (i) different varieties of crop, (ii) several feeds for animals, and (iii) different doses of a drug etc. are under investigation. In fact, *different levels of only one factor* form the treatment in varietal trials. The main purpose of such experiments is to compare the treatments in all possible pairs [2].

In factorial experiments, more than one factor with combination of *two or more* levels is considered. If the number of levels of each factor in an experiment is the same, the experiment is called as symmetrical factorial, otherwise the experiment is called asymmetrical factorial or mixed factorial. $\left(L^F \right)$

For example, consider an agricultural experiment with two factors namely Nitrogen at three levels and Irrigation with two denoted by N_0, N_1, and N_2 and I_0 and I_1, respectively. Using the above factors, the following six combinations taking one level from each factor namely I_0N_0, I_0N_1, I_0N_2, I_1N_1, I_1N_0, I_1N_2 can be formed. Such combinations form treatments in factorial experiment.

The comparisons required in this type of experiments are not the pair comparisons as in varietal trials but a special type of comparison called MAIN EFFECTS and INTERACTION EFFECTS.

Note 1: Factorial experiments are used to study individual effects of each factor and their interactions. Many biological and clinical trials used factorial experiments to give more insight and infer.

Note 2: To conduct an RBD with at least 10° of freedom for error variance, we require six replications for the experiment on nitrogen and 11 replications for the second experiment on irrigation. Therefore, a total of 40 experimental units are required. But for conducting a factorial experiment with six combinations of two factors as treatments, we require only 18 plots (units) for similar precision.

10.6.1 2^2 FACTORIAL EXPERIMENTS

Simple experiments involve only one factor at a time. If the effects of several factors are studied simultaneously, the experiment is called as FACTORAIAL EXPERIMENTS. In earlier experiments, we are primarily interested in comparing the effects of a single set of treatments. Such experiment which deals with 1 factor only is called simple experiments. In factorial experiments, the effects of several factors are studied simultaneously. Let us consider the design in which there are 2 factor search at 2 levels. Levels may be quite literally two quantitative levels or concentration of say fertilizers, or it may be two quantitative alternatives. Here, we have two factors each of level $(0,1)$, so that there are $2 \times 2 = 4$ treatment combination in all [3].

Let A and B be the names of the two factors and let "a" and "b" denote one of the two levels of each of the corresponding factors. It is called as a second level. The first level of A and B generally denote the absence of the corresponding factors in treatment combination. The four treatment combination can be enumerated as follows:

   i. a_0b_0 or "1": Factor A and B are at the first level
  ii. a_1b_0 or "a" : Factor A is at second level and B is of the first level
 iii. a_0b_1 or "b" : Factor B is at second level and A is at first level
  iv. a_1b_1 or "ab" : Both A and B are at the second level

The above treatment combination can be compared by laying out the experiment in

  a. RBD with r replicates, each contains 4 units or
  b. 4×4 LSD and ANOVA can be carried out accordingly.

Suppose these four treatments are applied to "r" blocks each containing four plots so that in each block all the four treatment combinations are applied. We define the main effects of A and B as follows:

$$\text{Main effect of } A = \left\{[ab]+[a]-[b]-[1]\right\}/2r = \frac{(a-1)(b+1)}{2r}$$

Where any letter enclosed in [ ] stands for the total yield of all the plots receiving that particular combination of treatment.

$$\text{Main effect of } B = \left\{[ab]-[a]+[b]-[1]\right\}/2r = \frac{(a+1)(b-1)}{2r}$$

$$\text{Interaction effect on } AB = \frac{(a-1)(b-1)}{2r}$$

$$SSA = \text{Sum of squares due to } A = \frac{\{[ab]+[a]-[b]-[1]\}^2}{4r}$$

$$SSB = \text{Sum of squares due to } B = \frac{\{[ab]-[a]+[b]-[1]\}^2}{4r}$$

$$SSAB = \text{Sum of squares due to } AB = \frac{\{[ab]-[a]-[b]+[1]\}^2}{4r}$$

Hypothesis:

H_{01}: the block effects are not significantly different.

H_{02}: the treatment effects are not significantly different.

Advantages:

- Provides estimates of interactions.
- Possible increase in precision due to so-called hidden replication.
- Experimental rates can be applied over a wider range of conditions.

Disadvantages:

- Some treatment combinations may be of little interest.
- Experimental error may become large with a large number of treatments.
- Interpretation may be difficult

Note: if the number of combinations in a full factorial design is too high to be logistically feasible, a fractional factorial design may be done, in which some of the

---

TABLE 10.17

ANOVA

Predictor	df	SS	Mean Square	$F_{\text{calculated}}$	$F_{\text{tabulated}}$
Block	$r-1$	BSS	BSS/$(r-1)$ = BMSS	BMSS/EMSS	$F_{(r-1, 3(r-1))}$
A	1	SSA	MSSA	MSSA/EMSS	$F_{(1, 3(r-1))}$
B	1	SSB	MSSB	MSSB/EMSS	$F_{(1, 3(r-1))}$
AB	1	SSAB	MSSAB	MSSAB/EMSS	$F_{(1, 3(r-1))}$
Error	$3(r-1)$	ESS[a]	ESS/$3(r-1)$ = EMSS		
Total	$4r-1$	TSS			

[a] ESS = TSS $-\{$SSA + SSB + SSAB$\}$

Inference: If the calculated F value exceeds the tabular F value in Table 10.17, we reject both H_{01} and H_{02}
5% level of significance.

possible combinations are omitted. In the ANOVA factorial design, if the interaction is significant, the main effects must be interpreted with care.

Illustration 10.12

An experiment was conducted to test the efficiency of nitrogen and potash in increasing the yield of corn. Four treatment combinations namely (i) no nitrogen and no potash, (ii) 10 lbs of N per acre, (iii) 20lbs of K per acre, and (iv) 10 lbs of nitrogen and 20 lbs of potash per acre were tested. The increases in yield are given in Table 10.18. Analyze the data.

Solution:

H_{01} : Blocks are same

H_{02}: Main effects of Nitrogen and Potash are same

H_{03}: There is no significant difference in the interaction effects

TABLE 10.18

Data on Efficiency of Nitrogen and Potash

Treatment Combinations	Blocks			
	1	2	3	4
(0,0)	0	33	17	10
(1,0)	30	50	41	25
(0,1)	23	14	12	33
(1,1)	64	75	76	73

TABLE 10.19

ANOVA

Variables	df	SS	Mean Squares	$F_{calculated}$	$F_{tabulated}$
Block	4 – 1 = 3	381.5	127.1667	1.2169	$F_{(3,9)} = 3.8625$
Nitrogen(main)	1	5329	5329	50.9952	$F_{(1,9)} = 5.1174$
Potash (main)	1	1681	1681	16.086	$F_{(1,9)} = 5.1174$
Interaction	1	900	900	8.6124	$F_{(1,9)} = 5.1174$
Error	9	940.5	104.5		
Total	15	9232			

Inference: Based on the comparison between critical value and table value in Table 10.19, it is inferred that the blocks are same; however, main effects of Nitrogen, Potash, and interaction effects are insignificant as critical F value is greater than tabulated F value. Since the table value is greater than the F calculated value, there is no sample evidence to reject H_{01}, i.e., there is no significant difference in blocks whereas the main effects and interaction effect are insignificant as F tabulated value is less than the F calculated value.

Level of significance $= 0.05$
Test statistics $= F$ test
Calculation:
Correction Factor $= 20{,}736$
Total sum of squares $= 29{,}968 - 20736 = 9232$
Block sum of squares $= 84{,}470/4 - 20736 = 381.5$
Sum of squares due to main effect Nitrogen $= (292)^2/16 = 5329$
Sum of squares due to main effect Potash $= (164)^2/16 = 1681$
Sum of squares due to interaction effect $= (120)^2/16 = 900$
Error sum of squares $= 9232 - 8291.5 = 940.5$

REFERENCES

1. Montgomery, D.C. (2007) *Design and Analysis of Experiments*, New York: John Wiley & Sons.
2. Paneerselvam, R. (2011) *Design of Experiments*, New Delhi: PHI Learning Pvt. Ltd.
3. Tamhane, A.C. (2009) *Single Factor Experiments: Completely Randomized Designs*, New York: John Wiley & Sons.
4. Antony, J. (2003) *Design of Experiments for Engineers and Scientists*, Amsterdam: Elsevier Ltd. DOI: 10.1016/B978-0-08-099417-8.00006-7.

11 Brain Computer Interfaces: The Basics, State of the Art, and Future

Muhamed Jishad. T. K and M. Sanjay
National Institute of Technology Calicut

CONTENTS

11.1   INTRODUCTION

In simple terms, Brain Computer Interface (BCI) is a system that enables mind control of devices. BCI is a technology that has the potential to transform the lives of people locked in due to spinal cord injury or other various neuromuscular diseases. Since BCI enables the user to execute certain actions by controlling the devices with his/her mind, BCI can be used as an assistive technology for such patients. The clinical use of BCI can be separately viewed as assistive and rehabilitation technologies. Assistive technology focuses on regaining communication ability or enabling the patient to perform daily life activities by controlling external devices, robotic hands, etc. Rehabilitation BCI helps the patients impaired due to stroke to regain motor functions such as movement by manipulating the brain reorganization. Recent developments show that neuro-prosthetic control can further improve network-based rehabilitation in severely impaired patients [1]. Additionally, noninvasive BCI can find applications in other areas like multimedia, gaming, and entertainment.

The history of BCI starts with the invention of electroencephalography (EEG) in the 1920s. In 1968, the possibility of controlling human EEG signals was experimented; eight of the ten subjects that participated in the experiment learned to control their alpha cycle durations after three to seven training sessions of 1-hour duration each [2]. The concept of direct brain computer communication was introduced at the beginning of the 1970s itself [3] and has started attracting the attention of researchers. Practical BCI experiments were first conducted on rodents and then gradually progressed to primates [4]. There are several examples for experiments with monkeys where multiple microelectrode arrays were surgically placed into different cortical areas to acquire brain signals and to establish BCI [5–7]. Signal quality is one of the key factors that affect the performance of the BCI system. Invasive methods of neuroelectric signal recording provide better quality signals but have limitations to apply in human subjects due to safety and ethical constraints. Also, the human brain is considered to be a very complex organ that contains about 100 billion neurons [8], and the exact explanation for how brain and mind works has not unravelled yet in its vivid sense. Despite the many challenges, several promising results have been produced by researchers around the globe to date. This chapter will look into the various aspects of the brain computer interface, its applications and future scope. Since the neuroelectric signals such as EEG are the widely used signals among other neuroimaging techniques for BCI applications, our main focus in this chapter will be on EEG-based BCI systems.

11.1.1   The Basic Architecture of a BCI System

The basic elements of any BCI system are a signal acquisition unit that captures the mental activity into signals; a signal processing unit that processes the signal, identifies the user intent and translates it into device commands; and an application unit that executes the intended task as per the command. Figure 11.1 illustrates a general outline of a BCI system. Brain signal types, brain signal patterns and stimulus modality are the three factors to consider when designing signal acquisition for BCI systems. Brain signal type is determined by the acquisition technique that is

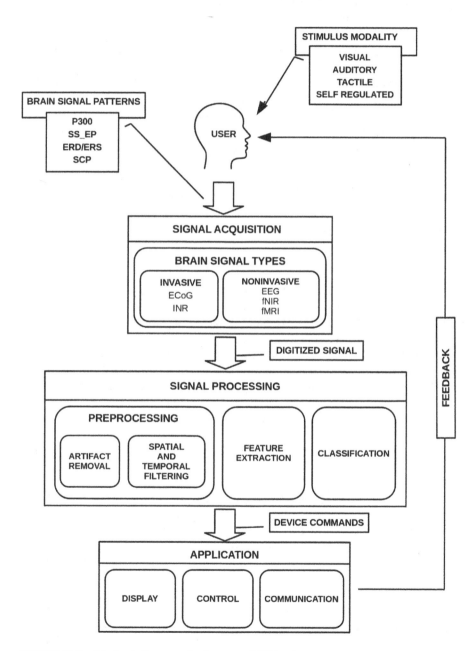

FIGURE 11.1   The basic framework of a general BCI system.

used to acquire neural activity. The signal acquisition can be either invasive (ECoG, INR, etc.) or noninvasive (EEG, MEG, fNIRS, fMRI, etc.) based on the electrode placement. Also, each BCI system will be based on certain standard brain signal patterns such as P300 Event-Related Potentials (ERP), Steady-State Evoked Potentials (SSEPs), Slow Cortical Potential (SCP) or Sensorimotor Rhythms (SMR). Some signal patterns such as P300 ERP and SSEPs require certain stimuli (visual, auditory, tactile, etc.) to elicit the response and other patterns such as SCP and SMR are self-regulated. The signal processing unit has three main functions; preprocessing, feature extraction and classification. In the preprocessing stage, the signal will first undergo artefact removal to eliminate the contaminations (electrical interference due to ocular, cardiac and muscle activities or other environmental factors) in the EEG signal. Then the signal undergoes filtering in the temporal and spatial domain to improve the signal to noise ratio. Temporal filtering is nothing but the filtering in the time domain typically using bandpass filters or notch filters. The purpose of spatial filtering is to combine the signals recorded from multiple sensors to form a single signal. Then the signal will be represented using the extracted features and the classifier classifies the features and translates it based on the user intent. Most of the time, BCI systems undergo several offline training trials where the filter, feature extraction and classifier parameters get calibrated before its final implementation. We will look into each of these blocks of BCI separately in the coming sections.

11.1.2 A General Classification of BCI Systems

The classification of BCI can be done in many ways based on several factors. A general classification based on some common parameters such as the method of signal acquisition, type of signal patterns, mode of presenting stimuli, mode of operation, strategy of operation, etc. is presented here [9].

Brain signal pattern: Brain signal can be viewed as a set of electrical impulses that are generated inside the brain during neuronal firing. Each BCI systems utilize different types of brain signal patterns, and they can be categorized accordingly. Some examples of brain signal patterns are P300 event-related potentials (ERPs), steady-state evoked potentials (SSEPs), slow cortical potentials (SCPs), sensorimotor rhythm (SMR), etc. They are explained separately in Section 11.3. Additionally, multiple signal patterns can be used together to form the so-called hybrid BCI systems [10]. For example, SSVEP and P300 ERP were used together in a speller and achieved improved spelling speed [11].

Stimulus modality: In some BCIs, it is required to administer some stimuli to the user for eliciting the required brain signal pattern. Visual stimuli are the most commonly used mode for BCI applications. Auditory and somatosensory stimuli are also in practice. For example, in the case of SSEPs, the potential can be evoked by steady-state stimuli in the form of visual (SSVEPs) [12] or auditory (SSAEPs) [13] or somatosensory (SSSEP) [14]. Similarly, in the case of P300 ERPs, the stimulus can be tactile [15], visual [16] or auditory [17].

Mode of operation: There are *synchronous* and *asynchronous* [18] BCIs based on the mode of operation. In synchronous BCI, the communication is cue-based where some indications are displayed to the user to evoke the brain signal response

[19], while in the asynchronous BCI, the user has the freedom to communicate to the system in his/her own pace and timing [20].

Operation strategy: This indicates the method by which the brain signal is evoked during the process. It can be *Selective attention* where visual, auditory or tactile stimuli are presented to the user to evoke brain signal response. This type of BCIs is termed as *exogenous* [18] since external stimuli are used to elicit the response. SSVEP and P300 ERP are some examples of such BCIs. Exogenous BCIs do not require training and usually possess a high information transfer rate. Another operation strategy is to train the user to maintain the desired level of brain wave frequency by using *cognitive effort*; such as *Motor imagery* where the user imagines muscle movement to stimulate spikes in neuronal activity in the motor cortex. This type of BCIs is termed as *endogenous* [18] since the brain signals are self-regulated without using any external stimuli. BCIs based on SCP or SMR comes under this category. The main advantage of this type of BCI is that it can be operated by the user's free will.

Recording methods: Recording can be *invasive* or *noninvasive*. Invasive methods such as Electrocorticography (ECoG) and Intracortical Neuron Recording (INR) require surgery to place electrodes inside the skull. On the other hand, the noninvasive signal recording is the safest and does not require surgery. EEG, fMRI, fNIRS, MEG, etc. are some examples of noninvasive recording method. They are explained separately in the next section.

11.2 SIGNAL ACQUISITION TECHNIQUES FOR BCI

The primary element of any BCI system is a signal acquisition setup that interprets neural activity into useful information which is then fed to the signal processing unit. Signal acquisition methods for BCI are generally classified based on the electrode placement; over the scalp (noninvasive) or inside the skull (invasive). These techniques can also be characterized by referring to their spatial and temporal resolution. *Spatial resolution* is an indication of how accurately the technique can locate the activity inside the brain. For example, the fMRI is said to have a spatial resolution in the millimetre range. On the other hand, the *temporal resolution* is the indication of how much quicker the technique is in recording the neural response. For instance, EEG can record responses with millisecond resolution. Hardware complexity is also a concern. In the order of increasing hardware complexity, it can be listed as EEG, fNIRS, fMRI, MEG, and PET; ie, EEG is the simplest of all, and PET is the most complex.

11.2.1 NONINVASIVE METHODS

Noninvasive recording methods are generally classified into two categories; *metabolic-based* and *electrophysiological*-based. Brain imaging techniques such as fMRI, fNIRS, PET, etc. can be listed under metabolic-based, and EEG and MEG come under electrophysiological recording technique. Generally, metabolic-based imaging techniques are considered to have a very good spatial resolution but poor temporal resolution. On the other hand, electrophysiological recordings exhibit excellent temporal resolution but with poor spatial resolution. Noninvasive techniques are the most used signal acquisition method for BCI applications. The common noninvasive recording techniques and their use in BCI application are described below.

11.2.1.1   Electroencephalography (EEG)

Electroencephalography (EEG) is a recording technique that utilises multiple electrodes that are placed over the scalp to measure the electrical activity of the brain. EEG provides measurement with an excellent temporal resolution. But it has a poor spatial resolution. Various head tissues in between the brain and scalp electrodes attenuate and blur the spatial distribution of the neural current, which makes it less accurate in terms of estimating the location of brain signal activation [21]. This mathematical difficulty in computing the current distribution within the brain using EEG signals is referred to as the *inverse problem* [22]. Also, EEG signals are likely to be noisy with the interference of other bioelectric activities like facial EMG, electrooculogram (EOG), etc which are commonly termed as EEG artefacts. Despite all the limitations, its better temporal resolution, noninvasiveness, low cost, simplicity and safe operation makes EEG a popular choice for BCI applications.

Normal EEG patterns (popularly known as *brainwaves*) are oscillatory and can be characterized based on their frequency and amplitude; frequency in the range of several hertz and amplitude in the range of up to few 100 µV. Table 11.1 indicates the general classification of brainwaves. Brain waves are considered to be the indication of mental state; for example, delta waves are generated during a deep dreamless sleep, theta occurs during light sleep, alpha indicates the awake resting state of the brain, etc.

Like many other biosignals, EEG is acquired using bipolar recording technique, requiring three electrodes – a recording electrode, a ground and a reference. Additionally, a signal processing unit comprising an amplifier, filter, analogue to digital converter, etc., and a display unit will be a part of the system. Either dry or wet electrodes can be used for EEG recording. Dry electrodes can be placed directly on the scalp, but wet electrodes need a conducting gel to be applied on the scalp before placing the electrode. The materials for electrodes are generally silver, silver chloride or gold.

The classical method for EEG signal acquisition involves an EEG cap with a universally defined arrangement of electrodes over the scalp. This will ensure a standard

TABLE 11.1
The General Classification of Brain Waves

Brain Wave	Frequency (Hz)	Amplitude (µV)
Slow Cortical Potentials (SCP)	<1	-
Delta	1–4	100–200
Theta	4–8	5–10
Alpha	8–12	20–80
Beta	12–30	1–5
Gamma	30–60	0.5–2

method for proper positioning of electrodes over the scalp according to the cortical areas of the brain which is vital for capturing reliable EEG data. The international 10/20 system which describes the internationally accepted electrode placement for EEG is illustrated in Figure 11.2. It is a 21-electrode system; electrodes are arranged in 10% and 20% of distances measured between inion and nasion as well as between the left and right preauricular points as shown in Figure 11.2 [23]. Additionally, extended versions of the 10/20 electrode system are available for a high-resolution EEG measurement; such as the 10/10 electrode system [24] (with up to 74 electrode locations), 10/5 electrode system (with up to 345 electrode locations) [25], etc. There are multiple EEG signal patterns that are used for implementing various BCI modalities, which we will discuss in Section 11.3.

11.2.1.2 Magnetoencephalography (MEG)

The electrical activity of the brain generates a tiny magnetic field around it. MEG measures this magnetic field using an extremely sensitive magnetometer. The presence

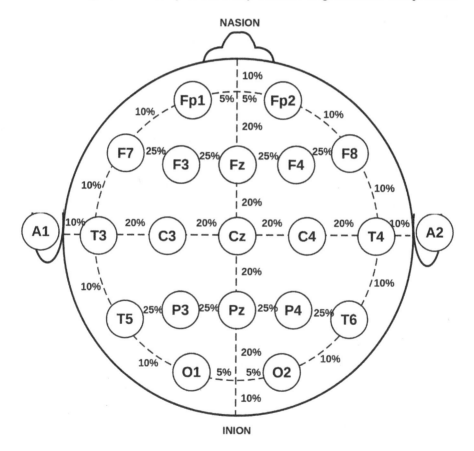

FIGURE 11.2 The 10/20 international system of electrode placement; indicates the location of the 21 electrodes for EEG recording. The spaces between electrodes are marked in percentage.

of weak alternating magnetic fields outside the human scalp was first demonstrated by the physicist David Cohen in 1968 [26]. Nowadays, MEG measurements largely depend on SQUID technology for detecting the magnetic field. The Superconducting Quantum Interference Device (SQUID) is an excellent detector of magnetic flux. MEG has a very good temporal resolution in the order of milliseconds as well as a better spatial resolution in the order of millimetres. Also, it is less affected by head tissues and less susceptible to muscle artefacts [27]. Despite all these advantages, MEG is very seldom used for BCI applications due to its higher operating cost, sensitive instrumentation, sophisticated methods and nonportability [9]. MEG measurements also require a shielded room for eliminating environmental magnetic interference.

11.2.1.3 Functional Magnetic Resonance Imaging (fMRI)

fMRI measures levels of blood oxygenation in the brain in terms of blood oxygen level-dependent (BOLD) contrast. The neuronal activity inside the brain is tightly coupled with the regional cerebral blood flow (rCBF) and cerebral blood oxygenation (rCBO); which is termed as neurovascular coupling [28]. Deoxygenated haemoglobin (deoxy-Hb) and oxygenated haemoglobin (oxy-Hb) differ in magnetic properties; oxy-Hb is diamagnetic, whereas deoxy-Hb is paramagnetic [29]. That means increased neuronal activity results in increased O_2 metabolism and increased cerebral blood flow. This finally results in decreased deoxy-Hb level and increased oxy-Hb level near the active neuron region. By making use of the difference in magnetic properties of oxy-Hb and deoxy-Hb, the increase in blood oxygen level can be detected using the powerful magnetic field produced by the MRI scanner. This difference is called blood oxygen level-dependent (BOLD) contrast [30]. In fact, functional MRI measures brain activity indirectly in terms of BOLD contrast.

Even though fMRI offers higher spatial resolution, which allows the system to monitor the entire brain activity at once, the temporal resolution is low when compared with other techniques like EEG or MEG. There are very limited number of BCI applications using fMRI.

11.2.1.4 Functional Near-Infrared Spectroscopy (fNIRS)

Similar to fMRI, fNIRS also makes use of the variation in the blood oxygen level inside the cerebral cortex to measure brain activity. But instead of magnetic property, fNIRS measurement is based on the differences in light absorption by oxygenated and deoxygenated haemoglobin. That is, light below 800 nm (roughly) gets absorbed relatively more in deoxygenated blood and light above 800 nm in oxygenated blood [31]. Electromagnetic waves in the near-infrared region (around 650–950 nm) is applied through the skull into the brain using an infrared light source. By measuring the reflected light intensity during the neural activity and comparing it with the light intensity during the normal state where no stimulus is present, the hemodynamic response can be obtained [32]. As stated earlier the hemodynamic response has a direct relation with neural activity.

fNIRS is not a popular signal acquisition method for BCI applications due to its slow temporal resolution. Also, it is sensitive to other physiological artefacts like the heartbeat, respiration effects, and physical movement [33]. Additionally, the measurement is limited to the surface of the brain due to the limitation of infrared light to penetrate deeper into the brain through the skull. But fNIRS offers a low-cost and portable option for BCI signal acquisition. Also, it has proven to be a better technique in the case of completely locked-in patients when compared to EEG-based BCIs [1].

11.2.1.5 Positron Emission Tomography (PET)

PET measures local blood flow by tracking the movement of radioactive water in the brain. The subject will be injected with water containing positron-emitting radio-active isotope. PET scanner will be able to track the flow of the radioactive water (which is in direct proportion to the blood flow itself) inside the brain for a minute after administering it. Since the active regions attract increased blood flow, the PET scanner indicates the regions of brain that are active during that particular activity [30]. Due to its high cost, hardware complexity and lower half-life of the radioactive isotope, the PET-based BCIs are usually limited to clinical research.

11.2.2 INVASIVE RECORDING METHODS

In invasive recording methods, microelectrode arrays are surgically placed inside the skull to acquire brain signal features. Either the electrodes are placed over the cortical surface (electrocorticography) or within the cortex (intracortical recording). ECoG and INR are two examples of invasive recording. Safety and ethical constraints are major concerns of this type of recording technique.

A communication device for ALS patients was the first implementation of invasive BCI in humans. The patient's brain was implanted with electrodes containing neuron growth factors such that it allows neurons to grow into it, and the user was able to control the cursor movement by controlling the axon firing [1].

11.2.2.1 Electrocorticography (ECoG)

Electrocorticography (ECoG) records the electrical activity of the brain using the electrodes placed beneath the skull but on the brain surface; either epidurally or subdurally. ECoG has a higher spatial resolution (1.25 mm for subdural recordings [34] and 1.4 mm for epidural recordings [35]) and better amplitude and is less susceptible to interference of artefacts when compared to EEG. Even though the electrodes are placed inside the brain surgically, they do not penetrate into the brain, which makes it safer than intracortical recording. Also, miniaturizing the electrode and increasing the electrode density will further improve the spatial resolution. Control of computer cursor [36–38], decoding finger movement [39], hand grasping movements [40], decoding hand gestures [41], P300 speller [42], etc. are some examples of ECoG-based BCI applications in human subjects.

11.2.2.2 Intracortical Neuron Recording (INR)

Intracortical neuron recording uses microscale recording electrodes implanted into the cortex for recording neuronal activity. This type of recording provides excellent spatial resolution, thanks to the high-density multielectrode arrays (MEAs). A high precision implant near the target neuron region is possible with this type of electrodes. The electrodes are usually made up of glass, platinum or tungsten. The signal resolution depends on two factors; the number of recording elements and their proximity to the target neuronal populations [43]. This technique is highly invasive and hence possesses high-risk factor among all the other methods. For this reason, the majority of BCI research involving INR is performed either on rodents or primates other than humans.

11.3 BCI TYPES AND BRAIN SIGNAL PATTERNS

As stated earlier, brain signals are a set of electrical impulses that are generated inside the brain during neuronal firing. Each BCI systems utilize different types of brain signal patterns. The brain signal pattern can be viewed as a time-locked response to a specific external or internal event; the event can be a sensory stimulus, thoughts, behavioral responses, or even emotional processes. The commonly used brain signal patterns like Steady-State Evoked Potentials (SSEP), P300 Event-Related Potentials, Sensorimotor Rhythms (SMR), Slow Cortical Potentials (SCP), etc. are usually extracted from the electrophysiological response of the neurons such as EEG, ECoG, etc. SSEP and P300 are based on selective attention where the response is elicited by applying external sensory stimuli, whereas SCP and SMR are self-regulated by the user.

11.3.1 P300 EVENT-RELATED POTENTIALS

The P300 is a component of the event-related potentials (ERP). ERPs are nothing but the potentials elicited in response to a particular event, typically when the user identifies an odd event. P300 potentials are evoked unexpectedly and provide task-relevant information [44]. For a BCI involving P300, the user is presented with two categories of sensory information (one rarely occurring event mixed with a stream of a commonly occurring irrelevant pattern) in a controlled fashion. Then the user is asked to observe and identify the odd event whenever it happens; popularly known as the *oddball* paradigm, in which a positive-going component in the ERP is elicited each time at around 300 ms after the occurrence of the odd event or stimulus [45] (Figure 11.3).

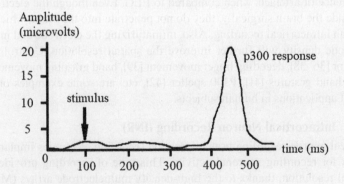

FIGURE 11.3 P300 ERP signal pattern.

They are termed as P300 event-related potentials, and P300-based BCI is one of the well researched BCI types. The first P300-based BCI (P300-based speller) was introduced by Farwell and Donchin in 1988 [45]. Even though EEG-based P300 BCIs are most common and practical, ECoG-based P300 BCI can provide a higher information transfer rate (ITR) [42].

It can be found that visual patterns are the most commonly used stimulus for P300 BCI applications. But in some cases, particularly where the user does not have sufficient eye movement control, auditory [17] or even tactile [15] stimuli can be used.

P300 patterns have been used for implementing BCI applications like spellers, web browsers, controlling external devices, paint applications, etc [46] (Figure 11.5). The increasing number of publications on the applications of the P300 patterns for BCI over the last 15 years is visible from Figure 11.4.

11.3.1.1 P300 BCIs Using Visual Stimulation
The first-ever and most common P300 BCI application is the P300 speller utilizing visual stimuli. P300 speller enhances the communication ability of patients suffering from motor neuron diseases such as ALS. It uses a character matrix displayed on a computer screen. And the rows and columns of that matrix flash or highlight rapidly one at a time. The row or column containing the character which the user wants to select induces a larger evoked potential [45]. Since the software controls the flashing, it can identify the user-intended character by detecting the P300 ERP. In addition to the so-called row-column paradigm [45] used in the first P300 speller, few other display-approaches also are in practice that improves the performance; such as the checkerboard flashing [47], region-based flashing [48], rapid serial visual

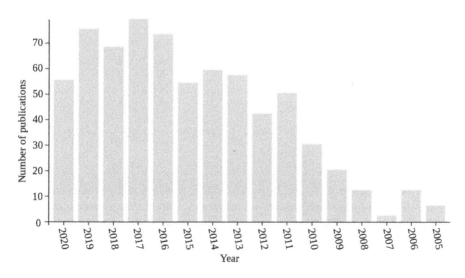

FIGURE 11.4 Year-wise statistics of P300 BCI research publications. (Generated from Web of Science search result analysis.)

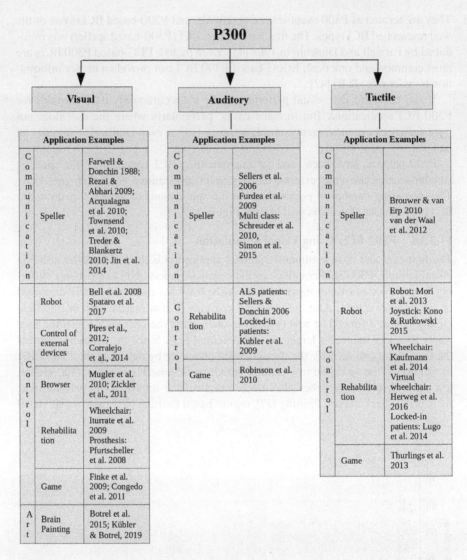

FIGURE 11.5   Example for P300-based BCI applications [15,45,47,52,48,49,53,54,60–62, 63–84].

presentation paradigm [49], etc. It was also found that different methods employed for the display of the stimuli produced different results. In the conventional P300 spellers, the rows or columns are highlighted by replacing the grey characters with white characters or a white box in a dark background [50]. Then it was noticed that by interchanging the character with an image of a face for a short duration [51] or using facial expression-change patterns [52] tends to improve the ERP response with better target identification. The main advantage of visual stimulus-based P300 BCI is that it can give better accuracy with lesser or no initial training.

11.3.1.2 P300 BCIs Using Auditory Stimulation

The first study with auditory stimulus used four randomly presented stimuli (YES, NO, PASS, END) to elicit P300 ERP [53]. The task involves an unambiguous question designed by the experimenter, and the user had to focus on the correct response to that question among the randomly presented words. In another experiment, a speller was implemented utilizing visual and auditory stimulus combined for eliciting the P300 ERP [54]. In this method, a character matrix is displayed to the user, and numbers were assigned (using row and column numbers) to represent each letter in the matrix. Instead of flashing the letter, the corresponding numbers were acoustically presented to them in order. But the information transfer rate and accuracy of this paradigm were turned out to be very low, and it did not lead to a practical BCI [50].

In another form of auditory BCI system, beeps of different frequencies were used to realize an auditory oddball paradigm; users were presented with a random sequence of beep sounds (80 dB, 1600 Hz (target) and 1500 Hz (nontarget) in intervals of 3–7 seconds [55]. Then the user was asked to count the beeps of 1600 Hz (target) tone. The loudness, pitch or direction of the target stimuli can affect the performance of an auditory-based BCI system [56]. Stimuli designed with tones of varying pitch and direction can improve the ITR better than that with varying loudness. In an experiment involving an eight-class auditory paradigm with eight pitches from eight directions produced an average ITR of 17.39 bits/min for an inter-stimulus interval (ISI) of 175 ms [57]. In later stages, familiar auditory stimuli such as drip-drop sound [58] or animal tones [59] were employed to improve the user-friendliness of the P300 BCI and achieved better accuracy and ITR. Unlike P300 BCI using visual stimulation, auditory-based P300 BCIs require more training for achieving a better result.

11.3.1.3 P300 BCIs Using Tactile Stimulation

The stimuli can also be presented using somatosensory modality. The usual practice is to use vibrating elements called tactors that are placed over specific areas of the subject body. By inducing vibrotactile stimulation patterns to simulate the oddball paradigm, P300 ERP can be elicited which is similar to that of other stimulation paradigms. The feasibility of a tactile P300 BCI was first demonstrated in the study using 12 equally spaced tactors placed around the user's waist [15] to induce somatosensory stimuli. Many studies have further validated the possibility of such systems using various stimulation approaches such as tactile speller using fingertip stimulation [60]; virtual wheelchair control by assigning four tactile stimulators that are placed at four different body parts to control the movement in four directions [61], etc. Also, studies suggest that training can improve the performance of somatosensory stimulus-based BCI [62].

11.3.2 STEADY-STATE EVOKED POTENTIAL (SSEP)

When a steady-state stimulus is presented to the brain, the brain will mimic the frequency of the stimulus, and it will generate a rhythmic brain activity in the associated cortical area [85]. In an SSEP-based BCI, many stimuli with different frequencies are applied simultaneously, and the user has to focus on any one of them according to the task user intents to perform. This will induce the steady-state evoked potentials

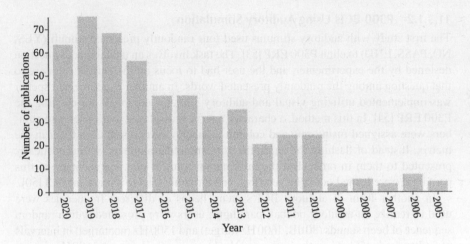

FIGURE 11.6 Year-wise statistics of SSEP-based BCI research publications. (Generated from Web of Science search result analysis.)

with a frequency similar to the target stimulus. In other words, in the EEG recorded from the corresponding cortical area (based on the stimuli type it will vary), the amplitude of the EEG component corresponding to target stimulus will be higher than that of the nontarget stimuli. By identifying the frequency component, the user intent can be interpreted. Similar to the P300 ERP discussed earlier, SSEPs can be induced by visual, auditory, or somatosensory stimuli. The SSEPs were originally observed for the first time in response to the visual stimuli. The increasing popularity of the SSEP patterns for BCI during the last 10 years is visible from Figure 11.6.

11.3.2.1 Steady-State Visual Evoked Potential (SSVEP)

SSVEPs are elicited by steady-state visual stimuli. They are also called the steady-state visual evoked response. In an SSVEP-based BCI, the user will be presented with multiple visual stimuli that flicker at different frequencies; each stimulus will be associated with a particular action. The focused attention on any one of the stimuli will induce neural activity in the occipital area which mimics the frequency of that particular stimulus [85] (Figure 11.7). The user intent can be identified by analyzing

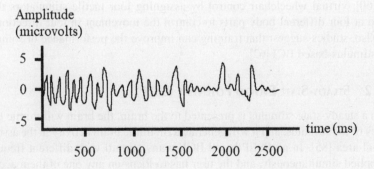

FIGURE 11.7 SSVEP brain signal pattern.

the spectral representation of the EEG activity taken from the occipital area. Since SSVEP in EEG matches with the frequencies of the stimulus or its harmonics or subharmonics [86], the stimulating frequencies should be as unique as possible. That is the stimulus should not be harmonic or subharmonic to each other.

Visual stimuli modulated with frequencies in the range 4–90 Hz can evoke SSVEP [86]. They are usually categorized into three frequency ranges; low (up to 12 Hz), medium (12–30), and high frequency (>30 Hz). The stimulus frequency has an impact on the amplitude of the induced response. The lower frequency (5–20 Hz) visual stimuli have been found to induce a higher amplitude response. The highest amplitude is observed for stimulus at 15 Hz. Also, the higher frequency stimuli induced responses were found to contain peaks at harmonic and subharmonic frequencies of the stimulus [86]. Low and medium-range frequency (below 30 Hz) stimuli are most common in SSVEP-based BCI applications. But the low-frequency flickering may cause discomfort to the user; even it may trigger an epileptic seizure in vulnerable subjects (15–25 Hz are most provocative) [87]. Even though the high-frequency SSVEP-based BCI can eliminate such difficulties, it has been very rarely used probably due to the technical difficulty in implementing the system without compromising the performance. An SSVEP BCI speller using stimuli frequency ranging from 30 to 39 Hz is reported in [88]. Another interesting fact to note that the human brain processes the stimuli that flicker at gamma frequency (particularly around 40 Hz) faster than any other frequency [89]. The characteristics of induced SSVEP depends also on phase, brightness and contrast of the visual stimuli.

SSVEP is usually considered as the most popular choice for noninvasive BCI in the current scenario; since they are quick, reliable and do not require training (Figure 11.8). The performance of the BCI mainly depends on the following parameters; speed, accuracy, and the number of targets [90]. The speed is often represented by using the information transfer rate (ITR). Average accuracy of 95.3% and an ITR of 58 ± 9.6 bits per minute were reported for a six target SSVEP base BCI experiment involving 12 subjects [91]. An SSVEP-based BCI speller using EEG recorded using wet electrodes achieved the highest ITR of 5.32 bits per second. This speller was implemented with a new frequency-phase modulation method to tag 40 characters using 0.5-second-long flickering signals and achieved a spelling rate of up to 60 characters [92].

11.3.2.2   Steady-State Auditory Evoked Potential (SSAEP)

SSAEPs are evoked by auditory stimulations, and they are sometimes referred to as auditory steady-state response (ASSR). SSEPs in humans can be induced by steady-state auditory stimuli presented at frequencies between 1 and 200 Hz [93]. In particular, the response was prominent at around 40 Hz stimuli. These recordings are best-taken from the vertex of the scalp. SSAEP is very rarely used for BCI applications due to their poor performance when compared to SSVEP. For this reason, SSAEP is often combined with other types of modalities; for example, combining SSAEP with SSVEP can reduce the discomfort due to the flickering light stimuli by replacing SSVEP by SSAEP during visual fatigue intervals [94]. Another advantage of combining SSVEP and SSAEP is that the BCI can work based on the EEG

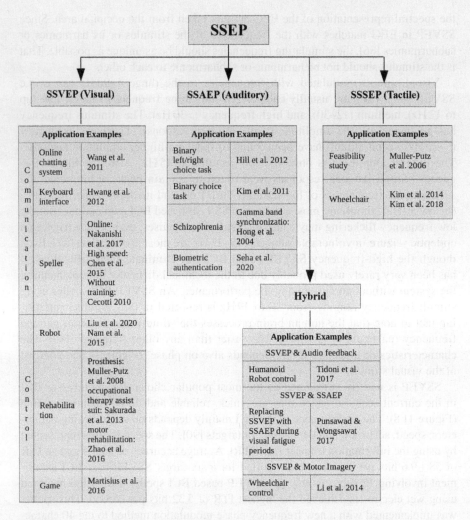

SSEP

SSVEP (Visual) | **SSAEP (Auditory)** | **SSSEP (Tactile)**

Application Examples		
C o m m u n i c a t i o n	Online chatting system	Wang et al. 2011
	Keyboard interface	Hwang et al. 2012
	Speller	Online: Nakanishi et al. 2017 High speed: Chen et al. 2015 Without training: Cecotti 2010
C o n t r o l	Robot	Liu et al. 2020 Nam et al. 2015
	Rehabilitation	Prosthesis: Muller-Putz et al. 2008 occupational therapy assist suit: Sakurada et al. 2013 motor rehabilitation: Zhao et al. 2016
	Game	Martisius et al. 2016

Application Examples	
Binary left/right choice task	Hill et al. 2012
Binary choice task	Kim et al. 2011
Schizophrenia	Gamma band synchronizatio: Hong et al. 2004
Biometric authentication	Seha et al. 2020

Application Examples	
Feasibility study	Muller-Putz et al. 2006
Wheelchair	Kim et al. 2014 Kim et al. 2018

Hybrid

Application Examples	
SSVEP & Audio feedback	
Humanoid Robot control	Tidoni et al. 2017
SSVEP & SSAEP	
Replacing SSVEP with SSAEP during visual fatigue periods	Punsawad & Wongsawat 2017
SSVEP & Motor Imagery	
Wheelchair control	Li et al. 2014

FIGURE 11.8 Examples for SSEP-based BCI applications [12–14,94, 96–111].

recorded from a more comfortable electrode location (behind-the-ears) without compromising the performance [95].

11.3.2.3 Steady-State Somatosensory Evoked Potential (SSSEP)

SSEPs can also be generated using vibrotactile stimulations. SSSEP was introduced in 1992 [112]; amplitude-modulated (AM) vibrations at 2–43 Hz frequencies were applied to the subject's fingers and palm and found out that it can induce study-state evoked potentials. Response with the highest signal to noise ratio was achieved at frequencies of stimuli at around 26 Hz. In the first feasibility study using SSSEP for BCI [14], vibratory stimulation was given simultaneously to both the index fingers. By concentrating attention on one of the two simultaneous stimulations, the user was able to modulate the SSSEP, which resembled the target stimulus. Based on the EEG response, the user's intent can be identified. Later studies indicate that stimulations applied to the foot areas induce efficient and stable SSSEP [113].

11.3.3 SLOW CORTICAL POTENTIAL (SCP)

SCPs are the electrical activity generated by the cortical neurons in the brain usually with a frequency of <1 Hz. They are the slow varying voltage changes in the brain and usually have a latency of around several seconds. The negative fluctuation of the SCP indicates behavioral preparation, or an increase in neuronal activity and positive fluctuation indicates behavioral inhibition or decrease in neuronal activity [114] (Figure 11.9).

A thought translation device (TTD) was introduced in [115], used for implementing an SCP-based spelling BCI for locked-in patients. TTD consists of a training device and a spelling program. This study indicates that with substantial training, the user can learn to self regulate the SCP through visual-auditory feedback and positive reinforcement. In another experiment, SCP was used to implement a web browser control for locked-in patients [116]. SCP takes around 0.5–10 seconds to develop and hence they provide a very low information transfer rate compared to P300 or SSVEP for BCI application [9]. Also, rigorous training with personalized cognitive and behavioral strategies need to be implemented. Currently, SCP is not a popular choice for BCI applications due to these reasons. Although, SCP-based BCIs do not require external stimuli to elicit the brainwave patterns. An overview of the number of publications happening each year is visible in Figure 11.10.

11.3.4 SENSORIMOTOR RHYTHMS (SMR)

Sensorimotor rhythms (SMR) are the brainwave patterns measured from the somatosensory and the motor cortices. They mainly consist of alpha-band frequencies (8–12 Hz) particularly around 10 Hz, and they are often called mu rhythms. SMR contains beta (around 20 Hz) and a gamma component (around 40 Hz) also. Movement or imagined movement results in changes in SMR patterns; these pattern changes are known as *event-related desynchronization (ERD)* or *event-related synchronization (ERS)*. ERD is the decrease in the oscillatory activity of

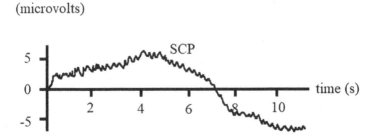

FIGURE 11.9   SCP signal pattern.

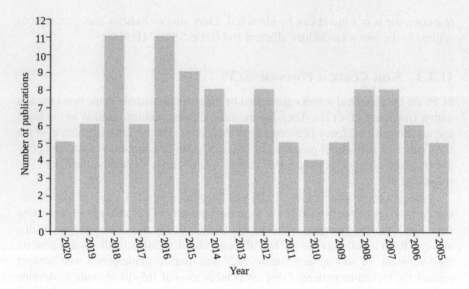

FIGURE 11.10 Year-wise statistics of SCP-based BCI research publications. (Generated from Web of Science search result analysis.)

SMR during movement, movement imagery or movement preparation. ERS is the increase in oscillatory activity of SMR during the period after movement or during relaxation [117]. In other words, SMR desynchronizes during the movement or imagined movement and synchronizes during the period after the movement. The ERD/ERS-based BCI works by motor imagery where the user imagines the movement without actually performing it to operate the BCI. On the other hand, SSEP and P300 BCIs were based on selective attention method where it needed external stimuli to be presented to the user by the BCI system. Imagined movement of feet, hands or tongue can be used to control an SMR-based BCI. Unlike selective attention-based BCI, this type of BCI needs extensive training for achieving maximum performance. SMR-based BCIs are gaining popularity among researchers and are visible from Figure 11.11.

One such example for an ERD/ERS-based BCI would be the left or right movement control of a cursor on a computer screen [118]. The user has to imagine the left or right-hand movement to induce different spatiotemporal EEG desynchronization patterns. A cursor movement control in two dimensions using motor imagery is demonstrated in [119]. A linear equation with weighted combinations of the amplitudes of mu (8–12 Hz) or beta (18–26 Hz) frequency band as the independent variables is used to control the vertical and horizontal movement of the curser. The weights were updated by an adaptive algorithm after each training trial. During the trial, the user has to move the cursor to the target locations on a computer screen by manipulating the EEG activity.

Another application of SMR is the "Graz-BCI" [120] developed for establishing communication ability for paralyzed patients. The experiment results by many researchers show that it is an easy to learn BCI for even the users with less exposure to

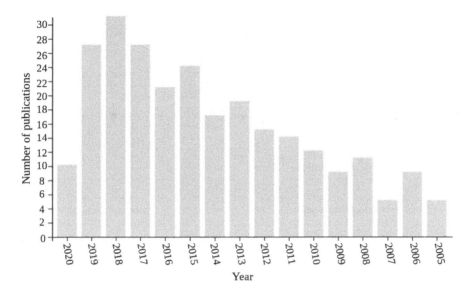

FIGURE 11.11 Year-wise statistics of SMR-based BCI research publications. (Generated from Web of Science search result analysis.)

this paradigm and delivers a better ITR and CA [46]. Additionally, SMR-based BCI has been used for game applications and control of external devices (Figure 11.12).

11.4 SIGNAL PROCESSING

Signal processing is an essential component of any BCI system. The acquired signal needs to be processed to identify the user's intent and generate a control signal to the final execution unit or application. This section will give a brief introduction to the process of signal preprocessing, feature extraction and classification for EEG-based BCI systems.

11.4.1 SIGNAL PREPROCESSING

The preprocessing stage prepares the acquired signal for further processing. EEG signals are often contaminated with ocular, muscular, cardiac artefacts and other noises originating from the power line or electrode impedance variations. Hence, preprocessing is necessary before feeding the signal to the classification algorithms. Preprocessing involves removing the artefacts and applying spatial and temporal filters for improving the EEG signal quality.

11.4.1.1 Artefact Handling

Artefacts are the unwanted electrical activities that interfere with the EEG signal. They can be generated by internal and external sources. Physiological activities like ECG, EOG, EMG or external factors like electrical and magnetic interference, electrode displacement, the subject's body movement, etc. can contribute to EEG

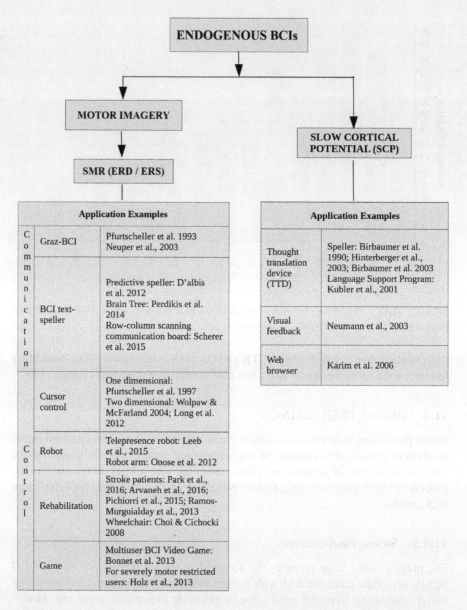

FIGURE 11.12　Application examples for endogenous BCIs [118,119,114–116,120,121–138].

artefacts. Artefacts can be minimized by taking precautionary measures such as instructing the subject to reduce body movement during measurement, avoid unnecessary eye blinking, or by ensuring proper instrument grounding and shielding, etc [139]. There are limitations to this approach and there will be some number of artefacts in the acquired signal; it has to be taken care while designing the signal processing algorithm.

Artefact detection: Artifacts interfere with the EEG in both temporal and spatial domains. Knowledge about both the artefact and the EEG characteristics are necessary for faster real-time artefact detection. Both supervised and unsupervised machine learning techniques are used for separating artefacts and EEG. Artificial neural networks (ANN) and support vector machine (SVM) are commonly used supervised learning-based techniques for artefact detection. Similarly, k-means clustering and outlier detection are most common among unsupervised learning-based detection techniques [139].

Artefact removal: Artifact removal is primarily done either by filtering and regression or by decomposing the signal to other domains. The purpose is to remove or correct the artefacts without distorting the brain signal. Currently, there are several methods employed for artefact removal, such as regression, blind source separation (BSS), time-frequency representation, wavelet transform, empirical mode decomposition (EMD), adaptive filtering, etc.

11.4.1.2 Spatial and Temporal Filtering

Before the feature extraction process, the signal needs to be filtered both in the temporal and spatial domain to improve the signal to noise ratio. Temporal filtering is nothing but the filtering in the time domain, typically using bandpass filters or notch filters designed according to the required EEG rhythms and signal patterns. Also, the optimal filter selection can be subject-specific, for example, in the case of motor imagery-based BCIs. Usually filtering with a broad frequency band is employed to get a better classification [140].

The purpose of spatial filtering is to combine the signals recorded from multiple sensors or channels to form a single signal. The process can be data dependant or data independent. Data independent filtering is based on physical considerations such as the process of the EEG propagation through skull bones and scalp tissues; Laplacian filter or inverse solution-based spatial filtering are a few examples. The data-driven spatial filtering uses either supervised or unsupervised learning methods. Data-driven spatial filters with unsupervised learning utilize methods such as principal component analysis (PCA) or independent component analysis (ICA) [141]. Data-driven spatial filters with supervised learning is now a popular approach due to the good classification performance obtained by them. Such supervised spatial filters include common spatial patterns (CSP) (used for band-power features and oscillatory activity BCI) [140], xDAWN (used to enhance P300 evoked potentials) [142], or Fisher spatial filters (used for ERP classification based on time point features) [143,144], etc.

11.4.2 Feature Extraction and Selection

Before proceeding to classification, the EEG signal should be represented by features that are extracted from the signal. Several features are available to extract from different channels and time segments of the signal [18]. Selection of the appropriate feature vector with feasible dimension from the available feature set is a challenging task. The most common EEG features used for BCI applications are frequency band power and time point features [144]. Frequency band power feature is extracted by calculating the EEG signal power in a given channel over a given period (typically 1 second). They are generally used in motor imagery and steady-state evoked potential-based BCIs where the oscillatory activity of the EEG is utilised. Features such as raw moment of first derivative of instantaneous frequency, area, spectral moment of power spectral density, peak value of power spectral density, etc. that are derived from analytic mode functions (AIMF) can also be used for motor imagery-based BCI tasks. AIMFS are obtained by applying empirical mode decomposition (EMD) and Hilbert transforms on EEG signals. Combining these features with the least squares support vector machine (LS-SVM) classifier can deliver good classification performance for MI-based tasks [145]. Timepoint features are generally used for BCIs based on Event-Related Potentials (ERP) such as P300 where the time-locked variations of EEG are utilised [144]. Recently, time-domain features such as the temporal moment of EEG sub-bands derived using flexible analytic wavelet transform are found to be effective for classification of motor imagery-based tasks [146].

Feature selection is the step that follows the feature extraction step during the training trails; the subset of the extracted features is selected for final use. Feature selection is important due to the following factors; some of the extracted features may not be relevant to the targeted mental state; with fewer features selected, it is easy to observe which feature is linked with the targeted mental state, reducing the number of features will lead to less number of parameters to optimize by the classifier, less number of parameters will improve the computational efficiency and reduced storage, etc. Mainly there are three feature selection approaches; filter, wrapper and embedded approaches [144].

11.4.3 Classification Algorithms

The function of the classification module in a BCI system is to translate the features extracted from the user-generated brain signal patterns into commands for the end device. The translation is based on the classification algorithm's ability to automatically identify the class of data presented as feature vectors [147]. There are several challenges associated with current classification methods; such as poor signal to noise ratio of EEG signals, EEG signal variations from trial to trial and subject to subject, the quantity of training data for calibrating classifiers, low reliability and performance of current BCIs, etc. From time to time, several approaches for implementing classification algorithms with improved performance have been emerged for tackling these challenges. Some of the existing classification algorithms are listed below [144,147].

Linear classifiers are one of the popular methods for classification which includes support vector machines (SVMs), linear discriminant analysis (LDA) and regularized LDA. Neural network classifiers are another family of classifiers which include multi-layer perceptron (MLP), gaussian classifier neural networks and learning vector quantization (LVQ) neural networks. Among the neural network classifiers, the multi-layer perceptron is the widely used method which is an assembly of several layers of artificial neurons with hidden layers. Another classifier family is the nonlinear

Bayesian classifiers that are based on the probability distribution and Bayes' rule. Hidden Markov models (HMMs) and Bayes quadratic classifiers are examples of such classifiers. The nearest neighbor classifier family which includes Mahalanobis distance classifiers or k-nearest neighbor algorithm [147] assign the class to a feature vector according to its nearest neighbor.

Adaptive classifiers can reestimate and update its parameters as new EEG data comes over time [148]. Such classifiers can deal with the EEG variation over time due to the ability to track EEG property changes during working. Also, they can work with minimum offline training data since they learn online. There are transfer learning techniques [149] that can transfer features or classifiers form on domain to another; for example, from one subject to another subject or from one session to another session of the same subject. Transfer learning techniques deal with challenges such as EEG property changes from subject to subject as well as training data limitations by transferring data from other domains.

There are new methods such as matrix and tensor classifiers and deep learning techniques where the processing (extraction and selection of features) and classification of signals are done in a single step. Riemannian geometry-based classification is one of the notable methods of matrix classifiers where the data is mapped directly into a geometric space such that the data can be easily manipulated for averaging, smoothing, interpolating, extrapolating and classifying, etc. [144]. Popular deep learning classifiers such as convolutional neural networks and restricted Boltzmann machines learn the features and classifiers together from the data.

11.5 SOFTWARE TOOLS FOR BCI

Various software tools have been developed since the time research on BCI began. Few of such tools are described below.

One of the software suites of present era, the freely downloadable BCI2000 (www.bci2000.org) started to be developed from the year 2000 in the United States of America with support from different laboratories. The first successful experimental implementation using this software was performed in 2001. Presently, BCI2000-based systems are used in experiments involving subjects with various disabilities mainly for communication purposes.

OpenVIBE (openvibe.inria.fr) originally developed in 2007 and having the latest release in 2018 is another software suite used for BCI research. This software can be used for real-time brain signal processing including signal classification. Both BCI2000 and OpenVIBE have been written in C++.

BCILAB (sccn.ucsd.edu/wiki/BCILAB), developed at the Swartz Center for Computational Neuroscience at California, is a MATLAB toolbox. This software has separate plugins for signal processing, feature extraction, machine learning and BCI paradigms along with a supportive graphical user interface.

BCI++ (http://www.sensibilab.lecco.polimi.it/bci) is a software tool for BCI research involving SSVEP and motor imagery. Many other software tools like Pygame, OpenEXP, Psychtoolbox, etc. are also available.

11.6   CONCLUSION AND FUTURE PERSPECTIVES

This chapter overviews the basic components of a brain computer interface system and its working; starting from signal acquisition to classification. BCI has found its major application in the medical field, either as an assistive and rehabilitation tool for patients with neuromuscular disorders or as a neuromodulation technique for treating diseases. Researchers are also exploring possibilities of noninvasive BCI for gaming and entertainment applications. For noninvasive BCI, signal acquisition is one of the major challenges; technology has not changed much compared to the advancements in the software and algorithms in the last decade. Apart from the EEG head cap that is traditionally used for clinical and laboratory applications, few companies are producing compact and portable EEG headbands which are commercially available in the market for domestic use.

In the last two decades, BCI research has shown significant advancements and developed as a promising technology that has the potential to improve people's lives. State of the art BCI paradigms has proven its viability for multiple clinical and nonclinical applications.

Noninvasive methods of brain interfacing sill remain as a promising and cost-effective method for assisting patients for regenerating abilities for day to day tasks. But for achieving improved degrees of freedom when treating motor impairment with neuroprosthetic devices and exoskeletons requires invasive BCI techniques. Recent trends indicate that invasive BCIs with implanted wireless electrode arrays will emerge as a safe and reliable technology for treating severe neurological disorders. Also, projects like Neuralink's scalable high bandwidth brain-machine interface [150] sheds light into the commercial viability of such surgical brain implants in the foreseeable future.

From recent research works, it is evident that BCI has potential scope for gaming and entertainment applications. Most of the existing works focus on smaller games which are customised for locked-in patients. By integrating BCI with the traditional gameplay, a better quality of game experience can be achieved. Virtual reality headset with built-in BCI can make gameplay more immersive and realistic. Additionally, in previous sections, we have seen many novel approaches for using BCI as a means of communication in disabled patients. This can be extended to healthy people for enhancing day to day human-machine interactions. Facebook's initiative to develop hands-free communication technology is one such example.

We have seen that BCI is a combination of software and hardware. Next-generation algorithms should be more adaptive and userfriendly. Also, more robust algorithms need to be developed in the future that can deliver improved and consistent signal classification accuracy. Moreover, the algorithms should be able to work online with minimal training data and noisy signals and yet deliver reliable performance.

Observing the current developments of this field and the rate at which it is progressing, one can hopefully anticipate the bright future of BCI. However, bringing out the technology into the real world from the laboratory remains a major challenge among researchers. Moreover, BCI research requires an interdisciplinary research environment with involvement of researchers from multiple fields such as engineering, computer science, neuroscience, material science, etc. among others.

ACKNOWLEDGMENTS

We gratefully acknowledge the technical and financial support offered by the National Institute of Technology Calicut for pursuing the research.

REFERENCES

1. Chaudhary, Ujwal, Niels Birbaumer, and Ander Ramos-Murguialday. "Brain–computer interfaces for communication and rehabilitation." *Nature Reviews Neurology* 12, no. 9 (2016): 513–525.
2. Kamiya, Joseph. "Conscious control of brain waves." *Psychology Today* 1, no. 11 (1968): 56–60.
3. Vidal, Jacques J. "Toward direct brain-computer communication." *Annual Review of Biophysics and Bioengineering* 2, no. 1 (1973): 157–180.
4. Lebedev, Mikhail A. and Miguel A. L. Nicolelis. "Brain-machine interfaces: From basic science to neuroprostheses and neurorehabilitation." *Physiological Reviews* 97, no. 2 (2017): 767–837.
5. Nicolelis, Miguel A. L., Asif A. Ghazanfar, Christopher R. Stambaugh, Laura M. O. Oliveira, Mark Laubach, John K. Chapin, Randall J. Nelson, and Jon H. Kaas. "Simultaneous encoding of tactile information by three primate cortical areas." *Nature Neuroscience* 1, no. 7 (1998): 621–630.
6. Wessberg, Johan, Christopher R. Stambaugh, Jerald D. Kralik, Pamela D. Beck, Mark Laubach, John K. Chapin, Jung Kim, Silmon James Biggs, Mandayam A. Srinivasan, and Miguel A.L. Nicolelis. "Real-time prediction of hand trajectory by ensembles of cortical neurons in primates." *Nature* 408, no. 6810 (2000): 361–365.
7. Carmena, Jose M., Mikhail A. Lebedev, Roy E. Crist, Joseph E. O'Doherty, David M. Santucci, Dragan F. Dimitrov, Parag G. Patil, Craig S. Henriquez, and Miguel A. L. Nicolelis. "Learning to control a brain–machine interface for reaching and grasping by primates." *PLoS Biology* 1, no. 2 (2003): 714.
8. Azevedo, Frederico A. C., Ludmila R. B. Carvalho, Lea T. Grinberg, José Marcelo Farfel, Renata E. L. Ferretti, Renata E. P. Leite, Wilson Jacob Filho, Roberto Lent, and Suzana Herculano-Houzel. "Equal numbers of neuronal and nonneuronal cells make the human brain an isometrically scaled-up primate brain." *Journal of Comparative Neurology* 513, no. 5 (2009): 532–541.
9. Nam, Chang S., Anton Nijholt, and Fabien Lotte, eds. *Brain–computer interfaces handbook: Technological and theoretical advances.* Boca Raton, FL: CRC Press (2018).
10. Pfurtscheller, Gert, Brendan Z. Allison, Günther Bauernfeind, Clemens Brunner, Teodoro Solis Escalante, Reinhold Scherer, Thorsten O. Zander, Gernot Mueller-Putz, Christa Neuper, and Niels Birbaumer. "The hybrid BCI." *Frontiers in Neuroscience* 4 (2010): 3.
11. Yin, Erwei, Zongtan Zhou, Jun Jiang, Fanglin Chen, Yadong Liu, and Dewen Hu. "A speedy hybrid BCI spelling approach combining P300 and SSVEP." *IEEE Transactions on Biomedical Engineering* 61, no. 2 (2013): 473–483.
12. Nam, Chang S., Matthew Moore, Inchul Choi, and Yueqing Li. "Designing better, cost-effective brain–computer interfaces." *Ergonomics in Design* 23, no. 4 (2015): 13–19.
13. Hill, N. Jeremy, and Bernhard Schölkopf. "An online brain–computer interface based on shifting attention to concurrent streams of auditory stimuli." *Journal of Neural Engineering* 9, no. 2 (2012): 026011.
14. Muller-Putz, Gernot R., Reinhold Scherer, Christa Neuper, and Gert Pfurtscheller. "Steady-state somatosensory evoked potentials: Suitable brain signals for brain-computer interfaces?" *IEEE Transactions on Neural Systems and Rehabilitation Engineering* 14, no. 1 (2006): 30–37.

15. Brouwer, Anne-Marie, and Jan B. F. Van Erp. "A tactile P300 brain-computer interface." *Frontiers in Neuroscience* 4 (2010): 19.

16. Bledowski, Christoph, David Prvulovic, Karsten Hoechstetter, Michael Scherg, Michael Wibral, Rainer Goebel, and David E. J. Linden. "Localizing P300 generators in visual target and distractor processing: A combined event-related potential and functional magnetic resonance imaging study." *Journal of Neuroscience* 24, no. 42 (2004): 9353–9360.

17. Musiek, Frank E., Robert Froke, and Jeffrey Weihing. "The auditory P300 at or near threshold." *Journal of the American Academy of Audiology* 16, no. 9 (2005): 698–707.

18. Nicolas-Alonso, Luis Fernando, and Jaime Gomez-Gil. "Brain computer interfaces, a review." *Sensors* 12, no. 2 (2012): 1211–1279.

19. Hazrati, Mehrnaz Kh, and Abbas Erfanian. "An online EEG-based brain–computer interface for controlling hand grasp using an adaptive probabilistic neural network." *Medical Engineering & Physics* 32, no. 7 (2010): 730–739.

20. Leeb, Robert, Doron Friedman, Gernot R. Müller-Putz, Reinhold Scherer, Mel Slater, and Gert Pfurtscheller. "Self-paced (asynchronous) BCI control of a wheelchair in virtual environments: A case study with a tetraplegic." *Computational Intelligence and Neuroscience* 2007 (2007): 79642.

21. Babiloni, Claudio, Vittorio Pizzella, Cosimo Del Gratta, Antonio Ferretti, and Gian Luca Romani. "Fundamentals of electroencefalography, magnetoencefalography, and functional magnetic resonance imaging." *International Review of Neurobiology* 86 (2009): 67–80.

22. Castaño-Candamil, Sebastián, Johannes Höhne, Juan-David Martínez-Vargas, Xing-Wei An, German Castellanos-Domínguez, and Stefan Haufe. "Solving the EEG inverse problem based on space–time–frequency structured sparsity constraints." *NeuroImage* 118 (2015): 598–612.

23. Klem, George H., Hans Otto Lüders, H. H. Jasper, and C. Elger. "The ten-twenty electrode system of the International Federation." *Electroencephalography and Clinical Neurophysiology* 52, no. 3 (1999): 3–6.

24. Gian Emilio Chatrian, Etorre Lettich, and Paula L Nelson. "Ten percent electrode system for topographic studies of spontaneous and evoked EEG activities." *American Journal of EEG Technology* 25, no. 2 (1985): 83–92.

25. Oostenveld, Robert, and Peter Praamstra. "The five percent electrode system for high-resolution EEG and ERP measurements." *Clinical Neurophysiology* 112, no. 4 (2001): 713–719.

26. Cohen, David. "Magnetoencephalography: Evidence of magnetic fields produced by alpha-rhythm currents." *Science* 161, no. 3843 (1968): 784–786.

27. Uhlhaas, Peter J., Peter Liddle, David E. J. Linden, Anna C. Nobre, Krish D. Singh, and Joachim Gross. "Magnetoencephalography as a tool in psychiatric research: Current status and perspective." *Biological Psychiatry: Cognitive Neuroscience and Neuroimaging* 2, no. 3 (2017): 235–244.

28. Lindauer, Ute, Ulrich Dirnagl, Martina Füchtemeier, Caroline Böttiger, Nikolas Offenhauser, Christoph Leithner, and Georg Royl. "Pathophysiological interference with neurovascular coupling–when imaging based on hemoglobin might go blind." *Frontiers in Neuroenergetics* 2 (2010): 25.

29. Pauling, Linus, and Charles D. Coryell. "The magnetic properties and structure of hemoglobin, oxyhemoglobin and carbonmonoxyhemoglobin." *Proceedings of the National Academy of Sciences* 22, no. 4 (1936): 210–216.

30. Bermúdez, José Luis. *Cognitive science: An introduction to the science of the mind.* Cambridge: Cambridge University Press (2014).

31. Jobsis, Frans F. "Noninvasive, infrared monitoring of cerebral and myocardial oxygen sufficiency and circulatory parameters." *Science* 198, no. 4323 (1977): 1264–1267.

32. Wilcox, Teresa, Heather Bortfeld, Rebecca Woods, Eric Wruck, and David A. Boas. "Hemodynamic response to featural changes in the occipital and inferior temporal cortex in infants: A preliminary methodological exploration." *Developmental Science* 11, no. 3 (2008): 361–370.

33. Matthews, Fiachra, Barak A. Pearlmutter, Tomas E. Wards, Christopher Soraghan, and Charles Markham. "Hemodynamics for brain-computer interfaces." *IEEE Signal Processing Magazine* 25, no. 1 (2007): 87–94.

34. Freeman, Walter J., Linda J. Rogers, Mark D. Holmes, and Daniel L. Silbergeld. "Spatial spectral analysis of human electrocorticograms including the alpha and gamma bands." *Journal of Neuroscience Methods* 95, no. 2 (2000): 111–121.

35. Slutzky, Marc W., Luke R. Jordan, Todd Krieg, Ming Chen, David J. Mogul, and Lee E. Miller. "Optimal spacing of surface electrode arrays for brain–machine interface applications." *Journal of Neural Engineering* 7, no. 2 (2010): 026004.

36. Leuthardt, Eric C., Gerwin Schalk, Jonathan R. Wolpaw, Jeffrey G. Ojemann, and Daniel W. Moran. "A brain–computer interface using electrocorticographic signals in humans." *Journal of Neural Engineering* 1, no. 2 (2004): 63.

37. Leuthardt, Eric C., Kai J. Miller, Gerwin Schalk, Rajesh P. N. Rao, and Jeffrey G. Ojemann. "Electrocorticography-based brain computer interface-the Seattle experience." *IEEE Transactions on Neural Systems and Rehabilitation Engineering* 14, no. 2 (2006): 194–198.

38. Schalk, Gerwin, Kai J. Miller, Nicholas R. Anderson, J. Adam Wilson, Matthew D. Smyth, Jeffrey G. Ojemann, Daniel W. Moran, Jonathan R. Wolpaw, and Eric C. Leuthardt. "Two-dimensional movement control using electrocorticographic signals in humans." *Journal of Neural Engineering* 5, no. 1 (2008): 75.

39. Kubanek, Jan, Kai J. Miller, Jeffery G. Ojemann. Jonathan R. Wolpaw, Gerwin Schalk. "Decoding flexion of individual fingers using electrocorticographic signals in humans." *Journal of Neural Engineering* 6, no. 6 (2009): 066001.

40. Pistohl, Tobias, Andreas Schulze-Bonhage, Ad Aertsen, Carsten Mehring, and Tonio Ball. "Decoding natural grasp types from human ECoG." *Neuroimage* 59, no. 1 (2012): 248–260.

41. Martin G Bleichner, Zachary V. Freudenburg, Johannus Martijn Jansma, Erik J. Aarnoutse, Mariska J Vansteensel, and Nick Franciscus Ramsey. "Give me a sign: Decoding four complex hand gestures based on high-density ECoG." *Brain Structure and Function* 221, no. 1 (2016): 203–216.

42. Brunner, Peter, Anthony L. Ritaccio, Joseph F. Emrich, Horst Bischof, and Gerwin Schalk. "Rapid communication with a "P300" matrix speller using electrocorticographic signals (ECoG)." *Frontiers in Neuroscience* 5 (2011): 5.

43. Gunasekera, Bhagya, Tarun Saxena, Ravi Bellamkonda, and Lohitash Karumbaiah. "Intracortical recording interfaces: Current challenges to chronic recording function." *ACS Chemical Neuroscience* 6, no. 1 (2015): 68–83.

44. Hillyard, Steven A., and Marta Kutas. "Electrophysiology of cognitive processing." *Annual Review of Psychology* 34, no. 1 (1983): 33–61.

45. Farwell, Lawrence Ashley, and Emanuel Donchin. "Talking off the top of your head: Toward a mental prosthesis utilizing event-related brain potentials." *Electroencephalography and Clinical Neurophysiology* 70, no. 6 (1988): 510–523.

46. Lazarou, Ioulietta, Spiros Nikolopoulos, Panagiotis C. Petrantonakis, Ioannis Kompatsiaris, and Magda Tsolaki. "EEG-based brain–computer interfaces for communication and rehabilitation of people with motor impairment: A novel approach of the 21st century." *Frontiers in Human Neuroscience* 12 (2018): 14.

47. Townsend, George, Brandon K. LaPallo, Chadwick B. Boulay, Dean J. Krusienski, G. E. Frye, Ckea Hauser, Nicholas Edward Schwartz, Theresa M. Vaughan, Jonathan R. Wolpaw, and Eric W. Sellers. "A novel P300-based brain–computer interface stimulus presentation paradigm: Moving beyond rows and columns." *Clinical Neurophysiology* 121, no. 7 (2010): 1109–1120.

48. Fazel-Rezai, Reza, and Kamyar Abhari. "A region-based P300 speller for brain-computer interface." *Canadian Journal of Electrical and Computer Engineering* 34, no. 3 (2009): 81–85.

49. Acqualagna, Laura, Matthias Sebastian Treder, Martijn Schreuder, and Benjamin Blankertz. "A novel brain-computer interface based on the rapid serial visual presentation paradigm." *In 2010 Annual International Conference of the IEEE Engineering in Medicine and Biology,* Buenos Aires, Argentina, pp. 2686–2689, IEEE (2010).

50. Allison, Brendan Z., Andrea Kübler, and Jing Jin. "30+ years of P300 brain–computer interfaces." *Psychophysiology* (2020): e13569.

51. Kaufmann, Tobias, Stefan M. Schulz, Claudia Grünzinger, and Andrea Kübler. "Flashing characters with famous faces improves ERP-based brain–computer interface performance." *Journal of Neural Engineering* 8, no. 5 (2011): 056016.

52. Jin, Jing, Ian Daly, Yu Zhang, Xingyu Wang, and Andrzej Cichocki. "An optimized ERP brain–computer interface based on facial expression changes." *Journal of Neural Engineering* 11, no. 3 (2014): 036004.

53. Sellers, Eric W., and Emanuel Donchin. "A P300-based brain–computer interface: Initial tests by ALS patients." *Clinical Neurophysiology* 117, no. 3 (2006): 538–548.

54. Kübler, Andrea, Adrian Furdea, Sebastian Halder, Eva Maria Hammer, Femke Nijboer, and Boris Kotchoubey. "A brain–computer interface controlled auditory event-related potential (P300) spelling system for locked-in patients." *Annals of the New York Academy of Sciences* 1157, no. 1 (2009): 90–100.

55. Güntekin, Bahar, and Erol Başar. "A new interpretation of P300 responses upon analysis of coherences." *Cognitive Neurodynamics* 4, no. 2 (2010): 107–118.

56. Halder, Sebastian, Massimiliano Rea, R. Andreoni, Femke Nijboer, Eva Maria Hammer, Sonja Claudia Kleih, Niels Birbaumer, and Andrea Kub'ler. "An auditory oddball brain–computer interface for binary choices." *Clinical Neurophysiology* 121, no. 4 (2010): 516–523.

57. Schreuder, Martijn, Benjamin Blankertz, and Michael Tangermann. "A new auditory multi-class brain-computer interface paradigm: Spatial hearing as an informative cue." *PLoS One* 5, no. 4 (2010): e9813.

58. Huang, Minqiang, Jing Jin, Yu Zhang, Dewen Hu, and Xingyu Wang. "Usage of drip drops as stimuli in an auditory P300 BCI paradigm." *Cognitive Neurodynamics* 12, no. 1 (2018): 85–94.

59. Simon, Nadine, Ivo Käthner, Carolin A. Ruf, Emanuele Pasqualotto, Andrea Kübler, and Sebastian Halder. "An auditory multiclass brain-computer interface with natural stimuli: Usability evaluation with healthy participants and a motor impaired end user." *Frontiers in Human Neuroscience* 8 (2015): 1039.

60. van der Waal, Marjolein, Marianne Severens, Jeroen Geuze, and Peter Desain. "Introducing the tactile speller: An ERP-based brain–computer interface for communication." *Journal of Neural Engineering* 9, no. 4 (2012): 045002.

61. Kaufmann, Tobias, Andreas Herweg, and Andrea Kübler. "Toward brain-computer interface based wheelchair control utilizing tactually-evoked event-related potentials." *Journal of Neuroengineering and Rehabilitation* 11, no. 1 (2014): 7.

62. Herweg, Andreas, Julian Gutzeit, Sonja Kleih, and Andrea Kübler. "Wheelchair control by elderly participants in a virtual environment with a brain-computer interface (BCI) and tactile stimulation." *Biological Psychology* 121 (2016): 117–124.

63. Treder, Matthias S., and Benjamin Blankertz. "(C) overt attention and visual speller design in an ERP-based brain-computer interface." *Behavioral and Brain Functions* 6, no. 1 (2010): 1–13.

64. Bell, Christian J., Pradeep Shenoy, Rawichote Chalodhorn, and Rajesh P. N. Rao. "Control of a humanoid robot by a noninvasive brain–computer interface in humans." *Journal of Neural Engineering* 5, no. 2 (2008): 214.

65. Spataro, Rossella, Antonio Chella, Brendan Allison, Marcello Giardina, Rosario Sorbello, Salvatore Tramonte, Christoph Guger, and Vincenzo La Bella. "Reaching and grasping a glass of water by locked-in ALS patients through a BCI-controlled humanoid robot." *Frontiers in Human Neuroscience* 11 (2017): 68.

66. Pires, Gabriel, Urbano Nunes, and Miguel Castelo-Branco. "Evaluation of brain-computer interfaces in accessing computer and other devices by people with severe motor impairments." *Procedia Computer Science* 14 (2012): 283–292.

67. Corralejo, Rebeca, Luis F. Nicolás-Alonso, Daniel Álvarez, and Roberto Hornero. "A P300-based brain–computer interface aimed at operating electronic devices at home for severely disabled people." *Medical and Biological Engineering and Computing* 52, no. 10 (2014): 861–872.

68. Mugler, Emily M., Carolin A. Ruf, Sebastian Halder, Michael Bensch, and Andrea Kubler. "Design and implementation of a P300-based brain-computer interface for controlling an internet browser." *IEEE Transactions on Neural Systems and Rehabilitation Engineering* 18, no. 6 (2010): 599–609.

69. Zickler, Claudia, Angela Riccio, Francesco Leotta, Sandra Hillian-Tress, Sebastian Halder, Elisa Holz, Pit Staiger-Sälzer et al. "A brain-computer interface as input channel for a standard assistive technology software." *Clinical EEG and Neuroscience* 42, no. 4 (2011): 236–244.

70. Iturrate, Iñaki, Javier M. Antelis, Andrea Kubler, and Javier Minguez. "A noninvasive brain-actuated wheelchair based on a P300 neurophysiological protocol and automated navigation." *IEEE Transactions on Robotics* 25, no. 3 (2009): 614–627.

71. Pfurtscheller, Gert, Gernot R. Müller-Putz, Reinhold Scherer, and Christa Neuper. "Rehabilitation with brain-computer interface systems." *Computer* 41, no. 10 (2008): 58–65.

72. Finke, Andrea, Alexander Lenhardt, and Helge Ritter. "The MindGame: A P300-based brain–computer interface game." *Neural Networks* 22, no. 9 (2009): 1329–1333.

73. Congedo, Marco, Matthieu Goyat, Nicolas Tarrin, Gelu Ionescu, Léo Varnet, Bertrand Rivet, Ronald Phlypo, Nisrine Jrad, Michael Acquadro, and Christian Jutten. "Brain invaders": A prototype of an open-source P300-based video game working with the OpenViBE platform." (2011).

74. Botrel, Loïc, Holz, Elisa Mira, and Andrea Kub'ler. "Brain painting V2: Evaluation of P300-based brain-computer interface for creative expression by an end-user following the user-centered design." *Brain-Computer Interfaces* 2, no. 2–3 (2015): 135–149.

75. Kübler, Andrea, and Loic Botrel. "The making of brain painting: From the idea to daily life use by people in the locked-in state." In Anton Nijholt, ed. *Brain art*, pp. 409–431. Cham: Springer (2019).

76. Sellers, Eric W., Andrea Kubler, and Emanuel Donchin. "Brain-computer interface research at the University of South Florida Cognitive Psychophysiology Laboratory: The P300 speller." *IEEE Transactions on Neural Systems and Rehabilitation Engineering* 14, no. 2 (2006): 221–224.

77. Furdea, Adrian, Sebastian Halder, D. J. Krusienski, Donald Bross, Femke Nijboer, Niels Birbaumer, and Andrea Kübler. "An auditory oddball (P300) spelling system for brain-computer interfaces." *Psychophysiology* 46, no. 3 (2009): 617–625.

78. Schreuder, Martijn, Benjamin Blankertz, and Michael Tangermann. "A new auditory multi-class brain-computer interface paradigm: Spatial hearing as an informative cue." *PLoS One* 5, no. 4 (2010): e9813.

79. Simon, Nadine, Ivo Käthner, Carolin A. Ruf, Emanuele Pasqualotto, Andrea Kübler, and Sebastian Halder. "An auditory multiclass brain-computer interface with natural stimuli: Usability evaluation with healthy participants and a motor impaired end user." *Frontiers in Human Neuroscience* 8 (2015): 1039.

80. Robinson, Chris, Nayef Ahmar, and Vladimir Sloutsky. "Evidence for auditory dominance in a passive oddball task." *Proceedings of the Annual Meeting of the Cognitive Science Society*, 32, no. 32 (2010): 2644–2649.

81. Mori, Hiromu, Shoji Makino, and Tomasz M. Rutkowski. "Multi–command chest tactile brain computer interface for small vehicle robot navigation." *In International Conference on Brain and Health Informatics*, pp. 469–478, Springer, Cham (2013).

82. Kono, Shota, and Tomasz M. Rutkowski. "Tactile-force brain-computer interface paradigm." *Multimedia Tools and Applications* 74, no. 19 (2015): 8655–8667.

83. Lugo, Zulay R., Javi Rodriguez, Alexander Lechner, Rupert Ortner, Ithabi S. Gantner, Steven Laureys, Quentin Noirhomme, and Christoph Guger. "A vibrotactile p300-based brain–computer interface for consciousness detection and communication." *Clinical EEG and Neuroscience* 45, no. 1 (2014): 14–21.

84. Thurlings, Marieke E., Jan B. F. Van Erp, Anne-Marie Brouwer, and Peter Werkhoven. "Controlling a tactile erp–bci in a dual task." *IEEE Transactions on Computational Intelligence and AI in Games* 5, no. 2 (2013): 129–140.

85. Regan, David. "Steady-state evoked potentials." *JOSA* 67, no. 11 (1977): 1475–1489.

86. Herrmann, Christoph S. "Human EEG responses to 1–100 Hz flicker: Resonance phenomena in visual cortex and their potential correlation to cognitive phenomena." *Experimental Brain Research* 137, no. 3–4 (2001): 346–353.

87. Fisher, Robert S., Graham Harding, Giuseppe Erba, Gregory L. Barkley, and Arnold Wilkins. "Photic-and pattern-induced seizures: A review for the Epilepsy Foundation of America Working Group." *Epilepsia* 46, no. 9 (2005): 1426–1441.

88. Chabuda, Anna, Piotr Durka, and Jarosław Żygierewicz. "High frequency SSVEP-BCI with hardware stimuli control and phase-synchronized comb filter." *IEEE Transactions on Neural Systems and Rehabilitation Engineering* 26, no. 2 (2017): 344–352.

89. Elliott, Mark A., and Hermann J. Müller. "Synchronous information presented in 40-Hz flicker enhances visual feature binding." *Psychological Science* 9, no. 4 (1998): 277–283.

90. Volosyak, Ivan. "SSVEP-based Bremen–BCI interface: Boosting information transfer rates." *Journal of Neural Engineering* 8, no. 3 (2011): 036020.

91. Bin, Guangyu, Xiaorong Gao, Zheng Yan, Bo Hong, and Shangkai Gao. "An online multi-channel SSVEP-based brain–computer interface using a canonical correlation analysis method." *Journal of Neural Engineering* 6, no. 4 (2009): 046002.

92. Chen, Xiaogang, Yijun Wang, Masaki Nakanishi, Xiaorong Gao, Tzyy-Ping Jung, and Shangkai Gao. "High-speed spelling with a noninvasive brain–computer interface." *Proceedings of the National Academy of Sciences* 112, no. 44 (2015): E6058–E6067.

93. Picton, Terence W., Michael Sasha John, Andrew Dimitrijevic, and David Purcell. "Human auditory steady-state responses: Respuestas auditivas de estado estable en humanos." *International Journal of Audiology* 42, no. 4 (2003): 177–219.

94. Punsawad, Yunyong, and Yodchanan Wongsawat. "Multi-command SSAEP-based BCI system with training sessions for SSVEP during an eye fatigue state." *IEEJ Transactions on Electrical and Electronic Engineering* 12 (2017): S72–S78.

95. Carmona, Luciano, Pablo F. Diez, Eric Laciar, and Vicente Mut. "Multisensory stimulation and EEG recording below the hair-line: A new paradigm on brain computer interfaces." *IEEE Transactions on Neural Systems and Rehabilitation Engineering* 28, no. 4 (2020): 825–831.

96. Wang, Xin, Teng Cao, Boyu Wang, Feng Wan, Peng Un Mak, Pui In Mak, Mang I. Vai, and Chaozheng Li. "An online SSVEP-based chatting system." *In Proceedings 2011 International Conference on System Science and Engineering,* Macau, China, pp. 536–539, IEEE (2011).

97. Hwang, Han-Jeong, Jeong-Hwan Lim, Young-Jin Jung, Han Choi, Sang Woo Lee, and Chang-Hwan Im. "Development of an SSVEP-based BCI spelling system adopting a QWERTY-style LED keyboard." *Journal of Neuroscience Methods* 208, no. 1 (2012): 59–65.

98. Nakanishi, Masaki, Yijun Wang, Xiaogang Chen, Yu-Te Wang, Xiaorong Gao, and Tzyy-Ping Jung. "Enhancing detection of SSVEPs for a high-speed brain speller using task-related component analysis." *IEEE Transactions on Biomedical Engineering* 65, no. 1 (2017): 104–112.

99. Cecotti, Hubert. "A self-paced and calibration-less SSVEP-based brain–computer interface speller." *IEEE Transactions on Neural Systems and Rehabilitation Engineering* 18, no. 2 (2010): 127–133.

100. Liu, Yiliang, Zhijun Li, Tong Zhang, and Suna Zhao. "Brain-robot interface-based navigation control of a mobile robot in corridor environments." *IEEE Transactions on Systems, Man, and Cybernetics: Systems* 50, no. 8 (2018): 3047–3058.

101. Muller-Putz, Gernot R., and Gert Pfurtscheller. "Control of an electrical prosthesis with an SSVEP-based BCI." *IEEE Transactions on Biomedical Engineering* 55, no. 1 (2007): 361–364.

102. Sakurada, Takeshi, Toshihiro Kawase, Kouji Takano, Tomoaki Komatsu, and Kenji Kansaku. "A BMI-based occupational therapy assist suit: Asynchronous control by SSVEP." *Frontiers in Neuroscience* 7 (2013): 172.

103. Zhao, Xingang, Yaqi Chu, Jianda Han, and Zhiqiang Zhang. "SSVEP-based brain–computer interface controlled functional electrical stimulation system for upper extremity rehabilitation." *IEEE Transactions on Systems, Man, and Cybernetics: Systems* 46, no. 7 (2016): 947–956.

104. Martišius, Ignas, and Robertas Damaševičius. "A prototype SSVEP based real time BCI gaming system." *Computational Intelligence and Neuroscience* 2016 (2016): 3861425.

105. Kim, Do-Won, Han-Jeong Hwang, Jeong-Hwan Lim, Yong-Ho Lee, Ki-Young Jung, and Chang-Hwan Im. "Classification of selective attention to auditory stimuli: Toward vision-free brain–computer interfacing." *Journal of Neuroscience Methods* 197, no. 1 (2011): 180–185.

106. Hong, L. Elliot, Ann Summerfelt, Robert McMahon, Helene Adami, Grace Francis, Amie Elliott, Robert W. Buchanan, and Gunvant K. Thaker. "Evoked gamma band synchronization and the liability for schizophrenia." *Schizophrenia Research* 70, no. 2–3 (2004): 293–302.

107. Seha, Sherif Nagib Abbas, and Dimitrios Hatzinakos. "EEG-based human recognition using steady-state AEPs and subject-unique spatial filters." *IEEE Transactions on Information Forensics and Security* 15 (2020): 3901–3910.

108. Kim, Keun-Tae, and Seong-Whan Lee. "Steady-state somatosensory evoked potentials for brain-controlled wheelchair." *In 2014 International Winter Workshop on Brain-Computer Interface (BCI)*, Gangwon, Korea (South), pp. 1–2, IEEE (2014).

109. Kim, Keun-Tae, Heung-Il Suk, and Seong-Whan Lee. "Commanding a brain-controlled wheelchair using steady-state somatosensory evoked potentials." *IEEE Transactions on Neural Systems and Rehabilitation Engineering* 26, no. 3 (2016): 654–665.

110. Tidoni, Emmanuele, Pierre Gergondet, Gabriele Fusco, Abderrahmane Kheddar, and Salvatore M. Aglioti. "The role of audio-visual feedback in a thought-based control of a humanoid robot: A BCI study in healthy and spinal cord injured people." *IEEE Transactions on Neural Systems and Rehabilitation Engineering* 25, no. 6 (2016): 772–781.

111. Li, Jie, Hongfei Ji, Lei Cao, Di Zang, Rong Gu, Bin Xia, and Qiang Wu. "Evaluation and application of a hybrid brain computer interface for real wheelchair parallel control with multi-degree of freedom." *International Journal of Neural Systems* 24, no. 04 (2014): 1450014.

112. Snyder, Abraham Z. "Steady-state vibration evoked potentials: Description of technique and characterization of responses." *Electroencephalography and Clinical Neurophysiology/Evoked Potentials Section* 84, no. 3 (1992): 257–268.

113. Tobimatsu, Shozo, You Min Zhang, Rie Suga, and Motohiro Kato. "Differential temporal coding of the vibratory sense in the hand and foot in man." *Clinical Neurophysiology* 111, no. 3 (2000): 398–404.

114. Birbaumer, Niels, Thomas Elbert, Anthony G. Canavan, and Brigitte Rockstroh. "Slow potentials of the cerebral cortex and behavior." *Physiological Reviews* 70, no. 1 (1990): 1–41.

115. Birbaumer, Niels, Thilo Hinterberger, Andrea Kubler, and Nicola Neumann. "The thought-translation device (TTD): Neurobehavioral mechanisms and clinical outcome." *IEEE Transactions on Neural Systems and Rehabilitation Engineering* 11, no. 2 (2003): 120–123.

116. Karim, Ahmed A., Thilo Hinterberger, Jürgen Richter, Jürgen Mellinger, Nicola Neumann, Herta Flor, Andrea Kübler, and Niels Birbaumer. "Neural internet: Web surfing with brain potentials for the completely paralyzed." *Neurorehabilitation and Neural Repair* 20, no. 4 (2006): 508–515.

117. Graimann, Bernhard, Brendan Allison, and Gert Pfurtscheller, eds. "Brain–computer interfaces: A gentle introduction." In *Brain-computer interfaces*, pp. 1–27. Berlin, Heidelberg: Springer (2009).

118. Pfurtscheller, Gert, Ch Neuper, Doris Flotzinger, and Martin Pregenzer. "EEG-based discrimination between imagination of right and left hand movement. "*Electroencephalography and Clinical Neurophysiology* 103, no. 6 (1997): 642–651.

119. Wolpaw, Jonathan R., and Dennis J. McFarland. "Control of a two-dimensional movement signal by a noninvasive brain-computer interface in humans." *Proceedings of the National Academy of Sciences* 101, no. 51 (2004): 17849–17854.

120. Pfurtscheller, Gert, Doris Flotzinger, and Joachim Kalcher. "Brain-computer interface: A new communication device for handicapped persons." *Journal of Microcomputer Applications* 16, no. 3 (1993): 293–299.

121. Neuper, Christa, Gernot R. Müller, Andrea Kübler, Niels Birbaumer, and Gert Pfurtscheller. "Clinical application of an EEG-based brain–computer interface: A case study in a patient with severe motor impairment." *Clinical Neurophysiology* 114, no. 3 (2003): 399–409.

122. Perdikis, Serafeim, Robert Leeb, John Williamson, Amy Ramsay, Michele Tavella, Lorenzo Desideri, Evert-Jan Hoogerwerf, Abdul Al-Khodairy, Roderick Murray-Smith, and J. d R Millán. "Clinical evaluation of BrainTree, a motor imagery hybrid BCI speller." *Journal of Neural Engineering* 11, no. 3 (2014): 036003.

123. D'albis, Tiziano, Rossella Blatt, Roberto Tedesco, Licia Sbattella, and Matteo Matteucci. "A predictive speller controlled by a brain-computer interface based on motor imagery." *ACM Transactions on Computer-Human Interaction (TOCHI)* 19, no. 3 (2012): 1–25.

124. Park, Wanjoo, Gyu Hyun Kwon, Yun-Hee Kim, Jong-Hwan Lee, and Laehyun Kim. "EEG response varies with lesion location in patients with chronic stroke." *Journal of Neuroengineering and Rehabilitation* 13, no. 1 (2016): 1–10.

125. Arvaneh, Mahnaz, Cuntai Guan, Kai Keng Ang, Tomas E. Ward, Karen S. G. Chua, Christopher Wee Keong Kuah, Gopal Joseph Ephraim Joseph, Kok Soon Phua, and Chuanchu Wang. "Facilitating motor imagery-based brain–computer interface for stroke patients using passive movement." *Neural Computing and Applications* 28, no. 11 (2017): 3259–3272.

126. Pichiorri, Floriana, Giovanni Morone, Manuela Petti, Jlenia Toppi, Iolanda Pisotta, Marco Molinari, Stefano Paolucci, et al. "Brain–computer interface boosts motor imagery practice during stroke recovery." *Annals of Neurology* 77, no. 5 (2015): 851–865.

127. Ramos-Murguialday, Ander, Doris Broetz, Massimiliano Rea, Leonhard Läer, Özge Yilmaz, Fabricio L. Brasil, Giulia Liberati et al. "Brain–machine interface in chronic stroke rehabilitation: A controlled study." *Annals of Neurology* 74, no. 1 (2013): 100–108.

128. Holz, Elisa Mira, Johannes Höhne, Pit Staiger-Sälzer, Michael Tangermann, and Andrea Kübler. "Brain–computer interface controlled gaming: Evaluation of usability by severely motor restricted end-users." *Artificial Intelligence in Medicine* 59, no. 2 (2013): 111–120.

129. Scherer, Reinhold, Martin Billinger, Johanna Wagner, Andreas Schwarz, Dirk Tassilo Hettich, Elaina Bolinger, Mariano Lloria Garcia, Juan Navarro, and Gernot Müller-Putz. "Thought-based row-column scanning communication board for individuals with cerebral palsy." *Annals of Physical and Rehabilitation Medicine* 58, no. 1 (2015): 14–22.

130. Bonnet, Laurent, Fabien Lotte, and Anatole Lécuyer. "Two brains, one game: Design and evaluation of a multiuser BCI video game based on motor imagery." *IEEE Transactions on Computational Intelligence and AI in Games* 5, no. 2 (2013): 185–198.

131. Long, Jinyi, Yuanqing Li, Tianyou Yu, and Zhenghui Gu. "Target selection with hybrid feature for BCI-based 2-D cursor control." *IEEE Transactions on Biomedical Engineering* 59, no. 1 (2011): 132–140.

132. Leeb, Robert, Luca Tonin, Martin Rohm, Lorenzo Desideri, Tom Carlson, and Jose del R. Millan. "Towards independence: A BCI telepresence robot for people with severe motor disabilities." *Proceedings of the IEEE* 103, no. 6 (2015): 969–982.

133. Onose, Gelu, Cristian Grozea, Aurelian Anghelescu, Cristina Daia, Crina Julieta Sinescu, Alexandru Vlad Ciurea, Tiberiu Spircu et al. "On the feasibility of using motor imagery EEG-based brain–computer interface in chronic tetraplegics for assistive robotic arm control: A clinical test and long-term post-trial follow-up." *Spinal Cord* 50, no. 8 (2012): 599–608.

134. Choi, Kyuwan, and Andrzej Cichocki. "Control of a wheelchair by motor imagery in real time." *In International Conference on Intelligent Data Engineering and Automated Learning*, pp. 330–337, Springer, Berlin, Heidelberg (2008).

135. Kübler, Andrea, Nicola Neumann, Jochen Kaiser, Boris Kotchoubey, Thilo Hinterberger, and Niels P. Birbaumer. "Brain-computer communication: Self-regulation of slow cortical potentials for verbal communication." *Archives of Physical Medicine and Rehabilitation* 82, no. 11 (2001): 1533–1539.

136. Hinterberger, Thilo, Andrea Kübler, Jochen Kaiser, Nicola Neumann, and Niels Birbaumer. "A brain–computer interface (BCI) for the locked-in: Comparison of different EEG classifications for the thought translation device." *Clinical Neurophysiology* 114, no. 3 (2003): 416–425.

137. Neumann, Nicola, Andrea Kübler, Jochen Kaiser, Thilo Hinterberger, and Niels Birbaumer. "Conscious perception of brain states: Mental strategies for brain–computer communication." *Neuropsychologia* 41, no. 8 (2003): 1028–1036.

138. Karim, Ahmed A., Thilo Hinterberger, Jürgen Richter, Jürgen Mellinger, Nicola Neumann, Herta Flor, Andrea Kübler, and Niels Birbaumer. "Neural internet: Web surfing with brain potentials for the completely paralyzed." *Neurorehabilitation and Neural Repair* 20, no. 4 (2006): 508–515.

139. Islam, Md Kafiul, Amir Rastegarnia, and Zhi Yang. "Methods for artifact detection and removal from scalp EEG: A review." *Neurophysiologie Clinique/Clinical Neurophysiology* 46, no. 4–5 (2016): 287–305.

140. Ramoser, Herbert, Johannes Muller-Gerking, and Gert Pfurtscheller. "Optimal spatial filtering of single trial EEG during imagined hand movement." *IEEE Transactions on Rehabilitation Engineering* 8, no. 4 (2000): 441–446.

141. Kachenoura, Amar, Laurent Albera, Lotfi Senhadji, and Pierre Comon. "ICA: A potential tool for BCI systems." *IEEE Signal Processing Magazine* 25, no. 1 (2007): 57–68.

142. Rivet, Bertrand, Antoine Souloumiac, Virginie Attina, and Guillaume Gibert. "xDAWN algorithm to enhance evoked potentials: Application to brain–computer interface." *IEEE Transactions on Biomedical Engineering* 56, no. 8 (2009): 2035–2043.

143. Hoffmann, Ulrich, Jean-Marc Vesin, and Touradj Ebrahimi. Spatial filters for the classification of event-related potentials. *European Symposium on Artificial Neural Networks,* Bruges, April 26–28 (2006).

144. Lotte, Fabien, Laurent Bougrain, Andrzej Cichocki, Maureen Clerc, Marco Congedo, Alain Rakotomamonjy, and Florian Yger. "A review of classification algorithms for EEG-based brain–computer interfaces: A 10 year update." *Journal of Neural Engineering* 15, no. 3 (2018): 031005.

145. Taran, Sachin, Varun Bajaj, Dheeraj Sharma, Siuly Siuly, and Abdulkadir Sengur. "Features based on analytic IMF for classifying motor imagery EEG signals in BCI applications." *Measurement* 116 (2018): 68–76.

146. Chaudhary, Shalu, Sachin Taran, Varun Bajaj, and Siuly Siuly. "A flexible analytic wavelet transform based approach for motor-imagery tasks classification in BCI applications." *Computer Methods and Programs in Biomedicine* 187 (2020): 105325.

147. Lotte, Fabien, Marco Congedo, Anatole Lécuyer, Fabrice Lamarche, and Bruno Arnaldi. "A review of classification algorithms for EEG-based brain–computer interfaces." *Journal of Neural Engineering* 4, no. 2 (2007): R1.

148. Buttfield, Anna, Pierre W. Ferrez, and Jd R. Millan. "Towards a robust BCI: Error potentials and online learning." *IEEE Transactions on Neural Systems and Rehabilitation Engineering* 14, no. 2 (2006): 164–168.

149. Pan, Sinno Jialin, and Qiang Yang. "A survey on transfer learning." *IEEE Transactions on Knowledge and Data Engineering* 22, no. 10 (2009): 1345–1359.

150. Musk, Elon. "An integrated brain-machine interface platform with thousands of channels." *Journal of Medical Internet Research* 21, no. 10 (2019): e16194.

12 Oriented Approaches for Brain Computing and Human Behavior Computing Using Machine Learning

Monali Gulhane and T. Sajana
Koneru Lakshmaiah Education Foundation

CONTENTS

12.1 OVERVIEW OF MACHINE LEARNING (DEFINITION APPROACHES)

Machine learning is a sub-category of artificial intelligence (AI). Machine learning can be defined by the aim of analyzing and understanding the data structure and fitting that data into modeling techniques that people can understand and use it for their application. Machine learning is termed as self-learning system, which learns from its inputs and performs task according to the application without being explicitly programmed by the programmer. Machine can individually learn from the data and can produce the accurate results related to it.

Machine learning collects different data from statistical tools or graphs to predict an output. The generated output is then combined or interfaced to the device or software to generate actionable insights. The machine learning is operated by taking data as input and then uses algorithm to generate the answer.

Example 12.1

User gets recommendation of movies or series on Netflix is completely based on the historical data or contents that has been watched previously by the user. Regardless of the fact that machine learning can be a field within informatics, it differs from conventional computational methods. A conventional computing includes computations that are sets of explicitly modified enlightening that computers used to calculate or issue fathom. Machine learning computations allow computers to prepare for inputs of information and use qualitative examination in order to yield values falling within a particular run. Machine learning thus encourages computers to construct. Machine learning therefore promotes computers to design things from test information so that decision-making aspects can be computerized based on the information inputs.

Nowadays, every innovation client has profited immensely with machine learning usage. Innovation in facial recognition allows stages of social media to help customers label and express companion photographs. Innovation in optical character recognition (OCR) shifts to mobile type over images of content. Proposal engines, fueled by machine learning, suggest what motion pictures or television appear to be observing based on client inclinations. Self-driving vehicles which rely on artificial intelligence (machine learning) to analyze might become available to customers before long [1].

12.1.1 MACHINE LEARNING IS PERFECT TO BE USED

1. To identify designs and patterns in big data
2. Complex issues for which employing a conventional approach yields no great solution
3. Apply to energetic circumstances where the machine learning can learn from the encompassing environment 4. Problems for which existing arrangements require a parcel of fine-tuning or long records of rules

12.1.2 EXAMPLES OF MACHINE LEARNING APPLICATIONS

This section briefly describes various machine learning applications.

1. To section news articles automatically
2. To abstract large and lengthy documents automatically
3. To detect tumors in human brain scans
4. To identify and eliminate offensive words in a public forum or website comments
5. To predict a company's economy graph for the following year
6. To create chat bots for specific user or personal assistants
7. To create gaming bots for one (signal) players
8. To divide customers based on their purchases
9. To recommend products to customers based on their pervious purchase patterns

Machine learning is a branch which continues to improve due to use work with machine learning methodologies there are stated considerations to keep in mind or study the effects of machine learning processes.

12.1.3 WHEN DO WE NEED MACHINE LEARNING?

In machine learning, there are generally two very basic components of any given problem: it uses of programs that learn and improve based on their "experience" and it entirely depends on the complexity of problem and its adaptively. Machine learning is implemented when: [3].

Tasks that are too complex to program. Tasks performed by humans: The human tasks in routine that are not able to do sufficiently, ML elaborates to extract task from a defined program. Examples: car driving, speech recognition, etc. These tasks can be automated using different machine learning algorithms. Tasks beyond human capabilities: tasks that are complex for humans to perform can benefit from machine learning techniques; these tasks are analysis of overly complex datasets: like astronomical data, medical applications like patient behavior analysis for disease prediction etc. and many more. The main difference in programmed tools is its rigidity – once the program is written and installed, it remains unchanged. However, tasks may change over time from user to user. Machine learning tools adapt to their input information [4], which will enable to predict output based on the previous data history. Machine learning implementation required study of different learning techniques and algorithms. They are as follows:

12.1.4 TYPES OF LEARNING

The board category of the types of learning is described in detail. The human behavior can be adaptive to different types of learning methods: All methods can be classified as follows [2]:

Learning Problems
1. Supervised learning
2. Unsupervised learning
3. Reinforcement learning

Hybrid Learning Problems
4. Semi-supervised learning
5. Self-supervised learning
6. Multi-instance learning

Statistical Inference
7. Inductive learning
8. Deductive inference
9. Transductive learning

Learning Techniques
10. Multi-task learning
11. Active learning
12. Online learning
13. Transfer learning
14. Ensemble learning

12.1.4.1 Learning Problems

This section briefly describes different learning problems and their various types.

12.1.4.1.1 *Supervised Learning*

Supervised learning describes a problem class involving multiple models to understand how to check (identify) and process information or examples between inputs and goals or corresponding outputs. Implementations wherein the training data set information contains examples of the feature vector such as their corresponding data vectors are recognized as supervised learning algorithms, according to Christopher Bishop [5].

In the areas of supervised learning, we can learn an example of inputs that have details of the required outputs that is given to the software system. The main objective of the SL technique is to be eligible to "learn" to identify the suitable answer or inaccuracies by evaluating its actual output with the "taught" outputs and then to reconfigure the method as per the outcomes. Supervised learning is using methods to identify label values on any sets of analysis that is unlabeled. Supervised learning is predicted to describe consistent reliability considering the input factors (x) and the output factors (Y), and the user includes an algorithm to evaluate and learn from input to output the mapping function $f(x)$ [2].

$$Y = f(x)$$

The main objective of supervised learning is to assume the mapping function correctly for machine to feed with any type of new input data (x) machine that can predict the output variable (*y*). Supervised learning problems are divided into two group 1. Classification problem and 2. Regression problems.

1. **Classification**: the classification problem arises when a classification or group is the output variable, such as "red" or "blue" or "disease," and "no disease."
2. **Regression**: a regression analysis is made when a true value, such as "dollars" or "weight" is the output variable.-+

A few prevalent types of issues associated to the regression and classification category above include suggestions made to the prediction of the user and time series including both. Some popular examples of supervised machine learning algorithms are:

1. **Linear regression problems**: used for applications computation of regression problems.
2. **Random forest**: used for application of classification problem and regression problems.
3. **Support vector machines**: used for application of classification problems.

The general configuration of supervised learning with usage of statistical information is to predict future events that are statistically likely. To anticipate upcoming fluctuations, it may use cultural share price information or be used to scan out spam emails. In training set of supervised learning, tagged dog pictures could be used to categorize untagged dog pictures as input data.

Systems operate properly on training information, which includes outputs and inputs that are used to predict various types of training set. Only information is provided throughout this method, and the output is matched to the variable's target value.

An example related to the classification task of problem can be the MNIST, which is hand-written datasets of digits; these datasets are situated wherein inputs are digits in hand-written format (pixel data) and final output of the dataset is the defined label of class for which the digit represents the image (numbers 0–9).

We illustrate some example of a regression problem which can be the data information (dataset) of Boston house price; in this case, inputs are changing and illustrating neighborhood, and final output is a house price. To explain more, Figure 12.1 shows the types of Supervised Machine Learning Algorithms with Applications.

For comparison purposes, to supervised learning, information with pictures of octopus classified as any species of water and images associated with oceans labeled under the tag water can be fed to an algorithm. The supervised algorithm should be able to identify(recognize) unidentifiable octopus images as aquaculture and unidentifiable ocean pictures under water, trained on the this type of information.

12.1.4.1.2 Unsupervised Learning

In this section of unsupervised learning, we explain a class of problems with the usage of a strategy to predict or trying to retrieve the accuracy of the data. In comparison to

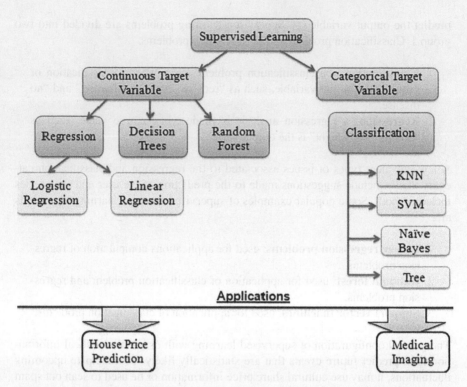

FIGURE 12.1 Types of supervised machine learning algorithms with applications.

supervised learning, unsupervised learning only works on input information without outcomes or desired output. Like supervised learning, unsupervised learning is not trained for correcting the model.

"As per the knowledge" by author in [6] there seems to be no teacher or trainer in unsupervised learning, as well as the algorithm should gain knowledge to make sense of the information without any training information.

The are several types of unsupervised learning but two of the major problems are that a practitioner usually encounters: Clustering that included the discovering of groups in enabling legislation and Density, which includes illustrating the dissemination of information.

- **Clustering:** For clustering unsupervised learning that included the problems for recognition of groups in data.
- **Density estimation**: Unsupervised learning issue involving summing up data distribution.

Example 12.2

The k-Means algorithm can be an instance of an algorithm for algorithm of clustering, where k corresponds with respect to quantity of clusters to have been found in data. The Kernel Density Estimation is an instance of an algorithm for density prediction that contains the use of small clusters of strongly linked data information's sets to approximate the allocation of recurring problem in the space of points.

Data are unlabeled in unsupervised learning; hence, the algorithm related to learning is used to identify the stylistic similarities in all of the input information. The methods of machine learning which involved unsupervised learning are specifically valuable; the unlabeled data are more abundant as compared to the labeled data.

The objective of unsupervised learning is stated that discover hidden patterns in data within the same set of data but may also include an objective of learning attribute that enables the numerical system to interactively identify the representations necessary for processes of raw data categorization.

In transactional information or data unsupervised learning is generally used. Users may have a wide range of consumer details and associated transactions, but one cannot create interpretation of similar characteristics could be produced from customer data in account and transaction types related to as human beings. The responsibilities pertaining in similar form of fed in unsupervised learning algorithm can be determined that no perfumed soaps can be pregnant for females in a particular age group scope, and thus an advertising campaign associated with pregnancy and baby items could be aimed to this community in the context of expanding the efficiency of associated sales.

To arrange it in a potentially valuable sequence, unsupervised learning approaches may evaluate complex information that is more expensive and apparently unrelated. Unsupervised algorithm is useful to the identification of anomalies containing irregular purchasing and recommendation mechanisms for credit cards that indicate which items to buy next. In the data input from a device, unsupervised learning, untagged dog-related images are used to find similarities to distinguish dog images. Figure 12.2 shows the types of Unsupervised Learning with Applications.

12.1.4.1.3 Reinforcement Learning

Reinforcement learning describes a type of problem in which agent functions in an atmosphere and can experience the process how to operate or function using input obtained from the system, software, or robot. To improve a numerical signal, reinforcement learning determines how to relate conditions to behaviors. Instead asking learner which actions to do, by attempting them, he must figure out which actions offer the most reward [7].

To use a framework implies that there is no predetermined database of training, but instead a target or collection of purpose the agent wants to accomplish, actions that they should execute, and execution feedback against the target. Many other algorithms for machine learning do not only undergo a fixed data set. For instance, reinforcement learning algorithms communicate with a scenario, to have a favorable loop for the learning environment and its perceptions [6].

It's comparable to supervised learning with the exception that the system has few reactions from which you want to learn, even though response may even be staggered and objectively noisy, making it more difficult for an agent or system model to relate a cause-and-effect. An instance of reinforcement issue is to play a game in which the participant strives to score a top mark and therefore can make a move throughout the game and receive input on punishments or bonuses.

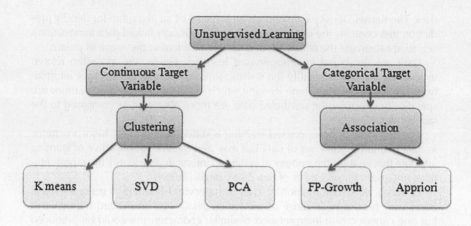

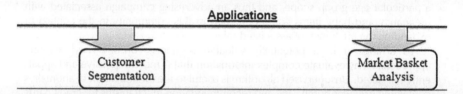

FIGURE 12.2 Types of unsupervised learning with applications.

Reinforcement learning is one of the feasible methods to carrying out a high-level curriculum in many diverse fields. For example, it is difficult for an individual to provide accurate and timely assessments of large numbers of overall game locations that would be specifically needed by examples for training and optimization technique. The only feasible way to employ a high-level implementation framework is reinforcement learning in several diverse domains. In games play, for example, it would be exceedingly difficult for individuals to provide reliable and timely evaluations of large numbers of locations that will be essential to explicitly train and optimization technique from observations. The system can be notified whether it has expected to win, and this data could be used to explain an iterative method that provides reasonably precise measurements of a probability of success from any scenario.

12.1.4.2 Hybrid Learning
There are fuzzy distinctions between unsupervised and supervised learning, and from each field of study, there are many hybrid approaches to draw. The different categories of Hybrid Learning discussed in this section are particularly useful.

12.1.4.2.1 Semi-Supervised Learning
Semi-supervised learning is collection of examples which are labeled and unlabeled in the data information set of training. Semi Supervised learning is as there are very less labeled examples and a numerous number in unlabeled examples in the training

data. The primary objective of a semi-supervised learning model is to use not only labeled data in supervised learning but also use all available data effectively.

We are given a couple of labeled examples in semi-supervised learning and must make what we can out of a large collection of unlabeled examples. Even the labels themselves are perhaps not the portentous truths humans hope for [9]. Implementing the unlabeled data effectively will need the usage methods of unsupervised leanings (i) Clustering (ii) Density Prediction (Density Estimation). In supervised approach or supervised learning, when groups or patterns are analyzed for instances, then it can be used to label all those instances which are unlabeled, or the process will be able to apply labels to all unlabeled instances used for prediction.

For representation of instances in the application, unsupervised learning can provide useful information as to how to group instances or examples. All adopted examples should be mapped to similar representations which cluster tightly in the input space [6]. Its general state that many problems of supervised learning problems are instance of all semi supervised learning problems. For examples, if distinguishing the photo, we may require data information of the photos or photo gallery that is already labeled or state by human worker or operators.

12.1.4.2.2 Self-Supervised Learning

The self-supervised learning process is stated to be a problem of unsupervised which is defined as a supervised problem for learning how to execute supervised learning algorithms for solving it. In this approach of learning, we can use an alternative or pretext task for solving problems which will give outcome in form of representation or model that is utilized to resolve the actual problem of modeling. The method of self-supervised learning provides a learning method, which allows unidentifiable data to produce a pretext learning method like interpretation prediction (estimation) or rotation of image; in this method, the target can be achieved without constant observation or analyzing [10].

Computer visualization in which a vernacular of unlabeled images is available is a notable example of self-supervised learning that can be used to train a supervised model, such as making visual information and making a model forecast a color representation (colorization) or eliminating image blocks, and providing a model determine its missing bits (painting). A classifier is built on a supplementary or 'pretext' task wherein floor-truth is easily accessible, in the self-supervised differential learning that will be the primary priority of this analysis. In certain situations, the pretext appropriate explanation step includes predicting certain secret part of the information from view (e.g., predicting color for grayscale scale) [11].

A classifier is based on an auxiliary or 'pretext' task in which floor-truth, the self-supervised discriminatory learning that will be the main subject of this analysis, is freely accessible. In certain situations, the pretext permissible explanation stage includes estimating such concealed sections of the data from view (e.g., predicting colored for gray-scale level) [6]. Most of these auto encoder classification models are centered on supplying both the source and the desired output of the model, requiring the model to reproduce that response by first converting it to an adhesive capsulated, then translating it back to its original. It is scrapped once the auto encoder is trained, then the encoder has been used to build compressed input descriptions, as necessary.

12.1.4.2.3   Multi-Instance Learning

Multi-instance learning is the part of supervised learning, in which individual instances are unlabeled. A supervised learning problem, where individual examples are unlabeled, is multi-instance learning; instead, bags or sample units are described. A whole set of examples is classified in multi-instance learning as including or not possessing an example of a class, but it does not mark the individual members of the array [6]. Occurrences are in "bags" instead of sets since one or perhaps more times an example given might be available, e.g., with multiple copies. Prediction involves using knowledge that a target mark is correlated with one or more of the situations in a bag and projecting the mark for new bags mostly in future, given several unlabeled instances being formulated [12].

12.1.4.3   Statistical Inference

Statistical inference is assumption applied to achieving an outcome or choice. Fitting a model and making a judgment are both forms of estimation in machine learning. There are numerous paradigms of inference that can be used as a foundation for analyzing how certain machine learning algorithms operate or how to solve certain learning problems. Instances of statistical inference can be states where inductive, deductive, and transduction learning and inference are ways of learning.

12.1.4.3.1   Inductive Learning

Inductive learning requires the use of proof to determine the outcome. Inductive methodology extends to the use of individual instances, e.g., specific to specific, to assess general performance. Many machine learning techniques apply a method of inductive inference or inductive reasoning to discover general rules (the model) from specific historical examples (data).

The problem of induction would be how to draw definite conclusions about the prospects from specific observations of the past [13]. It is an induction method to bring a machine learning model together. The model is a general statement of examples unique to the training dataset. A design or hypothesis is created about the issue using the data sets, and it is assumed to carry onto new unknown data later whenever the model is used.

12.1.4.3.2   Deductive Inference

Deduction inference involves the application of general principles to assess individual outcomes. We can understand inference better by comparing it with deduction. Deduction is the opposite of induction. If the induction continues from the specific observations to broader generalizations, the deduction goes from its general to the specific [14]. Thomas Mitchell's observation notes that induction is only the opposite of deduction. Deduction is often a top to down kind form of rationale goal of satisfying all hypotheses until concluding, where induction is a kind of bottom to up form of rationale that allocates data from resources as proof of a result. In the sense of machine learning, whenever we utilize induction to fit the model into a training data, the framework can also be used to make accurate predictions. As a sort of inference, the model is used.

12.1.4.3.3 Transductive Learning

In the field of statistical learning theory, given domain instances, transduction or transductive learning is used to contribute to interpreting concrete instances. It is distinct through induction, which includes the basic principles of learning, for example, unique to instances. Induction from which the feature is extracted from the information offered. Deduction deriving ideals for the given system function for points of interest. Transduction, extracting unidentified objective functions from the data provided for equity locations [15].

12.1.4.4 Learning Techniques

The numerous types of learning techniques can be stated. The below sections give a brief idea of some generally used methods. These techniques contain (i) Multi Tasking Learning, (ii) Active Learning, (iii) Online Learning, (iv) Transfer Learning, (v) Ensemble Learning.

12.1.4.4.1 Multi-Task Learning

The Multi-Tasking Learning is the approach of supervised learning which contains designing to frame a model in the one data information (dataset), which states the multiple related issues.

Multi-task training is a method of supervised learning that involves designed to fit a model into one dataset which addresses multiple related issues. The process includes development of such model which could be trained on various (multiple) allied task in such pattern that the given model enforcement is improved by training of a model that can be trained on multiple related tasks in such a way that model performance is enhanced by training evaluated to training on approximately particular mission, through the responsibilities. Multi-task training is a method of enhancing applicability by integrating the instances that come from multiple tasks (which could be weak limitations imposed on the variables) [6].

12.1.4.4.2 Active Learning

In Active Learning approach execution in learning method, the model will be enabling to enquire human operator (use operator) for resolving the ambiguity. In this learning approach learner will collect the entire training instance. Active learning is such methodology where, during the learning process, the model can query a human user operator in order to resolve ambiguity. Active learning: The learner collects training examples usually, by verifying an aegis to request labels for new points, adaptive or interactive [8].

12.1.4.4.3 Online Learning

Online learning includes using the existing evidence and immediately modifying the model until a prediction is required or since the last analysis has been produced.

Online approach of learning is ideal for issues where predictions are given across period and where in any given period the distribution of likelihood of findings is set to influence. Correspondingly, the model is expected to change constantly.

12.1.4.4.4 Transfer Learning

Transfer learning is such type of learning approach in which a set of models is provided with the specific training on the particular define single task; after this process is completed, some or the completed related model is consider to crosscheck (compare) the input state in the initial stage of the related task. The learner will have to make work of two or more than two different tasks in the transfer learning type of approach, but we can state the assumption that different factors can be used for explaining the difference in P1 task that are relevant to the difference that n will be needed to collect for learning P2 [6].

12.1.4.4.5 Ensemble Learning

Ensemble learning is a method in which several or more methods match a certain data and incorporate the assumptions stated from individual (each) model. Goodfellow, Ian et al. describe "The ensemble learning area offers many ways to combine the ensemble members' forecasts, involving uniform weight and these weights chosen on a validated range" [6].

Ensemble learning is aimed to achieve good performance with the desired models or system ensemble as compared to any desired single (individual model). The process of enhancing the performance includes two approaches: (i) Decide how to design the model which can be used in ensemble, (ii) Find the best way to combine the prediction of the ensemble members. The broad spectrum of the ensemble learning can be divided into two tasks: (i) to develop training set of data of population of base learners, (ii) To combine the develop data for forming the composite predictor. Other example of the learning algorithms includes weight average, stack generalization or stacking and aggregation of bootstraps or bagging.

12.2 MACHINE LEARNING ALGORITHM FOR BRAIN COMPUTING

This section briefly describes the working of brain with signal processing from brain to operate the BCI-based application in Figure 12.3.

12.2.1 WORKING OF BRAIN AND ITS DATA COMPUTING PROCESS

According to Alexandre Gonfalonier in A Beginner's Guide to Brain Computer Interface and Convolution Neural Networks [16], your brain has two main sections:

1. The limbic system
2. The neo cortex.

All our basic activities like eating and reproducing related to survival is done under responsibility of limbic system as shown in Section 12.2.1. The limbic system inside our brain is very advance and is responsible for the logical function that brings improvement in languages, technologies, different companies, and ideologies. The brain of human beings contains about 86 billion cells of the nerves called neurons, each one individually connected by connectors called axons and dendrites to other

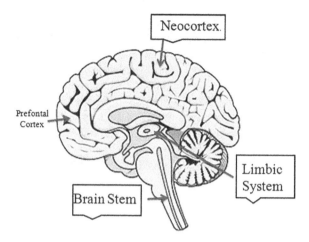

FIGURE 12.3   Major section of human brain.

neurons. Neurons are at work each time we think, move, or feel. Indeed, the brain produces enormous amount of neural activity. Basically, the work is done by small electrical signals which travel from neuron to neuron.

There are many signals generate by our complex brain system which can be captured in brain computer interface. These signals are divided into two main categories:

1. Spikes
2. Field Potential

By using machine learning algorithms, we can identify these signals, analyze and implement and can be used to different applications where interactive devices are used as shown in Figure 12.4 below.

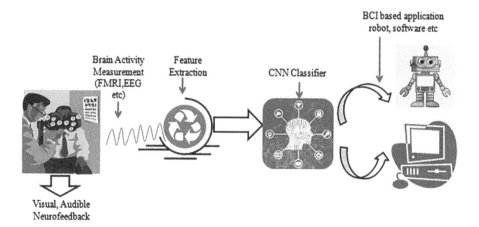

FIGURE 12.4   Signal processing from brain to operate the BCI-based application.

Boris Reuderink stated "in machine learning consultant based on cortex said that according to him the brain-computer interface issues have shown that signals obtained from the brain are unreliable and highly variable. Hence, we can face difficulties while training the classifier for using it every day" [17]. Measuring the signals from brain can be risky since to measure the signals small needle which contain rolled up in the skull, mesh is situated. Injected and unveiled, the mesh surrounds the brain.

The observed difficulties in development of BCI application can be overcome using artificial intelligence and machine learning; therefore, AI/MI in combination has become of the most trending tool for the development of BCI system, therefore for the implementation there are some considered aspects of requirement, they can be stated as follows.

12.2.1.1 Aspects of Requirement in Development of BCI

The following are the four aspects of requirements in development of BCI-

1. **Signal production**: To analyze the brain signal, production of signal is the first step for this requirement of HUMAN (Person). The signal production can be done in two ways: (i) By generation of signals by providing simulation in brain through sound or images or videos etc. (ii) By normal wave detection of already generated waves in the subject's brain
2. **Signal detection**: Second requirement is for detection of signal can be done through machine (e. g. EEG, MRI etc) the machine will help in the signal detection for signals received from brain.
3. **Signal processing**: Third requirement to process the signal define classifier, algorithms can be needed so that capture signals can be processed and send for task completion to the interfacing device or software.
4. **Output**: Last requirement aspect is need of interface that can be robot, device or software in which signals are provided as input to complete the task.

12.2.2 MACHINE LEARNING ALGORITHM FOR BRAIN COMPUTING INTERFACE

This section discusses different machine learning algorithms with their application to Brain-Computer Interfacing. A specific emphasis is placed on the types which can be applied in the BCI context.

12.2.2.1 Introduction to BCI

Machine learning area of recent trends has many useful algorithms for applications in the medical field, and most important and researched area is neuroengineering field; this area of application provides support to people with disabilities. The devices used in the neuroengineering field are interactive, which enable the disabled people to perform the specific task and activity.

One of the promising interaction tolls for both disable and healthy people with different application of VR (virtual reality), games, multimedia is BCI (Brain Computer Interface). Following Figure 12.5 shows the implementation of machine learning in any proposed BCI.

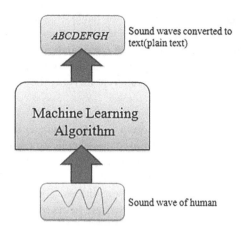

FIGURE 12.5   Machine learning and BCI.

The use of machine learning plays an incredibly significant role in BCI or BMI (Brain machine interface). Deep Learning algorithm of machine learning is one the efficient algorithm to perform the interfacing of brain and machine. The BCI is computer supported system which will analyze the brain signals and then translate them into the command which is dependent upon the significance of the output device to perform the given specified task. These signals are then used to study the actions according to the task stated about the movement of any body parts. The BCI do not use any brains normal signals like nerves or muscles signals, it used signals that are generated by the CNS (Central Nervous System) these signals can be voice also. In further studies we can conclude that EEG (electroencephalogram) is not BCI because it only records the signals and gives output in wave forms and do not perform any task or action related to the output. The BCI definition of human brain reading is misunderstood since any BCI and BMI are only used to research and interpret the signals provided to the brain via any parameter to predict human actions or intentions.

12.2.2.2   BCI Functions

BCI development and implementation are completely dependent upon the background of the functionality desired; hence, following are BCI operations in detail. The task of BCI starts from recording the brain signals; these signals are transmitted to computer system in final output to complete the task. The intention of the recorded signals or waves is to control the device or software; the process of these interlacing activities contains different operations given as follows.

12.2.2.2.1   Control and Communication of BCI

BCI defined as Brain Computing Interface is the intermediate system to link the connection of human brain and external environment. BCI can record the brain signals and decoding the signals for performing task. BCI designed for the disabled people to help them to give commands or write suggestions or opinions. There are several BCI applications like spelling [17], semantic categorization [18], or speech communication and recognition.

BCIs can also help to provide facility for hands-free applications help persons with ease and by using the mind-control of devices. These devices only allow brain signals to be integrated to execute a series of commands, and no muscle involvement is needed. BCI assistive robots will provide regular and professional assistance to disabled people, increasing their cooperation in merging the society [19].

12.2.2.2.2 User State Monitoring

In traditional times BCI application was target for disabled users who are having problem in speaking or mobility. The aim of these application was to provide alternative communication mode to disable user. But in recent application BCI is also used for healthy people and aim is to target for measuring physiological state of user or to get information about the user's emotional conditions. The goal of using brain notifications has been enlarged by controlling certain items or presenting a replacement for purposes in what is referred to as Passive or inactive BCI [20]. In many application BCI is also used for emotional and cognitive state [21], In developing intelligent improved reliability and software for experiences problems [22].

12.2.2.3 Phases of BCI

BCI phases are categorised in two phases:
   (1) Offline-Training Phase (2) Online-Operational Phase

1. In Offline-Training Phase system need calibration this calibration is the process which is carried out in offline training phase because the of Signal-to-noise ratios (SNR) is very variation parameter depending upon the subject to subject
2. In online operational phase is the activity where all the signals are filtered analyzed and preprocessing operations can be performed then operations are translated in computer commands i.e. in the form of application interface

The related two phases have different drawbacks of low accuracy of categorizations and limitation of implementations with real world. The study of BCI in this chapter assumes the working of phase operations. The above two phases can be technically categorised into four phase states in Section 12.2.2.3, 'Phase of BCI operations'.

1. **Signal capturing:** Brain signals are captured through scalp of the any patient or used using the electrodes, these electrodes are mounted in CAP EEG
2. **Preprocessing of signals:** The complex pattern of algorithms is applied to check the signal commands
3. **Analyses of the signals:** Using the study of algorithms the signals commands are filtered for processing
4. **Extraction of features from signals:** After analyzing the signals the features of commands are extracted and send for the categorization's unit of signal processing
5. **Categorization:** The type of commands is passed to the application used interface to perform the action, e.g., it might be action of clicking button; the

task of feature of clicking is extracted after categorization is passed to AUI (application user interface)

6. **Application user interface:** The task is performed accordingly stated in Figure12.6 if the task performed is correct, then feedback is recorded
7. **Feedback or output:** The feedback is the analyzed output from the signals received by electrodes from the scalp

12.2.2.4 Types of BCI

For ZazaZuilhof, it depends on the person and if you are willing to undertake operation or not. "For the intend of this theoretical model, let's suppose that only non-invasive BCIs that do not need surgery will be used by healthy individuals. There are mainly two methods developments in that situation, fMRI & EEG. First needs a huge computer, however the second has become open to a broader audience with user headsets like Emotiv and Neurosky" [20], according to Konstantinos J. Panoulas et. al author of [23]. BCI systems can be divided in three types:

1. Non-invasive during the usage of sensors which are place on the scalp, electroencephalography (EEG) is the measurement of electrical activity created by the brain or electromagnetic waves using a method called magneto encephalography (MEG).
2. Semi invasive when electrodes are imposed in an electro-corticography (ECoG) exercise to the outer portion of brain is called electrocorticography (ECoG).
3. Invasive while micro-electrode collections are located from inside cortex.

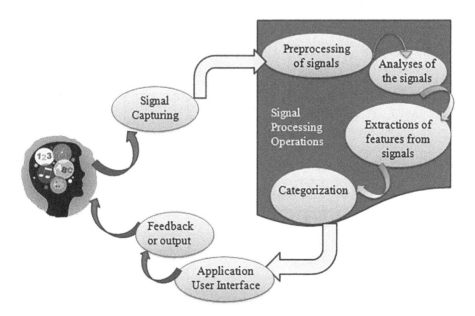

FIGURE 12.6 BCI operation in phase.

12.2.2.4.1   Non-Invasive

Non-invasive system is used in many applications of research primarily it used EEG recording as they are easily acquired at exceptionally low price by multitude of commercial off the shelf devices. The defined techniques based on EEG recording are addressed by the challenges inherent in EEG. The disadvantage of poor spatial resolution and it requires interaction from undesirable signals and its lower spectral noise ratio. MEG transcripts are often used to power BCI systems take advantage of MEG's high spatiotemporal resolution but require sensitive sensors and magnetically insulated rooms and have strictly restricted the significant potential of the academic research.

12.2.2.4.1.1   Magnetoencephalography (MEG)   Magnetoencephalography is defined under invasive type; it measures magnetic fields generated by electrical currents occurring in the brain. These currents are normally occurred inside the human brain. The magnetic signal outside of the brain is captured using the super-conducting quantum interference device (SQUID) stated by R. J. Ilmoniemi, R. J. Näätänen, in *International Encyclopedia of the Social & Behavioral Sciences*, 2001.

MEG signals compete with several other magnetic signals made up of the earth's magnetic field topic, so that laboratory requirements with covers and specialist equipment are included in this recording application.

12.2.2.4.1.2   Functional Magnetic Resonance Imaging System (fMRI)   The fMRI application is used to study the flow of blood which are responsible for the neural

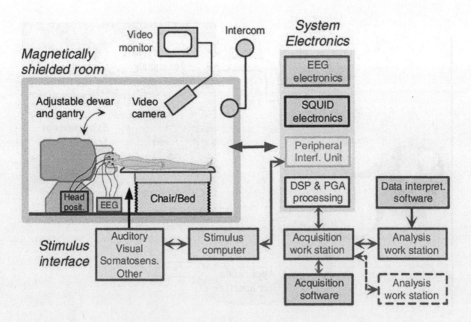

FIGURE 12.7   Block diagram of MEG. (Source: from Sternickel and Braginski [34].)

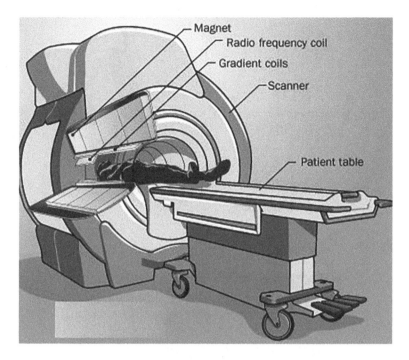

FIGURE 12.8   FMRI. (Source: https://science.HowStuffWorks.com/mri1.htm [35].)

activity carried out inside brain using the device shown in Figure 12.8. This study of blood flow helps in mapping the activities which are causally related to the part of brain which we used that part is defined as source localization problem. This denotes that part of brain that is used the flow of blood in that part of brain increases. The processes used contrast combination of blood-oxygen-level-dependent (BOLD), which is very susceptible to the output of hemodynamic [24].

12.2.2.4.1.3   Functional Near-Infrared Spectroscopy (fNIRS)   FNIRS is among the non-invasive methods used to identify neuronal activity by measuring blood fundamentals inside the brain. FNIRS works by using light within the near-infrasound range to identify blood flow. The merits in fNIRS are that its capable of it can provide high level signals of spatial resolution. Compared with the devices in temporal resolution based on electromagnetic signals, fNIRS recording is not effective. By comparing to the parameters of fMRI and fNIRS is more convenient and less expensive this is proofed in the Figure 12.9 Its advantages are that is reliable for medical study and for handy use [23].

12.2.2.4.1.4   Electroencephalogram System (EEG)   Electroencephalography system commonly called as EEGWorks by measuring electrical impulses from the scalp using voltage measurement volatility associated with neuronal activity within the brain. Electrodes throughout the cap form device are linked in features as shown in the Figure 12.10 [28].

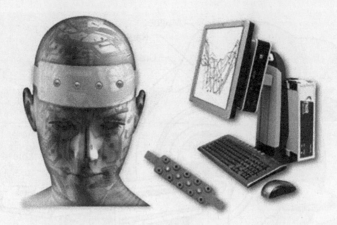

FIGURE 12.9   fNIR [24].

FIGURE 12.10   EEG. (Source: Ahmed Nazarmustafa University of Duhok/College of Medicine: slide share [36].)

Advantages of EEG:

a) It can be implemented for commercial use
b) Its portable and less expensive
c) It provides high temporal resolution.

Disadvantages: The signals conversions to noise ratio and spatial resolution determine the usage hence limiting it for development of BCI [25] as compared to other methods.

12.2.2.4.2 Invasive

ECoG semi-invasive system that offers to have greater spatial resolution and from signal to noise ratios than EEG at the expense of an invasive surgery called craniotomy. In several respects, the biophysical characteristics of EEG and ECoG measurements are identical and ECoG-based systems exploit a certain neurophysiological underlying cause as EEG-based systems, resulting in a signal processing process. According to Ref. [26] Invasive method the electrodes implanted under the scalp are used for measurement of brain waves either intracortial or cortical (electrocorticography (ECoG)) from inside the motor cortex. With their biggest benefit is that they have spatial and temporal resolution which increases the capacity of the received signal and the signal to noise ratio (switching frequency). But the techniques have demerits on many issues since the procedure involves surgical process, once completing the surgical procedure the planted electrodes cannot be shifted to another part of brain to measure the activity another major demerit observed was adaption of new object in the human body can fail causing the medical complication also maintaining the stability of the planted electrodes is also major concern. Hence, use of this invasive technique of recording brain activity is usually prohibited to BCI [25] medical application for usage of few user who are disabled [27].

12.2.2.4.2.1 Intracortical As shown in Figure 12.11 [28] intracortical acquisition technique represented as invasive method. Electrodes are installed in the brain beneath a cortex surface. The transmissions are managed to capture have one (single) electrode, or a lot or array of electrodes which are used to quantify the action of the individual neuron signals performed. In several applications, signals are not stable, but intracortial electrode points (tips) are planted near the initial position such that the arrays are secure for a long period of time to monitor the action.

Advantages: (i) Its implementation of source localization issue is generally recommended for use due to excessive spatial resolution. (ii) Intracortical procurement can long-term encounter signal changes.

12.2.2.4.2.2 Cortical Surface Method of recording electrocorticography (ECoG) is less invasive but its ability to overcome the benefits of invasive approach. It includes erecting electrode grids (electrode array) or spools inside the brain over the exterior of the cortex by medical operation shown in Figure 12.12 [29]. The electrodes are being used to record neuronal electrical impulses inside the brain at the accepting area, which has the highest number of electrodes used as unit of measure for the degree of invasiveness.

ECoG recording has partial advantages and disadvantages since its invasiveness accuracy and it is also safe for non-invasiveness application. Thus, it has high level spatial resolution and high-level signal amplitude as compared to the other non-invasive techniques e.g.,: as EEG. It is less affected by noise due to better amplitude generated from muscle engagement. Hence due to the stated ECoGa of the efficient solution for seizure localization problem. Widely application of this has been implemented for epilepsy patients before operations or surgery.

12.2.2.4.3 Semi-Invasive

Semi-invasive systems utilizing ECoG have greater spatial resolution with signal-to - noise rates than EEG at the expense of an invasive technique called craniotomy. In several perspectives, the biophysical characteristics of EEG and ECoG recordings are linked, and ECoG-based systems leverage a certain fundamental neurophysiologic

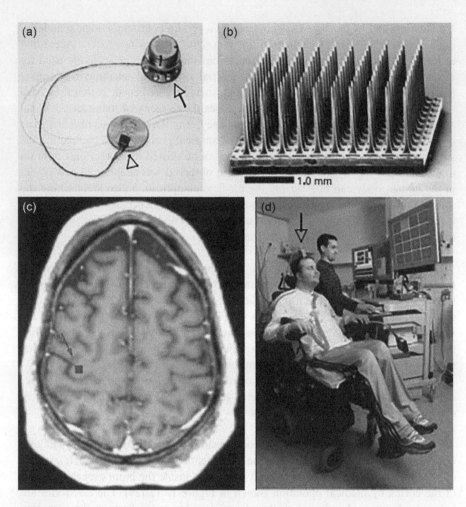

FIGURE 12.11   Intracortical acquisition technique [26].

mechanism as EEG-based systems, resulting in a reasonable signal processing technique from a viewpoint.

12.2.3 Why Machine Learning Algorithm Necessary in BCI

The applications of different machines used to record the brain signals has disadvantages of brain signals getting corrupted by different biological interference of the patient or user using the device. These biological reasons can be (e.g., blinking of eyes, muscle compression or contraction and relaxation, fatigue or drowsiness and concentration of human beings' measurement or problem of disabled people to interact with world through BCI) and environmental parameters (e.g., environmental disturbance like noise). The task of BCI is brain signal recognition, can be reliably done by deep learning models, these algorithms are the most popular algorithms.

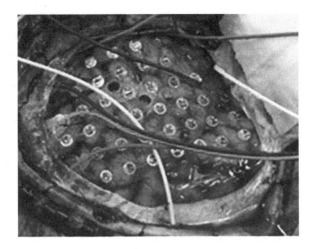

FIGURE 12.12 ECoG acquisition [26].

Deep learning provides better techniques to automatically extract the different features easily from the brain signals. Deep learning has been applied to the BCI applications and has shown success in addressing the challenges faced for capturing and processing of brain signals.

Deep learning major three advantages.

1. It takes less time to prepare and features engineering steps as it works directly on raw brain signals to learn conflicting data via back-propagation.
2. DNN (Deep Neural Network) is capable of capturing features of both high-level and latent connections through linguistic forms.
3. Deep learning methods have strong classifier groups such as Vector Machine assistance (SVM) and Linear Discriminant Analysis (LDA).

Shalu Chaudhary et al. [37] state that in many BCIs having problem of biological interference, the proposed methodology of "Motor Imagery" can help BCI to be implemented for the disabled people so that these people will also be able to communicate with real world like normal people. This methodology will act as an interface or supplementary way to interact between the user and the computer machine. It contains use of EEG (Electroencephalogram) for implementation of MI works or process in BCI. The operation of method is based on EEG signals which used flexible analytic wavelet transform i.e. FWAT, the author explains that the signals are divided into different sub band, these features are then process under k nearest classifier to complete the objective on MI (motor imagery). Hence this can be stated that implementation of machine learning techniques is very necessary to develop BCI for completion of all desire application operations.

Implementation of machine learning in BCI can improve the performance and accuracy in several application as the author Sachin Taran et al. [38] has given the methodology for the implementation of machine learning in BCI, author states that we can implement a feature based on analytic intrinsic mode functions i.e. AIMFs for categorization of EEG electroencephalogram signals of several Motor Imagery operations. The inputs to AIMFs are acquired by utilizing empirical mode decomposition (EMD) and Hilbert transform on all EEG signals. Implementation of different machine learning classifier we can normalize the feature so that the impact of biased nature of the classifier is reduced. The normalize features as input is applied to (LS-SVM) i.e. least square support vector classifier all the performance parameter is evaluated by using LS-SVM. IMF1's radial base kernel feature gives good MI task prediction performance of 97.56%, 96.45% sensitivity, 98.96% precision, 99.2% positive expected value, 95.2% negative predictive value, and 4.28% minimum error rate identification. The methodology stated that implementation of machine learning in BCI increases the performance of BCI in applications as compared to the other state-of-the-art methods.

12.3 MACHINE LEARNING ALGORITHM FOR HUMAN BEHAVIOR COMPUTING

You will also assume that humans are better at identifying fellow humans than machines are, or at least hope so. Yet a recent MIT research shows that an algorithm can predict somebody's actions more easily and accurately than humans can.

12.3.1 INTRODUCTION

Human behavior prediction is the recent trending application area for machine learning. Traditional consideration under human action focused primarily on worldwide features of digitizes image. The increment in human behavior computing demands in the applications different machine learning algorithms have been introduced in past many years. Deep Neural Network (DNN) are under very great chances for detecting and analyzing the human behavior. In many fields predicting human participants' behaviorin strategic settings is an important issue.

Deep learning could be used to educate overseen designs or unsupervised models [22]. These models can learn a hierarchical structure of features and extract high level features from lower level [27,29]. CNNs [30] are indeed a type of deep learning models has used skilled filters and fundamental elements on the input image in the local community; thus, an organizational structure with more complex features. Furthermore, CNNs could even perform extraordinary excellence on object detection after appropriate convolution during the training. Apart from traditional classifiers, CNNs were never required to categorized and commonly provide highest accuracy rates; in addition, it displays a wave equation to differences such as lighting, pose.

12.3.2 WHAT IS USER BEHAVIOR?

User behavior is nothing, but the way user reacts or interact to any product (software, device or robot). There are different machine learning algorithms are developed to analyze user behavior for implementation of these algorithms for analyzing user behavior there is need to set up various user parameter then these parameters are measure according to its usability and intuitive design. All user or person is different and similar is that with their behavior as customers or user. To know the user behavior following the pattern Machine learning is the most popular and efficient techniques for product implementation.

In machine learning clustering, the most powerful unsupervised learning problem, which can create the groups of object or individual having similar pattern of parameters (height, weight, color, shape, habits, social surfing, etc). These categories or groups are called clusters. A cluster is defined as a collection of points in a dataset. The points in one cluster are similar than the points in other clusters.

Example 12.3

1. Distance-based clustering groups can be defined as the group of clusters that has small distance between the object or points within the cluster, and distance between the clusters in group must be larger.
2. You managed to get student ID, age, gender, branch, mobile no, select sports. This last one is a selection of any sports completely depends on the student behavior, choice, interest, or pervious history data. You as a teacher want to motivate students to participate in new sports events and you want to target to only specific group of student so to cluster such kind of data we can use different distance calculating algorithm to clearly analyze the cluster and its dependency on client or use behavior.

12.3.3 ELEMENTS OF USER BEHAVIOR

User behavior depends on different parameters of human or user body these parameters are stated as elements of user behavior. The author in the paper [31] has described the Human Behavior Analysis (HBA) and its elements in details including the taxonomy of the human behavior which is stated as follows shown in Figure 12.13. The author Chaaraoui, A. A.; Climent-Pérez, P.; Flórez-Revuelta has reviewed the three basic elements of user behavior actions, activities, and behaviors. Further, the elements can be extended into inter-activities based on inter activity we can classify them into action [32].

Some elements stated for user behavior observed are as follows:

1. Used actions are temporarily short and the users make conscious muscular movements (e.g., picking up a cup, opening/closing fridge door, etc.).
2. The long activities which are contain many sub activities such as (making tea or dinner, taking shower, playing, etc.
3. Behaviors' describe the pattern how the user has completed the specific activities at varying times. Hence, it is analyzed from [20] that there are two types of behaviors identified. (i) The interactivity behaviors define how and why the user performs a given activity at different times (e.g., each user can have different playing techniques while playing, or while preparing dinner users can collect all the requirements and then start cooking, or some can start cooking and collect the required items simultaneously).

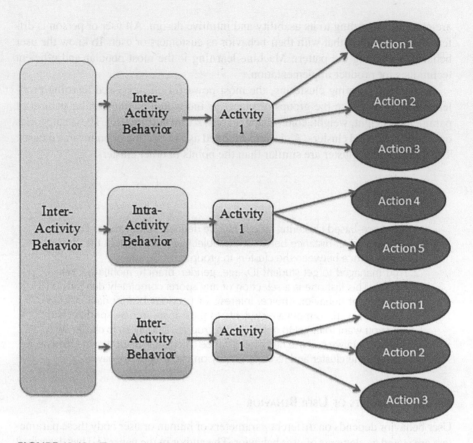

FIGURE 12.13 Taxonomy of the human behavior.

(ii) The interactive elements behaviors describe how and why the user co-related with various activities (e.g., getting up in the morning, having breakfast and going to work, and the sequence of activity might differ on no working day.

12.3.4 MACHINE LEARNING ALGORITHM FOR HUMAN BEHAVIOR COMPUTING

The elements of user or human behavior can be analyzed by using different algorithms these algorithms can be stated as [33]:

1. **Deep Neural Networks (DNNs):** Deep Neural Networks (DNNs) are typically **Feed Forward Networks (FFNNs)** in which data flows in sequential format starts input layer to the output layer. It does not support backtracking of any layer and the links connecting the layers are unidirectional i.e. one-way nodes cannot be visited again.
2. **Recurrent neural network (RNN):** A **Recurrent neural network (RNN)** addresses the issue stated in FFNN. This neural network has connections

between passes and connections using time as measuring constrain. These are categories of ANN, i.e. the artificial neural network from which the ties between nodes form a directed graph indicating the sequence called links from a layer to previous layers as characteristics, thus allowing information to flow backward and forth from the previous part of the network to the next part of the network. RNNs can thus utilize their internal state (memory) to analyze sequence data. RRN is widely used for applications like unsegmented, linking handwriting recognition or it can be connection of speech recognition. RNN is capable of processing data you feed and its related information from history of data'. **RNNs faces disadvantages of** vanishing (or exploding) which is also called as gradient/long-term dependency problem in which information rapidly gets lost over time. In simple terms its losses weight when value of node reaches to 0 or 1,000,000, not the neuron. Therefore, pervious node will not be able to get information from the previous state or stage but in this case, the previous state won't be very informative as it's the weight which stores the information from the past [33].

3. **Long Short-Term Memory (LSTM):** LSTM is one of the types of RNN but this method is having potential of learning dependencies which are based on long term through RNN gets capability to remember the history data, it's also beneficial in finding the patterns in the given time span to predict the next guesses [33].

4. **Convolutional Neural Network (CNN):** The CNN are one to the important class of Deep Neural Networks, this algorithm which is used to analyze visual imagery. Applications of CNN can include video understanding, speech recognition also natural language processing. Video interpretation, speech recognition and natural language processing are also included. The combined effect of LSTM and Convolutionary Neural Networks (CNNs) has improved instantaneous image capture. You could revise using and see RNN is reliable data to help us forecast data processing [33].

12.3.5   WHICH TECHNIQUES IS BETTER FOR HUMAN BEHAVIOR COMPUTING RNN OR CNN

RNN works on principal of sequences of vectors i.e. sequences in the input, the output. **Application Programming Interface (API)** generally helps in fixing the user interface or display problem of the device or software Therefore CNN is more efficient for such application than RNN. The disadvantage of RNN getting explode and data losses hence CNN overcome this disadvantage [33]. CNN also have demerits of not able to identify the specific location of object after recognizing the types of object. CNN can process the algorithm only on object and hence multiple objects remains visual field i.e. CNN cannot work by interference of multiple objects in visual. To overcome R-CNN (R standing for regional, for object detection) can implement the CNN in such way to focus on a single region at a time enhancing dominance of a particular object in each region. But Before implementing *CNN* for applications like classification and bounding box regression-CNN regions are divided into equal size this will help to specify a preferred object.

REFERENCES

1. Tagliaferri, L. (2017). *An Introduction to Machine Learning|DigitalOcean*. New York: Digital Ocean.
2. Brownlee, J. (2019). Supervised and unsupervised machine learning algorithms. Machine Learning Mastery Pty. Ltd., pp. 1–9.
3. Shalev-Shwartz, S., & Ben-David, S. (2013). *Understanding Machine Learning: From Theory to Algorithms*. doi: 10.1017/CBO9781107298019.
4. Brownlee, J. (2016). The {machine} {learning} {mastery} {method}. *Machine Learning Mastery*.
5. Bishop, C. M. (2006). Pattern recoginiton and machine learning. *Information Science and Statistics*.
6. Goodfellow, I., Bengio, Y., & Courville, A. (2016). Deep learning research. *Deep Learning*.
7. Sutton, R. S., & Barto, A. G. (2018). *Reinforcement Learning, Second Edition: An Introduction - Complete Draft*. Cambridge, MA: The MIT Press.
8. Mohri, M., Rostamizadeh, A., & Talwalkar, A. (2012). *Foundations of Machine Learning (Adaptive Computation and Machine Learning series)*. Cambridge, MA: The MIT Press. doi: 10.1007/978-3-642-34106-9_15.
9. Russell, S., & Norvig, P. (2002). *Artificial Intelligence: A Modern Approach* (2nd Edition). Upper Saddle River, NJ: Prentice Hall. doi: 10. 1017/S0269888900007724.
10. Kolesnikov, A., Zhai, X., & Beyer, L. (2019). Revisiting self-supervised visual representation learning. *Proceedings of the IEEE Computer Society Conference on Computer Vision and Pattern Recognition*. doi: 10.1109/CVPR.2019.00202.
11. Goyal, P., Mahajan, D., Gupta, A., & Misra, I. (2019). Scaling and benchmarking self-supervised visual representation learning. *Proceedings of the IEEE International Conference on Computer Vision*. doi: 10.1109/ICCV.2019.00649.
12. Witten, I. H., Frank, E., & Geller, J. (2002). Data mining: Practical machine learning tools and techniques with Java implementations. *SIGMOD Record*. doi: 10.1145/507338.50735.
13. Robert, C. (2014). Machine learning, a probabilistic perspective. *Chance*. doi: 10.1080/09332480.2014.914768.
14. Hierons, R. (1999). Machine learning, Tom M. Mitchell, Published by McGraw-Hill, Maidenhead, U.K., International Student Edition, 1997, 414 pages. Software Testing, Verification and Reliability. doi: 10.1002/(SICI)1099-1689(199909)9:3%3C191::AID-STVR184%3E3.0.CO;2-E.
15. Vapnik, V. N. (1995). *The Nature of Statistical Learning Theory*. doi: 10.1007/978-1-4757-2440-0.
16. Gonfalonieri, A. (2018). a-beginners-guide-to-brain-computer-interface-and-convolutional-neural-networks-9f35bd4af948@towardsdatascience.com. Towards Data Science.
17. Lelievre, Y., Washizawa, Y., & Rutkowski, T. M. (2013). Single trial BCI classification accuracy improvement for the novel virtual sound movement-based spatial auditory paradigm. *2013 Asia-Pacific Signal and Information Processing Association Annual Summit and Conference, APSIPA 2013*. doi: 10.1109/APSIPA.2013.6694317.
18. Wang, W., Degenhart, A. D., Sudre, G. P., Pomerleau, D. A., & Tyler-Kabara, E. C. (2011). Decoding semantic information from human electrocorticographic (ECoG) signals. *Proceedings of the Annual International Conference of the IEEE Engineering in Medicine and Biology Society, EMBS*. doi: 10.1109/IEMBS.2011.6091553.
19. Prataksita, N., Lin, Y. T., Chou, H. C., & Kuo, C. H. (2014). Brain-robot control interface: Development and application. *2014 IEEE International Symposium on Bioelectronics and Bioinformatics, IEEE ISBB 2014*. doi: 10.1109/ISBB.2014.6820928.

20. Zuilhof, Z. (2018). When brain-computer interfaces go mainstream, will dystopian Sci-Fi be our only guidance? - Core77. https://www.core77.com/posts/72957/When-Brain-Computer-Interfaces-Go-Mainstream-Will-Dystopian-Sci-Fi-Be-Our-Only-Guidance.

21. Tan, D., & Nijholt, A. (2010). *Brain-Computer Interfaces and Human-Computer Interaction*. doi: 10.1007/978-1-84996-272-8_1.

22. Fukushima, M., Inoue, A., & Niwa, T. (2010). Emotional evaluation of TV-CM using the fractal dimension and the largest lyapunov exponent. *Conference Proceedings - IEEE International Conference on Systems, Man and Cybernetics*. doi: 10.1109/ICSMC.2010.5642336.

23. Panoulas, K. J., Hadjileontiadis, L. J., & Panas, S. M. (2010). Brain-computer interface (BCI): Types, processing perspectives and applications. *Smart Innovation, Systems and Technologies*. doi: 10.1007/978-3-642-13396-1_14.

24. Ayaz, H., Shewokis, P. A., Bunce, S., & Onaral, B. (2011). An optical brain computer interface for environmental control. Proceedings of the Annual International Conference of the IEEE Engineering in Medicine and Biology Society, EMBS. doi: 10.1109/IEMBS.2011.6091561.

25. Wolpaw, J. R., & Wolpaw, E. W. (2012). Brain-Computer Interfaces: Principles and Practice. doi: 10.1093/acprof:oso/9780195388855.001.000.

26. Abdulkader, S. N., Atia, A., and Mostafa, M.-S.M. Brain computer interfacing: Applications and challenges. *Egyptian Informatics Journal* 16, no. 2 (2015): 213–230.

27. Rupp, R., Kleih, S. C., Leeb, R., del R. Millan, J., Kübler, A., & Müller-Putz, G. R. (2014). *Brain–Computer Interfaces and Assistive Technology*. doi: 10.1007/978-94-017-8996-7_2.

28. Hochberg, L. R., Serruya, M. D., Friehs, G. M., Mukand, J. A., Saleh, M., Caplan, A. H., Branner, A., Chen, D., Penn, R. D., & Donoghue, J. P. (2006). Neuronal ensemble control of prosthetic devices by a human with tetraplegia. *Nature*. doi: 10.1038/nature04970.

29. Roland, J. L., Hacker, C. D., Breshears, J. D., Gaona, C. M., Edward Hogan, R., Burton, H., Corbetta, M., & Leuthardt, E. C. (2013). Brain mapping in a patient with congenital blindness: A case for multimodal approaches. *Frontiers in Human Neuroscience*. doi: 10.3389/fnhum.2013.00431.

30. Brouwer, A.-M., van Erp, J., Heylen, D., Jensen, O., & Poel, M. (2013). *Effortless Passive BCIs for Healthy Users*. doi: 10.1007/978-3-642-39188-0_66.

31. Chaaraoui, A. A., Climent-Pérez, P., & Flórez-Revuelta, F. (2012). A review on vision techniques applied to Human Behaviour Analysis for Ambient-Assisted Living. *Expert Systems with Applications*. doi: 10.1016/j.eswa.2012.03.005.

32. Almeida, A., & Azkune, G. (2018). Predicting human behaviour with recurrent neural networks. *Applied Sciences (Switzerland)*. doi: 10.3390/app8020305.

33. Labs, S. (2019). Understanding deep learning: DNN, RNN, LSTM, CNN and R-CNN|by SPRH LABS|medium. https://medium.com/@sprhlabs/understanding-deep-learning-dnn-rnn-lstm-cnn-and-r-cnn-6602ed94dbff.

34. Migliorelli Falcone, C. M., Mañanas Villanueva, M. A. & Alonso López, J. F. (2017). Methods for noninvasive localization of focal epileptic activity with magnetoencephalography. https://dialnet.unirioja.es/servlet/tesis?codigo=230199&orden=0&info=link%0Ahttps://dialnet. unirioja. es/servlet/exttes?codigo=230199.

35. Gould, T. A. (2005). How MRI works. http://science.howstuffworks.com/mri.htm/printable.

36. Abdulkader, S. N., Atia, A., & Mostafa, M. S. M. (2015). Brain computer interfacing: Applications and challenges. *Egyptian Informatics Journal*, 16(2), 213–230. doi: 10.1016/j.eij.2015.06.002.

37. Chaudhary, S., Taran, S., Bajaj, V., & Siuly, S. (2020). A flexible analytic wavelet transform based approach for motor-imagery tasks classification in BCI applications. *Computer Methods and Programs in Biomedicine* 187, 105325.

38. Taran, S., Bajaj, V., Sharma, D., Siuly, S., & Sengur, A. (2018). Features based on analytic IMF for classifying motor imagery EEG signals in BCI applications. *Measurement* 116, 68–76.

13 An Automated Diagnosis System for Cardiac Arrhythmia Classification

Allam Jaya Prakash and Saunak Samantray
IEEE

C. H. Laxmi Bala
Rajiv Gandhi University of Knowledge Technologies

Y. V. Narayana
Tirumala Engineering College

CONTENTS

13.1 INTRODUCTION

Electrocardiogram (ECG) is a diagnostic device with the ability to measure the heart's electrical activity employing a skin electrode. It is a graphical time-related illustration of voltage obtained from the electrodes placed on the upper body region that records the electrical muscle activity. It provides the cardiac specialist with valuable knowledge about the rhythm and activity of the heart. A thorough examination is needed in irregular morphology and heart rate cases because those irregular heart rates can lead to life-threatening conditions [1].

13.2 THE HUMAN HEART

Cardiac cells exhibit slight variations in ion concentrations in the cell membrane at rest. The negative ion density is higher within the inner layer than the heart cell's outer layer, leading to a resting voltage of about 90 mV. Cardiac resuscitation results in a higher concentration of sodium (Na) positive ions, leading to a shift in polarity throughout the cell membrane. This shifting of the cell voltage from positive to negative and backward results in an electrical potential called nerve impulse [2]. The nerve impulse causes compression in the cardiac muscle. The action of depolarizing and repolarizing cardiac cells is recorded using ECG. The Sinoatrial (SA) node is regarded as the human heart pacemaker as a group of cells around the SA node is the source for the rapid depolarizing of the cardiac cells.

The human heart consists of two atriums and two ventricles left and right. The Sino Atria (SA) node is positioned within right atrium. The depolarization of cardiac cell within the SA area makes both atria contract simultaneously [3]. The atrioventricular valves attach the atria and ventricles. The Atrioventricular (AV) node is a collection of cells in the right atrium that causes the atria's depolarization by the mass of conducting fibres in the ventricles. Purkinje fibre depolarizes every part of the ventricles found on the ventricles' muscle walls. The activity of polarizing and depolarizing the cardiac muscle causes electrical energy to flow throughout the body. The electrical current difference is highest when a part of the heart is fully polarized while the opposite is fully depolarized. The electrocardiogram signal expresses the summary of the action potential of the heart. The right ventricle and the right atrium collectively create a pump to distribute blood into the lungs. The superior and inferior vena cava veins receive oxygen-deprived blood, which then flows inside the right atrium. The narrowing of the two atria shifts blood to circulate to the ventricles. Both ventricles are activated and start contracting when the atria conduct the signal, and blood is carried from the ventricles by the pulmonary and aortic arteries.

13.2.1 ECG RECORDING

ECG measurement comprises three chest (precordial), three bi-polar, and three augmented unipolar leads. The word lead relates to the electric voltage or potential variation, which develops between two electrodes placed on the skin. The ECG signals are measurement by fixing electrodes in designated anatomical locations [4]. Bipolar lead: It measures the electric potential variation between the two locations in which one electrode functions as +ve and the second one as −ve. Unipolar lead: It measures the electrical voltage with a single electrode in the designated area. Lead I, II, and III are usually indicated as a bipolar lead since they utilize precisely two electrodes to obtain feedback [1]. Lead-I: It registers the voltage variation within the left and right arm. Lead-II: It registers the voltage variation within the right arm and left leg. Lead-III: It registers the voltage variation within the left arm and the left leg.

The negative terminal is kept on the heart electric field centre, acting as a reference point while the other terminals are placed on the limbs act as positive. A single lead connected to the right arm for lead IV acts as +ve terminal simultaneously. The other two limbs left arm and left leg for lead V, VI, are attached to the −ve terminal and act as the reference terminal. The unipolar leg, when used in conjunction, provides a three-dimensional view of the critical vector in the frontal plane. The leads are: (i) Augmented Vector Right (AVR): It represents voltage variation within the right arm's potential and heart electrical field. (ii) Augmented Vector Left (AVL): It represents voltage variation within the left arm's potential and heart electrical field. (iii) Augmented Vector Foot (AVF): It represents voltage variation within the left foot's potential and heart electrical field. The following six leads are used to record the ECG signal effectively, they are $V-1$: It is positioned on the right side of the sternal edge, $V-2$: It is positioned on the left side of the sternal edge, $V-3$: It is positioned aced in the area amid the second and fourth electrodes, $V-4$: It is positioned on the midclavicular line, $V-5$: It is positioned on the 5th rib, anterior axillary line, $V-6$: It is positioned in the midaxillary line.

The chest lead measurement is performed when the three-limb leads are combined to create a reference electrode with large resistance. The chest leads essentially used to find vectors that may be directed back [1]. As the QRS vector is normally pointed towards the left side of the back area, the QRS vectors detected by the $V-1$ and $V-3$ leads are generally −ve, while the vectors obtained by the $V-5$ and $V-6$ are +ve [6]. The QRS wave is detected as −ve in leads $V-1$ and $V-2$ as the chest terminal at these leads are close to the heart bottom as represented in Figure 13.1, which is the most electrical negativity when the ventricular depolarization occurs. In leads, $V-4$, $V-5$, $V-6$, QRS wave is +ve since the thoracic terminal is close to the cardiac apex, which is most electrically positivity during depolarization [7].

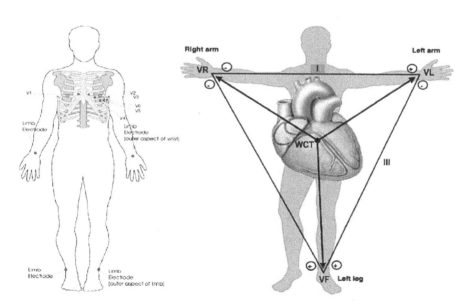

FIGURE 13.1 Different lead points to record the ECG signal [5].

13.3   DIFFERENT SEGMENTS OF ECG

The ECG is a continuous signal which contains a set of repetitive waves the collection of the set of waves is recognized as a cardiovascular cycle. These wave set are P, QRS, and T. P wave: The first rising pulse symbolizes the P wave, and it lasts for 0.04-seconds duration. It is formed by the contraction of the right and left atria, triggered by atria's depolarization. QRS complex wave: This wave is formed because of the ventricular depolarization. The descending pulse is a Q wave that is accompanied by a sharp, positive R wave. The negative S wave succeeds the R wave. The QRS complex is collectively formed by the Q, R and S wave as shown in Figure 13.2.

The QRS complex implies the time required for contraction of ventricles and has a duration of 80–120 ms [8]. T wave: The T-wave appears after the QRS complex signal. It is a rising pulse that shows the repolarization of ventricles. U wave: A slight deflection after T wave is called the U wave which suggests the repolarization of Purkinje fibers. PR segment: The segment between waves P and QRS is termed as PR segment. The electrical charge from the AV node passes through the atria into the ventricle during this time. The PR segment is shown as a horizontal signal since no heart muscle activity occurs during that time period. ST Segment: The section between the QRS complex and T wave is known as the ST segment. It reflects the ventricle depolarization action. ST section is approximately 80–120 ms.

13.3.1   DIFFERENT TYPES OF NOISES AFFECTING ECG SIGNAL

The ECG signal is often affected by noises and artifacts while recording, which may alter the signal's morphology. Therefore, extracting the required data for classification is a difficult task. The degradation of the ECG signal is due to the following significant noises:

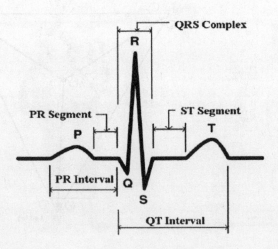

FIGURE 13.2   ECG signal.

i. **Baseline wander noise**: Baseline wandering leads to an upward and downward motion of the isometric line. It is triggered by the motion of electrodes attached through the chest during respiration, or by the activity of the leg or arm. The temperature variance and the amplifier device bias can also cause baseline drift. These types of noise have a spectrum of 0–0.5Hz [9].

ii. **Power line interference**: This comes in because of the bad insulation of ECG equipment attached to the power distribution [10]. The ECG recording device recognizes the 50–60Hz frequency ac signal making the ECG signal look dense.

iii. **Motion artifacts**: The contraction of muscles is caused by electrical activity. The resulting signals are Gaussian noise confined by a band with a mean distribution of zero. Interference due to Electromyogram (EMG) induces rapid variations that are quicker as compared to ECG signals [11]. It lasts for 50 ms with a frequency range of 0–10kHz [10].

13.3.2 Different Types of Noises Effected ECG Signal

A normal sinus rhythm (NSR) displays a cardiac disorder free ECG signal, with a 60–100 beats per minute (BPM). A heart rate differing from the normal rate puts stress on the internal organs. If the heart rate is above 100 BPM, it is known as tachycardia, and if it is below 60 BPM, it is known as bradycardia. Different types of arrhythmias are explained below.

a. **Sinus node arrhythmias:** This kind of arrhythmias originates in the Sinoatrial node. This kind of arrhythmias' distinctive characteristic is the presence of normal P wave morphology in the ECG signal. The different types of arrhythmias that come under the class of sinus node arrhythmias are Sinus arrhythmia, Sinus arrest, and Sinus bradycardia.

b. **Atrial arrhythmias:** The source of this kind of arrhythmias is from within the atria but from the outside of the SA node. The various forms are:
   • **Atrial Tachycardia:** It is identified with a 160–240 BPM heart rate. The physical signs displayed by the body are palpitations, agitation, and nervousness.
   • **Premature Atrial Contractions (PAC):** In this, the P wave has an irregular morphology, but normal morphology is present in the QRS complex as well as in T wave. This problem occurs as a result of the ectopic pacemaker's early firing before the SA node.
   • **Atrial flutter:** It causes the heart to beat faster, exceeding the standard heart rate with a pretty high margin of 240–360 beats per minute. The P wave appears at a rapid pace and forms a sawed tooth.
   • **Atrial fibrillation:** Due to the random excitation of various parts of the atria, the heart rate surpasses 350 BPM. Due to which the ventricles do not get filled with blood.

c. **Junction arrhythmia:** Such forms of arrhythmias arise due to the excitation of the atrioventricular (AV) node, or the excitation of its bundles. Due to the depolarization within the AV node into the atria [1], the P wave has an unusual morphology of reverse polarity as compared with typical sinus rhythm P waves.

  b. **Ventricular arrhythmia:** This kind of arrhythmias is identified with a large, bizarre shaped QRS wave. The impulse signal originally comes from ventricles and then travels to the rest of the heart.

  e. **Bundle branch blocks:** Bundle branch blocks can cause cardiac infraction as the flow of the impulse signal from the AV node to the whole transmission structure is restricted. There are two kinds of Bundle Branch Blocks (BBB) groups. The right BBB stops the depolarization of the right ventricle from the impulse signal originating from the AV node where the left BBB stops depolarization of left ventricular muscles. Such blocks induce an infarction of the myocardial [1].

  f. **Premature ventricular contraction (PVC):** The heartbeat is caused by the Purkinje fibers inside the ventricles, which results in earlier ventricular constriction, and the atria cannot become sufficiently depolarized. Oxygen deficiency in heart muscles can induce PVC beats, and it can happen at a random place in the ECG beat cycle.

In general, the diagnosis of ECG arrhythmia follows the following three: (i) pre-processing, (ii) feature extraction, and (iii) classification. In Pre-processing phase, noise is removed from the ECG signals. Using various digital signal processing methods such as Fourier transform (FT) [12], wavelet transform [13], Hilbert-Huang transform (HHT) [14], discrete wavelet transform (DWT) [15] and S-transform (ST) [13], different temporal and morphological features are derived after pre-processing. One of the popular techniques for extracting the features from the ECG signal is wavelet transform. In Ref. [12], detailed and estimated wavelet coefficients are calculated to form a feature vector from the ECG signals. Defined ECG segment features are derived using principal component analysis and dynamic time warping in Ref. [16]. Traditionally, efficient extraction techniques improve the classifier's classification accuracy.

After feature extraction of ECG signals, selected features are regarded as input to the automated classifiers. In literature, many algorithms are proposed for automatic classification based on signal processing techniques such as artificial neural networks (ANNs) [17], support vector machine (SVM) [18], multi-layer perceptron neural network (MLP-NN) [12], CNN [19], Alex net [20], modular neural network model [21], random forest [22], long short-term memory (LSTM) network [23] and hidden Markov models [24]. In Ref. [25], classified five types of ECG beats based on beat interims and morphological characteristics of the ECG signals applying linear discriminant statistical classifier. In Ref. [26], detected five types of different ECG beats using SVM and increased the generalization capability of SVM using particle swarm optimization (PSO) with an accuracy of 90.52%.

SVM and MLP were used in the study carried out by [18] as these two classifiers gave the most effective result. Computation time is critical for operations like extracting and categorizing of the features. The classifier performance measured depends on time and other performance factors. This research objective was to employ certain wave-based transformation systems such as DWT, wavelet transformations on ECG to improve the classification ability.

In Ref. [27], the objective was to improve the diagnosis of arrhythmia via implementing a novel trait termed heartbeat amplitude variation. The classification used two methods: (i) beat recognition and extraction characteristic and (ii) random forest for classification of heartbeat characteristics. Thorough investigations exploring the results of adding a new beat classification feature employing the MIT-BIH dataset indicate that taking into account the amplitude variation characteristic will boost classification efficiency by decreasing the frequencies of false-positive and false-negative.

In Ref. [28], Deep Neural Network is used for the classification. It is visible that the recording from the patient suffering from arrhythmia or sinus rhythm is classified with high accuracy using the DNN. The classifier's specificity, precision, specificity, and error rate clearly demonstrate the DNN superior performance. The DNN classifier acquires an accuracy of 94%.

In Ref. [29], the arrhythmia is classified utilizing a form of a higher-order neural unit with error Backpropagation by Levenberg-Marquardt method in batch optimization. The paper's purpose is to offer a procedure utilizing the classifier to assist the doctors in classifying arrhythmia.

In Ref. [30], DNN techniques such as MLP and CNN are used to differentiate between various arrhythmias. The accessible ECG databases available at PhysioBank.com were utilized. The DNN techinques where trained, tested, and validated on this database. In this paper, the implemented algorithm constitutes of four hidden layers consisting of weights and biases in MLP. It furthermore comprises a four-layered CNN applied to classify ECG representations into various types of arrhythmia.

In Ref. [31], a unique method is adopted in which coarse-to-fine arrhythmia classification is used for effective processing ECG databases of significant quantity. By decreasing the dimension of the heartbeat and quantizing the number of beats utilizing Multi-Section Vector Quantization, it decreases the time required for classification without losing the classification accuracy.

In Ref. [32] an area of the research field is dedicated for considering various neural networks to assess their performance in the category or class recognition and separation. Feed Forward backpropagation was picked for classification between all other neural systems. To train the system, a perfectly stable input is utilized, taking the equivalent aggregate of exemplars from each class. The trained system is then employed to identify an entirely different dataset.

In paper [33] proposes a method for predicting eight different arrhythmias employing the radial basis function neural network. Predictions of NSR, left bundle LBBB, Sinus bradycardia, Atrial fibrillation, RBBB, Atrial flutter, Second-degree block, and PVC are carried out using the proposed algorithm.

In recent trends, due to the impressive success in pattern recognition applications, deep learning techniques have been very common. There is no separate feature extraction stage in deep learning. The input layers are used in deep learning approaches to derive the features from raw data. Therefore, many researchers turned their focus to classification methods focused on deep learning. Kiranyaz proposed patient-specific ECG beat classification based on 1D CNN that is used to classify long-term Holter ECG records [19]. In Ref. [34], 1D CNN with active learning was implemented to classify SVEB, and VEB beats and the active learning phase is used to fine-tune the model to detect arrhythmias more efficiently. U.R. Acharya et al. implemented 16 layers deep CNN to classify 13, 15, and 17 types of arrhythmias, including normal sinus rhythm (NSR) with 95.20%, 92.51%, and 91.33% of accuracy, respectively [35]. Recently, Tae et al. proposed an efficient ECG arrhythmia classification which is based on two dimensional CNN that shows good accuracy in pattern recognition and compare the obtained results with AlexNet and VGGNet also [36]. Segmented ECG beats are converted into images during a pre-processing stage, and transformed imageries are used as input to the classifier for eight categories of ECG beats, i.e., normal beat (NOR), PVC beat, paced beat (PAB), RBBB, atrial premature contraction beat (APC), ventricular flutter wave beat (VFB), ventricular escape beat (VEB), and LBBB in Ref. [36]. In Ref. [37], reported an automatic ECG arrhythmia classification using transfer learning method with the combination of 2D deep CNN to classify four types of ECG arrhythmia beats. Reported literature mentioned above shows that deep learning techniques can automatically extract unique and efficient features from the input data adaptively, which reduces the dependencies on manual feature extraction techniques.

The remaining book chapter is structured as follows: Database used in the experimental results are explained in Section 13.4, proposed method is described in Section 13.5, and experimental results, conclusion is explained in Sections 13.6 and 13.7 respectively.

13.4 DATABASE

The Massachusetts Institute of Technology (MIT), in partnership with the Boston's Beth Israel Hospital (BIH), established a conventional database for the classification of different arrhythmia. It was used as a conventional database by research scholars to verify the performance of their own system and equate their findings with previous systems. The data is collected from 25 males aged between 32–89 years and 22 females aged between 23–89. There are two-channel recordings for each ECG record. The recording of the first channel is done using altered lead limb II (MLII), and the next channel recording is done using lead $V-1(V-2,\ V-4$ or $V-5)$. The database consists of totals 48 records, each with an ECG segment of 30 minutes selected from a 24-hour record of 47 different subjects as the record 200 and 201 are taken from the same subject [38]. The first 23 (100–124) recordings are regular clinical records, while the remaining of the records (200–234) comprise the intricate arrhythmias. The ECG recordings were sampled at 360 Hz, and a 0.1–100 Hz bandpass filter was used [39].

13.5 PROPOSED METHODOLOGY

The general proposed methodology for arrhythmia detection is as shown in Figure 13.3 In this proposed methodology section, four important stages are described to detect

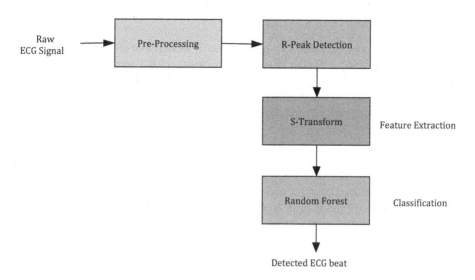

FIGURE 13.3 Proposed methodology for detecting ECG beat.

the different types of cardiac arrhythmia, they are (i) pre-processing, (ii) R-peak Detection, (iii) feature extraction, and (iv) classification.

The sampling frequency of the ECG signal is 360 Hz, and the band-pass filter used has a cutoff frequency range of 0.1–100 Hz which removes the unwanted disturbances in the signal. The majority of information about the heart is located at the QRS complex of the ECG signal. Therefore, to obtain the ORS complex the Pan-Tompkins algorithm is utilized to identify the R-peak in the ECG signal, which helps to extract the vital information about the cardiac arrhythmias. Based on the location of R-peak, ECG signal is segmented by recognizing +128 to −128 samples about the position of the R-peak [20]. A total of 256 samples are collected and applied to Stockwell transform (ST) [14] on each ECG segment to extract the features from the ECG signal. ST transform is used to collect the frequency domain characteristics from the ECG signal. Time-frequency domain illustration of the ECG signal $E(t)$ at time $t = \tau$ given below:

$$S(\tau, f) = \int_{-\infty}^{\infty} E(t) \frac{|f|}{\sqrt{2\pi}} e^{-\frac{(\tau-t)^2 f^2}{2}} e^{-j\omega t} dt \qquad (13.1)$$

Random forest is used for regression and classification [22]. It is the most effective machine learning algorithm. A random forest classifier is a combination of different random trees. Extracted features using ST transform are used as input to the random forest classifier. Classification algorithm performance is always depending upon the extracted features. A subgroup of the feature vectors is used in different random trees of the random forest classifier. This subgroup of the feature vector is called bootstrap. Detected arrhythmia is dependent upon the majority voting system which is an aggregate of all predicted values by the random decision trees.

13.6   EXPERIMENTAL RESULTS

In this work, the proposed classifier performance is assessed on the MIT-BIH arrhythmia database to detect different types of arrhythmias. In this work, the following 17 types of ECG beats are classified: normal sinus rhythm, 15 cardiac disorders, and pacemaker rhythm. Random forest is used to classify all the 17 types of different beats.

A performance matrix of the classifier is shown in Table 13.1.

13.6.1   THE RANDOM FOREST METHOD PERFORMANCE IN CLASSIFICATION

In this work, 500 decision trees are used to design a random forest classifier, where individual decision trees access sub-sample of feature set among 28 features extracted by the ST transform. The random forest counts all individual decision trees votes and declared the output class based on majority votes. From the confusion matrix of the random forest, it is clear that 97 normal beats (N), 12 pacemaker beats (PM), 467 arrhythmia beats (O) are misclassified. The proposed random forest classifier displays the overall classification performance of 98.75% and sensitivity of 97.46% which is relatively higher than the earlier described methods. Figure 13.4

TABLE 13.1
Performance Matrix of Random Forest

Method	Class	Acc(%)	Sen(%)	Spe(%)	Ppr(%)	F-Score
Random forest	N	99.63	98.73	98.71	99.87	0.99
	PM	98.72	97.87	97.23	98.34	0.99
	O	97.68	97.72	98.67	98.46	0.98

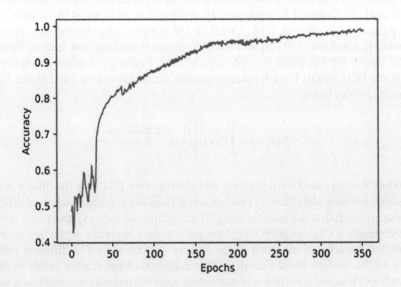

FIGURE 13.4   Accuracy curve of the classifier.

TABLE 13.2
Classification Performance of Random Forest

Class	TPR	FPR	AUC
N	0.984	0.041	0.999
P	0.996	0.004	1.000
L	0.968	0.001	0.999

represents the accuracy curve of the random forest while its running curve for 350 epochs. From this curve, it is observed that accuracy almost reaches more than 98% after 320 epochs.

The random forest-based method with ST features produces a mean sensitivity of 97.05%, a mean specificity of 98.69%, and mean positive predictivity 99.78%. The overall F-score of the classifier is 0.99.

For all 17 types of classes Area under curve (AUC) is more than 0.9, which means the classifier performance extremely well in detection of different ECG beats. AUC is also treated as receiver operating characteristic (ROC). Table 13.2 shows that the AUC, True Positive rate (TPR), False Positive Rate (FPR) of individual classes.

13.7   CONCLUSION

In this work, an automated ECG beat classification method using a machine learning algorithm is proposed. For QRS complex detection in the ECG signal Pan-Tompkins, algorithm is applied. ST transform is used to extract the frequency domain information from the QRS complex of the respective ECG beat. ST transform features are used as input to the random forest classifier which produces excellent results in the detection of 17 types of arrhythmia beats. MIT-BIH arrhythmia database is utilized to conduct all the experiments in this chapter. Random forest classifier detects all 17 types of ECG beats with an accuracy of 98.75, which is extremely greater than the earlier reported techniques. The proposed technique can be incorporated in the remote patient monitoring system.

REFERENCES

1. R. U. Acharya, J. S. Suri, J. A. E. Spaan, and S. M. Krishnan, *Advances in Cardiac Signal Processing*. Berlin: Springer, 2007.
2. W. J. Germann and C. L. Standield, *Principles of Human Physiology*. San Francisco, CA: Benjamin Cummings, 2002.
3. J. P. Fisher, *Heart: Cardiac Function & ECGs*, Textbook of Medical Physiology, 11th edition, Pennsylvania: Elsevier, 2012.
4. M. Gabriel Khan, *Rapid ECG Interpretation*, Third edition, NJ: Humana Press, Springer, 2003.
5. F. Morris, J. Edhouse, W. J Brady, and J. Camm, *ABC of Clinical Electrocardiography*, BMJ Books, 2003.
6. Z. Ihara, Design and performance of lead systems for the analysis of atrial signal components in the ECG, 2006. doi: 10.5075/epfl-thesis-3565.

7. J. Moss and S. Stern., *Noninvasive Electro Cardiology: Clinical Aspects of Holter*, London, Philadelphia: W.B. Saunders, 1996.
8. O. Malgina, J. Milenkovic, E. Plesnik, M. Zajc, and J. F. Tasic, ECG signal feature extraction and classification based on R peaks detection in the phase space, *2011 IEEE GCC Conference and Exhibition (GCC)*, Dubai, UAE, 2011, pp. 381–384.
9. X. Hu, Z. Xiao, and N. Zhang, Removal of baseline wander from ECG signal based on a statistical weighted moving average filter, *Journal of Zhejiang University-SCIENCE C (Computers & Electronics)*, vol. 12, pp. 397–403, 2011.
10. G. M. Friesen, T. C. Jannett, M. A. Jadallah, S. L. Yates, S. R. Quint, and H. T. Nagle, A comparison of the noise sensitivity of nine QRS detection algorithm, *IEEE Transactions on Biomedical Engineering*, vol. 37, pp. 85–98, 1990.
11. Lippincott Williams & Wilkins, *ECG Interpretation Made Incredibly Easy*, 5th edition, Cbspd, 2011.
12. S. Jain, V. Bajaj, and A. Kumar, Effective denoising of ECG by optimized adaptive thresholding on noisy modes, *IET Science, Measurement Technology*, vol. 12(5), pp. 640–644, 2018. doi: 10.1049/iet-smt.2017.0203.
13. S. Chaudhary, S. Taran, V. Bajaj, and S. Siuly, A flexible analytic wavelet transform based approach for motor-imagery tasks classification in BCI applications, *Computer Methods and Programs in Biomedicine*, vol. 187, p. 105325, 2020. doi: 10.1016/j.cmpb.2020.105325.
14. X. Yang and J. Tang, Hilbert-huang transform and wavelet transform for ecg detection, *In 2008 4th International Conference on Wireless Communications, Networking and Mobile Computing*, Hunan, China, October 2008, pp. 1–4.
15. N. K. Dewangan and S. P. Shukla, Ecg arrhythmia classification using discrete wavelet transform and artificial neural network, *In 2016 IEEE International Conference on Recent Trends in Electronics, Information Communication Technology (RTEICT)*, Bangalore, India, May 2016, pp. 1892–1896.
16. S. Jain, V. Bajaj, and A. Kumar, An efficient algorithm for classification of ECG beats based on ABC-LSSVM classifier, *IET Electronics Letters*, vol. 52(14), pp. 1198–1200, 2016. doi: 10.1049/el.2016.1171.
17. A. S. Alvarado, C. Lakshminarayan, and J. C. Principe, Time-based compression and classification of heartbeats, *IEEE Transactions on Biomedical Engineering*, vol. 59(6), pp. 1641–1648, 2012.
18. H. I. Bulbul, N. Usta and M. Yildiz, Classification of ECG Arrhythmia with machine learning techniques, *In 2017 16th IEEE International Conference on Machine Learning and Applications (ICMLA)*, Cancun, 2017, pp. 546–549. doi: 10.1109/ICMLA.2017.0-104.
19. S. Kiranyaz, T. Ince, and M. Gabbouj, Real-time patient-specific ecg classification by 1-d convolutional neural networks, *IEEE Transactions on Biomedical Engineering*, vol. 63(3), pp. 664–675, 2016.
20. A. Isin and S. Ozdalili, Cardiac arrhythmia detection using deep learning, *Procedia Computer Science,* vol. 120, pp. 268–275, 2017, *9th International Conference on Theory and Application of Soft Computing, Computing with Words and Perception, ICSCCW 2017*, Budapest, Hungary, 22–23 August 2017. Available: http://www.sciencedi- rect.com/science/article/pii/S187705091732450X.
21. F. I. Alarsan and M. Younes, Analysis and classification of heart diseases using heartbeat features and machine learning algorithms. *Journal of Big Data*, vol. 6, p. 81, 2019. doi: 10.1186/s40537-019-0244-x.
22. S. M. Jadhav, S. L. Nalbalwar, and A. A. Ghatol, Ecg arrhythmia classification using modular neural network model, *In 2010 IEEE EMBS Conference on Biomedical Engineering and Sciences (IECBES)*, November 2010, pp. 62–66.

23. S. Jain, V. Bajaj, and A. Kumar, Riemann Liouvelle fractional integral based empirical mode decomposition for ECG denoising, *IEEE Journal of Biomedical and Health Informatics*, vol. 22(4), pp. 1133–1139, 2018. doi: 10.1109/JBHI.2017.2753321.

24. D. A. Coast, R. M. Stern, G. G. Cano, and S. A. Briller, An approach to cardiac arrhythmia analysis using hidden markov models, *IEEE Transactions on Biomedical Engineering*, vol. 37(9), pp. 826–836, 1990.

25. P. de Chazal and R. B. Reilly, A patient-adapting heartbeat classifier using ecg morphology and heartbeat interval features, *IEEE Transactions on Biomedical Engineering*, vol. 53(12), pp. 2535–2543, 2006.

26. F. Melgani and Y. Bazi, Evolutionary computation approach to ECG signal classification, pp. 19–24, 2008.

27. J. Park, S. Lee, and K. Kang, Arrhythmia detection using amplitude difference features based on random forest, IEEE, 2015.

28. T. Paul, A. Chakraborty, and S. Kundu, Hybrid shallow and deep learned feature mixture model for arrhythmia classification, IEEE, 2018.

29. R. Rodriguez, J. Bila, O. O. Vergara Villegas, V. G. Cruz Sánchez, and A. Mexicano, Arrhythmia disease classification using a higher-order neural unit, *The Fourth International Conference on Future Generation Communication Technologies (FGCT 2015)*, IEEE, 2015.

30. S. Savalia and V. Emamian, Cardiac arrhythmia classification by multi-layer perceptron and convolution neural networks, *Bioengineering (Basel)* vol. 5(2), pp. 35, 2018. doi: 10.3390/bi-oengineering5020035. PMID: 29734666; PMCID: PMC6027502.

31. S. Chakroborty and M. A. Patil, Real-time arrhythmia classification for large databases, IEEE, 2014.

32. S. K. Dash and G. S. Rao, Robust multiclass ECG arrhythmia detection using balanced trained neural network, *International Conference on Electrical, Electronics, and Optimization Techniques (ICEEOT) – 2016*, IEEE, 2016.

33. J. P. Kelwade and S. S. Salankar, Radial basis function neural network for pre- diction of cardiac arrhythmias based on heart rate time series, *2016 IEEE First International Conference on Control, Measurement and Instrumentation (CMI)*, IEEE, Kolkatha, India, 2016.

34. D. Li, J. Zhang, Q. Zhang, and X. Wei, Classification of ecg signals based on 1d convolution neural network, *In 2017 IEEE 19th International Conference on e-Health Networking, Applications and Services (Healthcom)*, Dalian, China, October 2017, pp. 1–6.

35. Ö. Yıldırım, P. Pławiak, R.-S. Tan, and U. Rajendra Acharya, Arrhythmia detection using deep convolutional neural network with long duration ecg signals, *Computers in Biology and Medicine*, vol. 102, pp. 411–420, 2018. Available: http://www.sciencedirect.com/science/article/pii/S0010482518302713.

36. T. J. Jun, H. M. Nguyen, D. Kang, D. Kim, D. Kim, and Y. Kim, ECG arrhythmia classification using a 2-d convolutional neural network, *Computer Science*, 2018. Available: http://arxiv.org/abs/1804.06812.

37. M. Salem, S. Taheri, and J. Yuan, ECG arrhythmia classification using transfer learning from 2- dimensional deep CNN features, *2018 IEEE Biomedical Circuits and Systems Conference (BioCAS)*, 2018. doi: 10.1109/biocas.2018.8584808.

38. MIT-BIH Database Distribution, Massachusetts Institute of Technology, Cambridge, MA, 1998. http:// www.physionet.org/physiobank/data- base/mitdb.

39. G. B. Moody, and R. G. Mark, The impact of the MIT-BIH arrhythmia database, *IEEE Engineering in Medicine and Biology Magazine*, vol. 20(3), pp. 45–50, 2001.

Index

Note: **Bold** page numbers refer to tables and *Italic* page numbers refer to figures.